ZOOLOGY

Edited by **María-Dolores Garcia**

Zoology
Edited by María-Dolores Garcia

Published by InTech

Janeza Trdine 9, 51000 Rijeka, Croatia

Publishing Process Manager Danijela Duric

Technical Editor Teodora Smiljanic

Cover Designer InTech Design Team

Additional hard copies can be obtained from orders@intechopen.com

Zoology, Edited by María-Dolores Garcia
p. cm.
ISBN 978-953-51-0360-8

For EU product safety concerns:
IN TECH d.o.o., Prolaz Marije Krucifikse Kozulić 3, 51000 Rijeka, Croatia,
info@intechopen.com or visit our website at intechopen.com.

Meet the editor

María-Dolores García is professor at the Zoology and Physical Anthropology Department of the University of Murcia (Spain) where she has taught different topics on Zoology since 1979, including a Master in Forensic Sciences. She graduated in Biological Sciences, specialty Zoology, from the University Complutense of Madrid and got a PhD in Biology from the University of Murcia. Since her graduation, her research interests have included taxonomy, diversity and acoustic behaviour of Orthoptera and, more recently, also Forensic Entomology. She is an author in numerous publications and contributions to scientific meetings covering these fields of research. Lately, she has made contributions in the area of the European Credit Transfer System implemented in the Spanish Universities.

Contents

Preface

Despite the title the present book is not a classical manual. For instance, the reader should not expect to find the usual treatment of animal groups, or an evolutionary approach to animal diversity. The book does not deal with the study of the different groups of animals known to date from a traditional point of view, that is, mainly their morphological and anatomical aspects and in some cases with notes on their physiology, ecology... Some people may feel disappointed when consulting the index, mainly if searching for something that is considered standard. But the reader, if interested in Zoology, should not be disappointed when trying to find novelties on different topics that will help to improve the knowledge on animals.

When thinking about Zoology, the first thought that comes to mind is animals, although this term is very often restricted to only vertebrates. Yet, from an etymological point of view, Zoology refers to the knowledge of animals. Without any doubt, this is such a wide concept that it is hardly understood without some type of meditation. What does it mean the knowledge of animals? Does it mean their identification or recognition? Does it mean to understand how they are constructed and how they function? As a matter of fact, Zoology means all those and many other things. Zoology is one of those "natural sciences" as old as the human being. It is the result of the humans' tireless curiosity and desire for knowledge that from the beginning of humanity interacted with the nearby environment trying to control it and, in order to do so, they first needed to understand it.

Thus, Zoology, as a "science", existed even before the name was assigned. It is easy to imagine that hunting was one of the principal incentives behind animal knowledge, in particular those aspects that concern the anatomy of the species of cynegetic value. However, in a hunter-gathered society, hunting was not its only fundamental aspect. In addition, knowing all those things that could be gathered and be taken advantage of and all those organisms that took advantage of what humans had gathered and stored is important. These organisms competed with the humans themselves, causing a clear prejudice which, undoubtedly, should be avoided. This empirical knowledge, driven by those aspects that concern the human beings (food, coat, and even magic and religious fundamentals (Grassé, 1963)) led to the study of animals in all their imaginable aspects. As the University of Cambridge proclaims (http://www.zoo.cam.ac.uk/) *"Zoology is central to our understanding of the world.*

Zoologists seek to discover the fundamental principles that underpin animal life focusing on the diversity, function and evolution of animals and thus providing the scientific basis for our knowledge both of the creatures with whom we share this planet and of ourselves". In fact, Zoology is devoted to any aspect related to all living animals, such as their morphology, structure, embryology and development, evolution, behavior, distribution..., since the extinct animals are the subject of study of a different, but related, discipline, Paleontology.

The idea that Zoology has been an object of interest to human beings is obvious when considering all the findings dedicated to animals. Independently of the prehistoric paintings that exist all over the world that represent animals of large size, which are usually subject of hunting, and activities related to other animals of utility, such as bees, there are a number of documents that concern different and varied aspects of animals. In addition to several passages in the Bible, such as those dedicated to different animals in the books of Leviticus 11 and Proverbs 30, it is worth mentioning the works of Aristotle (*Historia Animalium* and *De Partibus Animalium*), Plinio (*Naturalis Historia*), Claudio Eliano (*De Natura Animalium*) and Opiano (*Cynegetica* and *Halieutica*). All these authors already attempted the study of animals from a scientific point of view, even if, in some cases, they gathered the information from tradition, whether it was right or not. On the other hand, the constant presence of animals appears to be fundamental in the symbolic and transcendental thinking of the different human cultures, including old (i.e. ancient Egypt) as well as current ones (i.e. Aymara from northern Chile (Grebe, 1984)). This aspect of animal knowledge forms part, of Ethnozoology (i.e. Alves & Souto, 2011).

Zoology, as a scientific discipline, is significantly disconnected from other sciences dedicated to particular animal species that are included in the Health Sciences. These sciences refer to Veterinary (the stable animals, or those who offer a clear profit to the human beings), and Medicine (the man, himself, as the animal species that he is). Though these sciences contemplate methodological aspects also used in Zoology, their studies have a different purpose, since they devote themselves to supporting, re-setting, and improving the health of animals and humans.

Zoology has been the "mother" science of many scientific disciplines that, nowadays, have their own identity. Animal Physiology, for example, derived from Zoology and remained together until it became an independent discipline with methods and specific fundaments of its own. The same can be said about animal Cytology and Histology, Ethology, Parasitology, Pathology..., and even Genetics, among others, when applied to animals. The rapid technological development that started in the last century has enabled further specialization of these disciplines, separating them conceptual and methodologically of the mother science, i.e. the most traditional Zoology. To many scientists that study other disciplines, Zoology is thought to focus on the knowledge of the morphology and the anatomy of the animals from a descriptive point of view and, clearly, in the taxonomy. It is true that the knowledge of animal diversity is critical because if we do not know which animals exist, then how will be learn more of them? How will we know which animals became extinct? In addition, once animal diversity is understood, we must use some method(s) to classify them in order to understand, even in a simple way, this enormous diversity. However, despite the importance of these facts, there are others of interest in Zoology. As it concerns biodiversity, there exists great interest in the precise knowledge of the geographical distributions of species, in the development of predictive elements of the distribution of species, mainly the harmful ones, in behavioural, physiological, genetic ... aspects, all of which somehow contribute to the overall knowledge of the animals thus contributing, among others, to their hierarchic arrangement.

Thus, at the present time, when considering Zoology in its global term, the knowledge of the animals begins, without any doubts, with knowledge of their biodiversity. Once the species are known, or rather recognized (morphology), biodiversity continues with the knowledge of the internal organization (anatomy), stage forced for the ultra-structural knowledge (cytology and histology) which, in turn, will allow knowledge of how such structures work (physiology, biochemistry). All these studies run in parallel with the information, descriptive or analytical, of how the animals act and behave (ethology), and without losing perspective of which origin such an animal could have and with which other animals is related (phylogeny, genetics). All this information allows determining its location in the complex tree of the animal life and, somehow, closes the cycle of the knowledge of biodiversity since all this knowledge allows to conclude if what it was once considered to be a species it is a true species or not. But zoological knowledge also includes other scientific and applied aspects; it is not only a question of knowing about an animal in an isolated space but, also, about its relationships with other forms of life, with its environment (Ecology) where humans exist, and how influences on the environment, or on the own human species, contributing benefits or causing injuries (applied Zoology). Currently, biodiversity research requires a multidisciplinary approach (Boero, 2009) and the study of biodiversity should proceed with the contribution of integrative taxonomy (Boero, 2010).

Despite the fact that the modern concept of Zoology is clearly multidisciplinary, the consideration of this science is different between academic and research environments. In the Universities, there are Faculties of Zoology, Institutes of Zoology and many Departments of Zoology with different academic contents, some multidisciplinary, others exclusively focused on the strictest zoological and traditional theme, but all with a common purpose, the integrated knowledge of animals. In the area of research, the different fields of knowledge clearly affect this science. In the ISI Web of Knowledge one can find, among others, categories such as Behavioural sciences, Biodiversity conservation, Ecology, Entomology, Evolutionary biology, Marine and freshwater biology, Parasitology or Zoology which gather reviews publishing very diverse aspects related to Zoology. In most cases, the final aim of many articles is not the knowledge of the animals themselves but to be able to better deriv the mechanisms or fundamentals of a specific discipline (Biochemistry, Genetics, Physiology ...). For example, studying the DNA structure of many insect species in order to know the recombination mechanisms is not Zoology but Genetics of animals. In this area, there exists a notable bias of the major indexes of impact towards the topics of molecular nature to the detriment of zoological topics of more traditional court. It is also worth noting the large asymmetry on the number of magazines gathered in every category, with those concerning zoological aspects being much less than those dedicated to Biochemistry and Molecular Biology. Some aspects of Zoology are, in fact, reviled.

For example, although politicians have supported efforts to increase the knowledge of global biodiversity, and there is an understanding for a need to learn and name the existing species, Taxonomy, which is the discipline in charge of that mission, has fallen in disregard for not obvious reasons. This is not just the result of recent science policies, based on certain "objective" indicators (i.e., impact index of the Institute for Scientific Information) that ignore, in general, the publications devoted to taxonomy, and in particular to traditional taxonomy (Boero 2010). The result of this policy is a continuous drop of traditional taxonomic practices for the benefit of other studies which provide more advantages and benefits to the authors, but probably contribute less to scientific knowledge. But Taxonomy is the basis of scientific knowledge in regard to Zoology, and to Biology in general. Even in relation to more technologically advanced biological disciplines (Biochemistry, Genetics, Physiology...), Taxonomy is critical. For example, there are databases such as Gene Bank, which may contain sequences attached to misidentified taxa (Boero 2009). Just for determining whether the reference database is appropriate for a particular research work, it would be necessary to know in detail the specific species that are involved (Wells & Stevens 2010). Thus, the recognition of species, through the study of phenotypes (Boero 2009), remains central to all levels of knowledge of biodiversity.

I wish at this time, to highlight the work of taxonomists for their scientific and public recognition. Without a taxonomical basis, there will not be any applied work that is based on species. And we should not consider that the existing knowledge is sufficient. As Boero (2009, 2010) points out, about 2 million species are known and have been named so far, but there should be up to about 10 to 15 million more species that we need to identify and, of course, to name. And, once known and named, their study will proceed on other levels of depth, until their overall knowledge is achieved.

Though many people may think that almost everything is already known in Zoology and that, at most, some species remain to be discovered, which may not be considered a critical issue, the fact is that a lot of things still remain to be discovered. In fact, not a long time ago a new large group was described, the Micrognathozoa (Kristensen & Funch, 2000), nowadays with level of phylum, that continues contributing with anatomical innovations (Sørensen, 2003) and, more recently, a new organizational type of sponges has been described (Cavalcanti & Klautau, 2011) as well as some new species and news in the systematic and the evolution of that group (Dohrmann et al., 2011). But, even if it is only a question of learning about a few species still not discovered or described, this would be already important, because basic knowledge is fundamental for developing applied knowledge, which is the most useful aspect from the human point of view. One can imagine that when Meigen, in 1830, described *Drosophila melanogaster*, initiating the knowledge of this species, he would not do it thinking about the usefulness for allowing the development of Genetics, as some years later occurred. Nevertheless, it is not possible to imagine modern Genetics without any of the advances achieved thanks to this species. In another order of things, we must have in mind that, for example, the first multicellular organism genome that was completely sequenced (The *C. elegans* Sequencing Consortium, 1998), was a small Nematode worm, *Caenorhabditis elegans* Maupas, 1900, an animal frequently used nowadays as a model for studying the aging process (i.e. Cypser et al., 2006, Johnson, 2006; Luo & Murphy, 2011) or pathologies, such as diabetes or Alzheimer's disease (i.e. Morcos & Hutter, 2009; McColl et al., 2009), or in studies on genetics of the development (Gent et al., 2009), among others, and which is the preferential object of study of, nobody else, but two Nobel prizes (Sydney Brenner and Martin Chalfie). There are many examples of animals that are facilitating enormously research in other fields. Besides those already mentioned, and among many others, it is worth mentioning *Rhodnius prolixus* Stål, 1859 that, in addition to being important as a vector of Chagas's disease (i.e. Spurling et to., 2005), it is an animal frequently used as model for biochemical (i.e. Cricket et al., 2007), physiological (i.e. O'Donnel et al., 1983) and, even behavioural studies (i.e. Abramson et al., 2005, Ferreira et al., 2007). It would be impossible, because of the amount of work , to try to make a relation of the animals directly or indirectly involved in the vegetable or in the human health, including in the latter one aspects of forensic interest, though their importance as species themselves is huge to global scale. For instance, one can think about the vectors of diseases so relevant as the malaria (*Anopheles* sp.), the contagious disease that causes more deaths with the exception of the tuberculosis, the trypanosomiasis (*Glossina* sp.), the onchocerciasis (*Simulium* sp.) and the schistosomiasis (*Bulinus* sp., *Biomphalaria* sp., *Neotricula* sp. among others)... Undoubtedly, research on these animals is completely active in all the fronts, including more traditional aspects, as the morphologic ones (i.e. Reis dos Santos-Mallet et al., 2005), the ecological ones (i.e. Morgan et al., 2005) or those related to the distribution (i.e. Antonio-Nkondjio et al., 2011, El-Badry & Al-Ali, 2010, Kengluecha et al., 2005), relevant aspects to bear in mind when managing the control of their populations. But, as mentioned above, the studies on animal species of not obvious utility and that, therefore, are only an object of study by themselves are also active, both in still little explored environments, as the Antarctica (i.e. Eakin et al., 2009, Matallanas, 2009, Kuhn et al., 2011) as in other more explored environments (i.e. Marin, 2011, Pilato et al., 2011), so much in relation with the description of species as in their behaviour (i.e. Lorier et al., 2010) or distribution (i.e. Thomas, 2005).

It would not be possible to report here the immense possibilities of knowledge offered by the animals, both because of the space and the time that would be necessary to accomplish this task. The intention that has guided me has been, meanly, to expose the modern concept of Zoology that, besides its technologies and own methods, is nourished of and complemented by those developed for other scientific disciplines, some of which proceed from it.

In the current background of the science, no aspect must be disregarded; all them have value themselves, individualy and in synergy with the others. And, in this background, Zoology turns out to be a paradigm of multidisciplinary science.

This book is a compendium of contributions to some of the many different topics that can be considered in relation to the knowledge of animals. Individual chapters have been written by authors concerned with different scientific questions. Thus, contributions are quite heterogeneous. Some of them deal with more or less general aspects, such as those concerning biodiversity which include molecular aspects related to animal variation. Others are related to applied zoology and technologies under development not only to study animals but also to control them when acting as pests. Others, finally, deal with ultra-structural and physiological topics that surely will help to understand other overall processes within animals. In this way, this issue includes recent contributions to Zoology illustrating the diversity of research conducted in this discipline and providing new data to be considered in future overall publications. It would have been desirable to include more diverse contributions, in order to provide a wider outlook of what is being investigated about animals and the novelties it provides, but I think the contributions included are a good outlook of that research. Because of the diversity and number of contributions, this book is not divided in "Sections"; the chapters have been arranged in a "logical" manner, from the most overall questions to the most specific ones. Of course, this arrangement could have been the opposite, in the same way as the scientific achievement are usually pursued, that is, from the specific to the global aspects, but I have considered the current arrangement as the most pedagogical, which would guide the readers into progressively deeper questions related to animals.

At this point, I wish to thank InTech - Open Access Publisher for the initiative of publishing such a book on Zoology and for having trusted me with the editing task. Thanks are also due to all the contributors for the efforts they made to accomplish their chapters and others who helped me to review some of the chapter proposals which, in some cases, are far removed from my own research field. I wish all of them successful future research careers, which will help to maintain this science, Zoology, a fundamental and still alive one. Thanks are due especially to the Publishing Process Manager assigned to this task for their patience with all of us, especially me. She has been very helpful during the time of writing and editing this book.

I hope this initiative will encourage other scientists to share some of their achievements and engage in a more complete work on this exciting science in the future.

Dr. María-Dolores Garcia
Department of Zoology, Faculty of Biology, University of Murcia,
Spain

References

Abramson, C.I.; Sulbaran Romero, E.; Frasca, J.; Fehr, R.; Lizano, E. & Aldana, E. (2005). Psychology of learning: a new approach to study behavior of Rhodnius prolixus Stal under laboratory conditions. *Psychological reports,* 97, 3: 721-31. ISSN: 0033-2941.

Alves, R.R.N. & Souto, W.M.S. (2011). Ethnozoology in Brazil: current status and perspectives. Journal of Ethnobiology and Ethnomedicine, 7: 22. doi: 10.1186/1746-4269-7-22. ISSN: 1746-4269.

Antonio-Nkondjio, Ch., Tene Fossog, B., Ndo, C., Menze Djantio, B., Zebaze Togouet, S., Awono-Ambene, P., Costantini, C., Wondji, Ch.S. & Ranson, H. (2011). *Anopheles gambiae* distribution and insecticide resistance in the cities of Douala and Yaoundé (Cameroon): influence of urban agriculture and pollution. *Malaria Journal,* 10: 154. Disposable at http://www.malariajournal.com/content/pdf/1475-2875-10-154.pdf (1 December 2011) ISSN: 14752875.

Boero, F. (2009). Zoology in the era of biodiversity. *Italian Journal of Zoology,* Vol. 76, No. 3, pp. 239, ISSN 1125-0003.

Boero, F. (2010). The study of species in the era of biodiversity: a tale of stupidity. *Diversity,* Vol. 2, pp. 115-126, ISSN 1424-2818.

Cavalcanti, F.F. & Klautau, M. (2011). Solenoid: a new aquiferous system to Porifera. *Zoomorphology,* 130: 255–260. ISSN: 0720-213X.

Cypser, J.R., Tedesco, P. & Johnson, T.E. (2006). Hormesis and aging in *Caenorhabditis elegans.* Experimental Gerontology, 41, 10: 935-939. ISSN: 0531-5565.

Dohrmann, M., Göcke, Ch., Janussen, D., Reitner, J., Lüter, C. & Wörheide, G. (2011). Systematics and spicule evolution in dictyonal sponges (Hexactinellida: Sceptrulophora) with description of two new species. *Zoological Journal of the Linnean Society,* 163, 1003–1025. ISSN: 0024-4082

Eakin, R.R., Eastman, J.T. & Near, T.J. (2009). A new species and a molecular phylogenetic analysis of the Antarctic fish genus Pogonophryne (Notothenioidei: Artedidraconidae). *Copeia,* 4:705–713. ISSN: 0045-8511.

El-Badry, A.A. & Al-Ali, K.H. (2010). Prevalence and seasonal distribution of dengue mosquito, *Aedes aegypti* (Diptera: Culicidae) in Al-Madinah Al-Munawwarah, Saudi Arabia. *Journal of Entomology,* 7, 2: 80-88. ISSN: 1812-5670.

Ferreira, R.A.; Lazzari, C.R., Lorenzo, M.C. & Pereira, M.H. (2007). Do Haematophagous Bugs Assess Skin Surface Temperature to Detect Blood Vessels? *PLoS ONE,* 2(9): e932. ISSN: 1932-6203.

Gent, J.I., Schvarzstein, M., Villeneuve, A.M., Gu, S.G., Jantsch, V., Fire, A.Z. & Baudrimont, A. (2009). A *Caenorhabditis elegans* RNA-Directed RNA Polymerase in Sperm Development and Endogenous RNA Interference. *Genetics,* 183(4): 1297–1314. ISSN: 0016-6731.

Grassé, P.P. (dir.) *Zoologie 1. Généralités-Protozoaires-Métazoaires I.* Encyclopédie de La Pléiade XIV. Librairie Gallimard. Paris.

Grebe, M.E. (1984). Etnozoología andina: Concepciones e interacciones del hombre andino con la fauna altiplánica. *Estudios Atacameños,* 7: 335-347. ISSN: 0716-0925.

Grillo, L.A.M., Majerowicz, D. & Gondim, K.C. (2007). Lipid metabolism in Rhodnius prolixus (Hemiptera: Reduviidae): role of a midgut triacylglycerol-lipase. *Insect Biochemistry and Molecular Biology,* 37: 579-588. ISSN: 0965-1748.

Johnson, T.E. (2006). For the special issue: The nematode Caenorhabditis elegans in aging research. Experimental Gerontology, 41, 10: 887-889. ISSN: 0531-5565.

Kengluecha, A., Rongnoparut, P., Boonsuepsakul, S., Sithiprasasna, R., Rodpradit, P. and Baimai, V. (2005). Geographical distribution of *Anopheles minimus* species A and C in western Thailand. *Journal of Vector Ecology,* 30 (2): 225-230. ISSN: 1081-1710.

Kristensen, R.M. & Funch, P. (2000). Micrognathozoa: A new class with complicated jaws like those of Rotifera and Gnathostomulida. *Journal of Morphology* 246: 1-49. ISSN: 0362-2525.

Kuhn, K.L., Near, T.J., H.W Detrich III, H.W. & Eastman, J.T. (2011). Biology of the Antarctic dragonfish Vomeridens infuscipinnis (Notothenioidei: Bathydraconidae). *Antarctic Science,* 23:18-26. ISSN: 0954-1020.

Lorier, E., Clemente, M.E.; García, M.D. & Presa, J.J. (2010). El comportamiento acústico de *Fenestra bohlsii* Giglio-Tos (Orthoptera: Acrididae: Gomphocerinae). *Neotropical Entomology,* 39, 6: 839-853. ISSN 1519-566X.

Luo, S. & Murphy, C.T. (2011). *Caenorhabditis elegans* reproductive aging: Regulation and underlying mechanisms. *Genesis,* 49, 2: 53-65. ISSN: 1526-954X.

Marin, I. (2011). Two new species of alcyonacean-associated shrimp genus *Alcyonohippolyte* Marin, Okuno & Chan, 2010 (Crustacea: Decapoda: Hippolytidae) from the Great Barrier Reef of Australia. *Zootaxa* 3123: 49–59. ISSN: 1175-5326.

Matallanas, J. (2009). Description of *Gosztonyia antarctica,* a new genus and species of Zoarcidae (Teleostei: Perciformes) from the Antarctic Ocean". Polar Biology 32, 1: 15-19. ISSN: 0722-4060.

McColl, G., B. R. Roberts, A. P. Gunn, K. A. Perez, D. J. Tew, C. L. Masters, K. J. Barnham, R. A. Cherny & Ashley I. Bush. (2009). The *Caenorhabditis elegans* Aβ1–42 Model of Alzheimer Disease Predominantly Expresses Aβ3–42. *Journal of Biological Chemistry,* 284 (34): 22697-22702. ISSN 0021-9258.

Morcos, M. & Hutter, H. (2009). The model *Caenorhabditis elegans* in diabetes mellitus and Alzheimer's disease. Journal of Alzheimer's Disease, 16(4): 897-908. ISSN: 1387-2877.

Morgan, J.A.T., Dejong, R.J., Adeoye, G.O., Ansa, E.D.O., Barbosa,C.S., Brémond, Ph., Cesari, I.M., Charbonnel, N., Corrêa, L.R., Coulibaly, G., d'Andrea, P.S., Pereira de Souza, C., Doenhoff, M.J., File, S., Idris, M.A., Incani, R.N., Jarne, Ph., Karanja, D.M.S., Kazibwe, F., Kpikpi, J.; Lwambo, N.J.S., Mabaye, A., Magalhães, L.A., Makundi, A., Moné, H., Mouahid, G., Muchemi, G.M., Mungai, B.N., Séne, M., Southgate, V., Tchuem Tchuenté, L-A., Théron, A., Yousif, F., Zanotti-Magalhães, E.M., Mkoji, G.M. & Locker, E.S. (2005). Origin and diversification of the human parasite *Schistosoma mansoni.* Molecular Ecology, 14, 2: 3889-3902. ISSN: 1365-294X.

O'Donnell, M.J., Maddrell, S.H.P.& Gardiner, B.O.C. (1983). Transport of uric acid by the Malpighian tubules of *Rhodnius prolixus* and other insects. *Journal of experimental Biology*, 103: 169-184. ISSN: 0022-0949.

Pilato, G., Kiosya, Y., Lisi, O., Inshina, V. & Biserov, V. (2011). Annotated list of Tardigrada records from Ukraine with the description of three new species. *Zootaxa*, 3123: 01–31. ISSN: 1175-5326.

Reis dos Santos-Mallet, J., Angela Cristina Verissimo Junqueira, A.C., de Carvalho Moreira, C.J., Andrade, Z., Rodrigues Coura, J. & Monte Gonçalves, T.C. (2005). Morphobiological aspects of *Rhodnius brethesi* Matta, 1919 (Hemiptera: Reduviidae) from the Upper and Middle Negro River,Amazon region of Brazil. I - Scanning electron microscopy. *Memorias do Instituo Oswaldo Cruz*, Rio de Janeiro, 100, 8: 915-923. ISSN: 0074-0276.

Sørensen, M.V. (2003). Further structures in the jaw apparatus of Limnognathia maerski (Micrognathozoa), with notes on the phylogeny of the Gnathifera. Journal of Morphology, 255,2: 131-45. ISSN: 0362-2525.

Spurling, G., Lucas, R. & Doust, J. (2005). Identifying health centers in Honduras infested with *Rhodnius prolixus* using the seroprevalence of Chagas disease in children younger than 13 years. *American Journal of Tropical Medicine and Hygiene*, 73, 2: 307-308. ISSN: 0002-9637.

The *C. elegans* Sequencing Consortium (1998). Genome sequence of the nematode *C. elegans*: a platform for investigating biology. *Science* 282: 2012-2018. ISSN: 0036-8075.

Thomas, J.A. (2005). Monitoring change in the abundance and distribution of insects using butterflies and other indicator groups. *Philosophical Transactions of the Royal Society of London Series B Biological Sciences*, 360(1454): 339–357. ISSN: 0962-8436.

Waters, K.A. & Reinke, V. (2011). Extrinsic and intrinsic control of germ cell proliferation in *Caenorhabditis elegans. Molecular Reproduction and Development*, 78, 3: 151–160. ISSN: 1040-452X

1

Mapping a Future for Southeast Asian Biodiversity

Alice C. Hughes
Department of Biology, Faculty of Science, Prince of Songkla University, Hat Yai, Thailand

1. Introduction

1.1 Global conservation priorities

Globally, biodiversity levels are currently changing at an unprecedented rate due to a myriad of anthropogenically induced factors (Sala et al., 2000). Over the next century these negative trends in biodiversity are set to continue, and therefore the identification of areas for conservation prioritisation are necessary in order to best protect areas of greatest diversity (Brook et al., 2006). Though studies have used different criteria in prioritisation of areas, some studies have combined a number of criteria (Myers et al., 2000) which have led to the identification of 25 global hotspots of biodiversity and species endemicity, which only comprise 1.4% of the global land surface, but contain 44% of all known plant species and 35% of currently described vertebrates.

In this chapter I will principally dwell on three of these biodiversity hotspots, which join to form Southeast Asia (SEA). The following section details the biodiversity present through the region, followed by a brief discussion of the threats to biodiversity. To effectively conserve species present, knowledge of distributions and identification of species is essential, and thus appropriate techniques will be discussed and demonstrated. This will be followed by an analysis of methods to quantify the impacts of such threats, and thus develop the most suitable strategies to effectively conserve the maximum number of species throughout the region.

Though this chapter focuses predominantly on Southeast Asia many regions round the world currently face similar situations. The techniques and approaches discussed here will be broadly applicable to other regions, and species, than those discussed here.

1.2 The biodiversity of Southeast Asia

Southeast Asia (SEA) contains a number of the biodiversity hotspots identified by Myers et al. (2000) and has some of the richest biodiversity and endemicity on the planet (Gaston, 1995a). The area consists of a number of biotas including the Indo-Burmese region, Wallacea, Sundaland and the Philippines. When considering the number of endemic plants and vertebrates, three Southeast Asian regions rank in the global top ten (Sundaland-2nd,

Indo-Burma-8th, Philippines-9th) and when the ratio of endemic species relative to area are considered these three are in the top 5 (Phillipines-2nd, Sundaland-3rd, Indo-Burma-5th) (Myers et al., 2000). SEA also contains high endemic evolutionary diversity at species, family and clade levels. On a global ranking Sundaland is in 2nd place, Wallacea 3rd and Indo-Burma 5th in terms of unique evolutionary history, with between 65 - 40 My (million years) of unique evolutionary history in each region (Sechrest, et al., 2002). Therefore SEA contains irreplaceable biodiversity and thus represents a priority area for conservation. Indeed, the forests of SEA have been deemed among the highest of all conservation priorities for biologists (Laurance, 2007).

The landscape of SEA is also diverse and varied and comprises a large number of ecoregions (Olson et al., 2001). Stibig (2007) categorised sixteen native forest types, in addition to woodland, savannah, two types of thorn scrub and forest, alpine grassland and cold desert among the native vegetation types. Such diversity in vegetation cover also creates very varied ecosystems with very different animal and plant communities. Karsts (limestone outcrops) make up around 400,000 km^2 of SEA, and though they only make up one percent of the land area, around two percent of Malaysian species are endemic to karst landscapes (Clements et al., 2006). Globally karsts also harbour a great proportion of endemic species, and therefore contribute significantly to landscape diversity and heterogeneity throughout SEA.

One reason for the high levels of diversity and endemicity in SEA is the dynamic and complex geo-physical history of the region, which has been described as a biogeographic theatre (Woodruff, 2003). Some of the landmasses that form SEA only joined as little as 15 Mya (Million years ago), and the addition of new landmasses caused faults and regional instability in many regions (Hall, 2002), which in turn contributed to the formation of unique biotas. Even at only five Mya SEA had not taken its present shape and landmasses within it were still subject to small but significant movements (Hall, 2002). Since this time glacial cycles have periodically transformed SEA, both in terms of shape and vegetation cover (Woodruff, 2003). During successive glaciations mainland and insular areas of SEA have been joined, and glaciers existed as recently as 10 Kya (thousand years ago) in Borneo and Sumatra (Morley & Flenley, 1987). This dynamic geophysical history has led to a highly complex pattern of species distributions and the area contains no less than three zoo/floro-geographic boundaries: Wallace's line, the Kangar-Pattani line and the Isthmus of Kra (Whitmore, 1981; Baltzer, 2008; Cox & Moore, 2010; A.C.Hughes et al., 2011). Therefore the region has a rich and highly varied biota, and thus represents a priority region for conservation.

2. Threats to biodiversity

Southeast Asia has been stated by many to be facing a crisis in terms of biodiversity loss (Laurance, 2007). SEA has the highest global rate of deforestation, with rates over double those documented elsewhere (Laurance, 2007). Despite possessing extensive biodiversity, Thailand only has around 17.6% of its potential forest remaining, and Peninsula Malaysia around half (Witmer, 2005). Rates of change in vegetation cover in SEA between 1981 and 2000 were the highest globally (Lepers et al., 2005) and what is more these rates of change are accelerating (Hansen & DeFries, 2004).

Loss of habitat and deforestation are not the only threats to the biodiversity of SEA. The Convention on International Trade in Endangered Species (CITES) listed that at least 35 million animals in addition to 18 million pieces of coral (and 2 million kg live coral) were exported from SEA between 1998 and 2007 (Nijman, 2010). Many species are also hunted for recreation (Epstein et al., 2009) in addition to bushmeat (Brodie et al., 2009). Furthermore the Chinese medicine trade is stated to be the "single major threat" for some species (EIA, 2004; Ellis, 2005). These problems are not limited to "unprotected areas" as even National Parks fail to offer protection from either illegal logging (Sodhi et al., 2010) or high levels of hunting (Brodie et al., 2009).

The above mentioned factors affecting biodiversity loss are further complicated by the effects of climate change (Figs.1-2), which may act to amplify other threats, and which itself may be amplified by other threat factors (such as wood burning and subsequent release of greenhouse gases-Brook et al., 2006). Fires present a major threat to biodiversity in the region, and during the past decade major fires have started progressively further north in response to climate change (Taylor et al., 1999). Even without considering of many of these factors, projections of the number of extinctions have been made, which project the extinction of 43% of endemic Indo-Burmese fauna within the next century (Malcolm et al., 2006). Thus despite harbouring considerable biodiversity, few areas in SEA have sufficient levels of protection, and with many new species still to be found (as demonstrated by the rapid rate of discovery (Giam et al., 2010)) it is currently almost impossible to determine the most effective means of conservation prioritisation within SEA given the level of knowledge of much of the fauna, and high levels of corruption (Global Witness, 2007).

Some conservation biologists have advocated the use of "indicator species" to monitor more general threats to biodiversity (Carignan & Villard, 2002). Chosen species must obviously be sensitive to the potential threats in the area, and such species must be possible to monitor in a standardised and repeatable way to generate meaningful and comparable data over large spatial and temporal scales. Indicator species can also be used to indicate trends in overall biodiversity (Mace & Baillie, 2007) and therefore provide a gauge of biodiversity change at large regional scales over time. Bats provide an ideal indicator group (G. Jones et al., 2009), and their diversity means that species can be susceptible to a wide variety of different threats. Bats form a large component of bush-meat through SEA (Mickleburgh et al., 2009), and many of these species often perform vital roles within ecosystems and their loss could have negative implications for a wide range of interacting taxa (Mohd-Azlan et al., 2001). A number of ecosystem services are provided by bat species, including pollination, seed dispersal and insect control, and therefore bats are frequently keystone species (Myers, 1987; Fujita and Tuttle, 1991; Hodgkison et al., 2004). Effective conservation of these keystone species is crucial not only for their survival, but for the ecosystems dependent upon them. Furthermore many bat species are either dependent on forests or caves for foraging and roosting, and some species have limited dispersal ability (Kingston et al., 2003), suggesting that their status may be indicative of destruction and consequent fragmentation of both karst and forest areas.

To try to reduce impacts of the Southeast Asian biodiversity crisis requires a number of steps: quantification of how species are distributed and their distribution changes, analysis of the threats each species faces and determination of the probable impact of threats they are likely to face. Only once these initial steps have been achieved is it possible to formulate effective impact mitigation strategies. Though in this chapter bats will provide the main case study (due to their potential as indicator species) most of what will be discussed here is broadly applicable for the conservation of biodiversity throughout SEA, and in developing strategies for mitigating species loss in other regions of the world which faces similar issues to those discussed here.

3. Identifying species and distributions

Although over 320 species of bat are currently described from SEA (Simmons, 2005; Kingston, 2010) research in the area has been sporadic and the rate of species discovery is now high for not only bats (Bumrungsri et al., 2006), but across many other taxa (Duckworth & Hedges, 1998; Bain et al., 2003; Giam et al., 2010). Recent research has revealed that many bats previously regarded as one species are in reality complexes, comprising a number of cryptic species (Soisook et al., 2008, 2010; Francis et al., 2010). Therefore before any conservation measures can be put in place the distribution and status of current species must first be established. SEA has some of the highest diversity of bats on the planet in addition rate of species discovery (Simmons & Wetterer, 2011). A projection of the species richness of 171 species throughout SEA (Fig.1) shows that most forested regions still retain high species richness, and therefore present priority regions for research.

However recent research has clearly demonstrated that currently known SEA bat species only represent a fraction of total species numbers (Francis et al., 2010; Giam et al., 2010; A.C.Hughes et al., in prep a). Both recent taxonomic and genetic research show that much further work is needed in order to identify all species in the region, and similar trends are liable to exist across biotic groups. Species identification is clearly a priority, because it is impossible to try to develop effective conservation strategies when there is little understanding of the true ranges of many species; and when species currently classified as showing large distributions are in actuality made up of a number of cryptic species with small ranges and much smaller populations (A.C.Hughes et al., in prep a). Both taxonomic (Soisook et al., 2008, 2010) and genetic work (Francis et al., 2010) demonstrate that there are many currently undescribed and potentially cryptic species throughout SEA.

Methods used to determine species present obviously involve detailed taxonomic surveys (as advocated by Webb et al., 2010), in addition to genetic analyses where possible. However other protocols for species identification and monitoring may also be valuable components of species discovery in some taxa, such as the use of call analysis to identify cryptic bat species (e.g. G. Jones & Van Parijs, 1993). In such cases the identification of potentially cryptic species may begin with call analysis, as was recently found to be the case in Hipposideros bicolor, (Douangboubpha et al., 2010). Acoustic monitoring also provides a means of potentially monitoring population trends as well as identifying possible cryptic species (K. E. Jones et al., 2011). Two protocols have recently been developed which describe the potential for using localised call libraries for identifying bat species in SEA (A.C.Hughes et al., 2010, in press). Once acoustic identification libraries have been developed then acoustic surveys and inventories of surrounding regions (e.g. 1° of the areas used to develop the library) can be made to identify species present (using discriminant function analysis) and the presence of species outside their known range. The presence of novel call variants could cue and promote further research to determine if sub-species or cryptic species are present, and the spatial distributions of call variants of some species suggests spatial segregation which could denote cryptic species (A.C.Hughes et al., in prep a). Monitoring surveys are also essential to determine distribution and population trends, however funds and specialists are not always available to carry out this valuable work when it requires repeated taxonomic surveys and specialist knowledge. Acoustic analysis and monitoring only requires specialists initially, during the creation of acoustic libraries, and surveys can then be carried out by non-specialists or automated software programs (K.E. Jones et al., 2011). Thus protocols such as these provide a viable means of both identifying species present and subsequently monitoring trends, and may be able to detect variation over shorter periods than in trapping-based monitoring which has been previously been advocated (Meyer et al., 2010). Acoustic surveys are currently limited in species coverage, and are biased towards bat taxa that use high-intensity echolocation calls. Acoustic surveys are therefore best used side-by-side with conventional survey techniques such as using mist-nets and harp traps in a standardised manner (MacSwiney G et al., 2008). However invasive trapping techniques are expensive and require highly trained experts, whereas acoustic surveys can be carried out with little training and recordings can then be forwarded to highly trained researchers for analysis, or analysed by software to provide standardised and comparable data for any region. If initially surveys combine both trapping and acoustic techniques to establish acoustic libraries within a given area then those libraries can subsequently be employed to monitor trends in many species across wide areas. The use of common species as indicators for abundance and distribution of rarer species has been found to be accurate in previous studies, as correlations have been found in the trends of common species with other species present (Pearman et al., 2010). Therefore even if acoustic surveys cannot cover all species, the trends in the distributions and populations of common species may still be more widely applicable.

Logistical constraints also mean that it is not always possible to survey all areas in a region, and thus methods which determine range based on limited spatial knowledge of an organism's total distribution provides a valuable tool when applied properly (i.e. predictive modelling approaches, Box 1, Fig. 1). Former distributions of species and zoogeographic constraints must also be considered and included in analyses of species distributions. Within SEA the geophysical history is to a large extent responsible for the current patterns of diversity and species' distributions, and thus analyses of present species distributions cannot be conducted without by making reference to the past (Woodruff, 2003). The connections and separations of the various parts of SEA during past time periods not only influence current distributions but further constrain possible responses to future change. A zoogeographic transition in the distributions of some animal groups centred around the Isthmus of Kra has persisted for over a million years (De Bruyn et al., 2004). Recent analyses (A.C. Hughes et al., 2011) show that although breaks in the distribution patterns of bats are apparent along the Thai peninsula, they occur not only at the Isthmus of Kra and are influenced by climatic discontinuities in conjunction with biogeographic consequences associated with the narrow breadth of the peninsula; and it is probable that these circumstances have also caused divisions known to occur in the distributions of other taxa in the region (J.B. Hughes, et al., 2003). Zoogeographic transitions have persisted over long time periods along the peninsula because the position of climatic boundaries appears remarkably constant. Climatic discontinuities continue to affect the distributions of species, and will also affect how effectively species can respond to climatic change in the future.

Identification of species present, their ranges and trends in distribution and population form an important first step in the development of effective conservation plans. Once these steps have been fulfilled then threats to current distributions and diversity can be analysed (Fig. 1) and necessary conservation actions planned.

3.1 Assessing and quantifying threats to current diversity, and determining impacts

Analyses have previously shown that species richness is negatively related to human population density (A.C. Hughes et al., in prep b), and therefore further increase in human population size is likely to have detrimental effects on bat biodiversity. Projections suggest that human populations will continue to increase until at least 2050 and further urbanisation is likely throughout SEA (CIESEN, 2002; Gaffin et al., 2003; United Nations Population Division, 2008; Seto et al., 2010). Larger human populations impinge on biodiversity in a number of ways: through increased demand for wild-sourced products and via higher pollution (Corlett, 2009; Peh, 2010). Urbanisation and increasing deforestation also increase the potential for invasive species to spread throughout SEA (Riley et al., 2005) and further work is necessary to determine the effects of invasive species on the native fauna.

Forest fragment size correlates positively with bat species richness (Struebig et al., 2008; A.C. Hughes et al., in prep b). As deforestation is projected to increase throughout most of SEA, including in "protected areas" (Fuller et al., 2003), this trend is likely to lead to progressive loss of species richness in many areas due to the increased fragmentation of large forest patches. Currently many protected areas fail to offer protection, and are subject to both high hunting pressure (Steinmetz, et al., 2006) and deforestation (Fuller et al., 2003). Heightened accessibility of parks and involvement of rangers may indeed lead to greater pressures within National Parks than in other forested regions. Many regions were predicted to have high species richness during this study, however many forests have been described as showing "empty forest syndrome" (Redford, 1992; Tungittiplakorn & Dearden 2002). Therefore although many areas may be suitable for certain species, they are overexploited by humans, and do not contain the native fauna previously held. Empty forest syndrome and overexploitation have serious implications for a wide range of species: rodents are the most "harvested" taxa, followed by bats, and almost all bat species in SEA are eaten (Mickleburgh et al., 2009). The loss of species due to hunting has implications for the entire ecosystem. Frugivorous and nectarivorous bats, large bodied mammals and birds all have essential functions in seed dispersal and pollination and fulfil vital ecosystem services, yet such species are often the most threatened by human hunting activities. If such species are lost, there may be negative consequences for the entire ecosystem. Yet these animals are among the most hunted organisms in the region (Wright, et al., 2007; Corlett, 2008; Brodie et al., 2009).

When projections of the distribution of bats under future climatic scenarios are made, three broad outcomes can be noted (Fig. 2) (A.C. Hughes et al., in review). First almost all species are projected to show reductions in original range under future scenarios and second, most species are projected to move north. The third probable outcome is the large projected loss of species (up to 44) from areas currently predicted to have the highest levels of species richness (figs 1-2). Though some species were projected to show expansions in original range, this is unlikely to be logistically possible due to the limited dispersal abilities of many species (Struebig et al., 2008). This loss in species richness is based on climate change alone, and therefore is a conservative projection, and though it is possible to prevent the loss of species due to deforestation in protected areas it is not possible to prevent species loss due to climatic change. Forest is becoming increasingly fragmented even within "protected areas", and mining rates in SEA are the highest in the tropics (Day & Ulrich, 2000). Mining not only destroys important roost sites (Clements et al., 2006), but also degrades areas and increases accessibility to previously remote areas (which in turn facilitates deforestation, McMahon, et al., 2000; Laurance, 2008a). Therefore not only are current suitable habitat and roosts being destroyed, but the distance between suitable areas may actually be increasing for the same reasons. Other factors such as fires are also prevalent through SEA, and fires have increasingly been found to move north in response to climatic change, therefore posing an increasing threat to the biodiversity of SEA (Taylor et al., 1999). Projections of total biodiversity loss currently estimate the extinction of up to 85% of current biodiversity in SEA within this century (Sodhi et al., 2010). However the estimates of undiscovered species show that we may potentially have only discovered around half of the species in many orders (Giam et al., 2010) and only around 40% of bat species (A.C. Hughes et al., in prep a). Groups containing cryptic species are likely to have particularly high numbers of undiscovered species, and this is highlighted in bats by recent genetic work (Francis et al., 2010). Species with smaller distributions are more likely to have specialist requirements (limiting overall distribution), and will be more susceptible to loss of range and therefore have a higher probability of extinction (Kotiaho et al., 2005). Hence many species currently regarded as widespread, and thus of "Least Concern" by the IUCN may comprise a complex of cryptic species each of which will show higher categories of threat. As almost all species analysed here (fig. 2) showed a loss in original habitat in all scenarios, and many of those species may be species complexes it is likely that impacts for many of the species will be worse than estimated during this study (fig. 2). Projections here (Fig. 2) only account for climate change, but cannot consider hunting, fires, mining and the plethora of other threats. Fungal diseases have recently devastated populations of North American bats (Blehert et al., 2009), in addition to South American frogs (Berger et al., 1999). Moreover the spread of pathogens has been associated with temperature change, for example the spread of chytrid fungus is believed to be related to global warming (Pounds et al., 2006; Boyles & Willis, 2010). Therefore the effect of climate change on species is dynamic and complex, as it has both direct and indirect implications for distributions and populations of all species. Furthermore climatic changes have already been shown to cause changes to the distribution of different biomes (Salazar et al., 2007), and hence has profound implications for species within those biomes.

SEA is currently in the midst of a biodiversity crisis which has been described as a 6th mass extinction (Myers, 1988). There are some undeniable implications of the current threats, and others such as the possibility of 'no-analogue' communities (Stralberg et al., 2009) and the effect of invasive species, which are less certain. However native species are likely to attempt to either migrate north spatially, or move to higher altitudes (Malcolm et al., 2006). Continued decreases in the patch sizes of rainforest will decrease species richness, and increasing accessibility for humans will increase the probability of hunting within areas. Increases in human population will negatively affect biodiversity, if current unsustainable practices continue. Not only is the modification of human activities necessary to decrease further species loss, but human intervention is necessary to allow species any opportunity to respond effectively to climatic changes. The methods to mitigate possible threats require detailed evaluation to try to curb species extinctions.

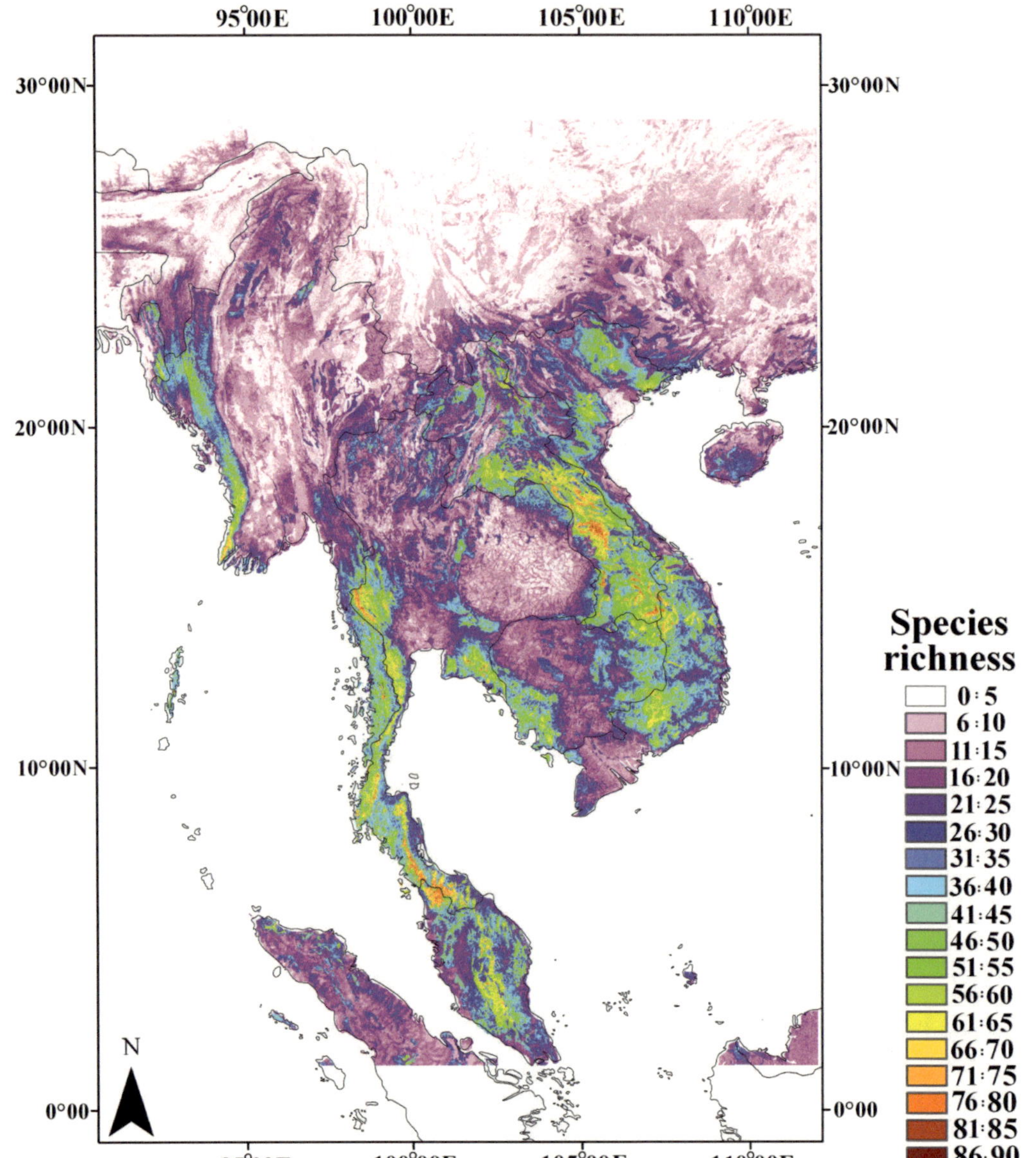

Fig. 1. The current projected species richness for 171 Southeast Asian bat species on a km² basis. Projections were generated using Maxent, methods are shown in Box 1. Environmental variables used in projections are included in Appendix 1.

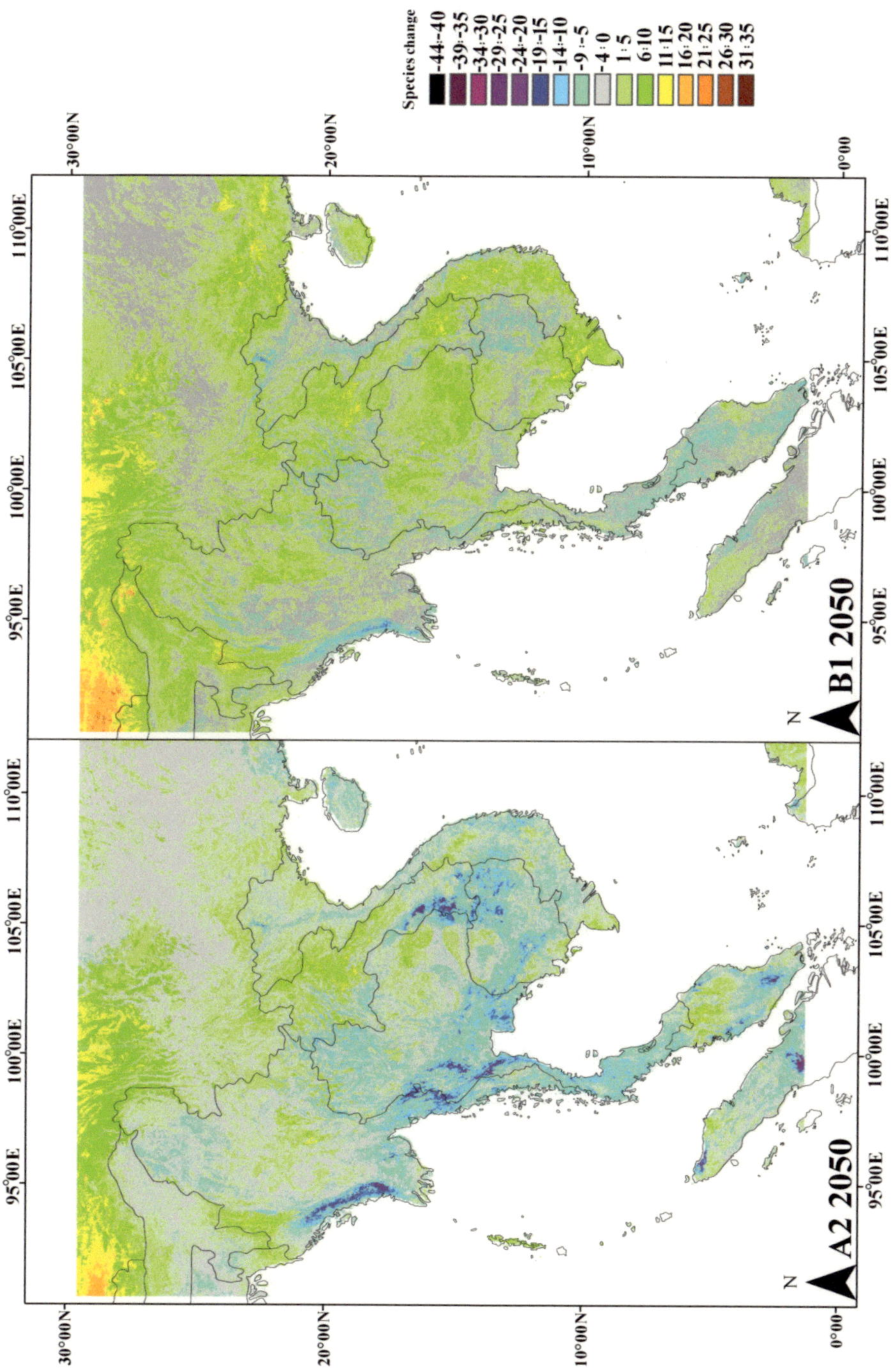
Species change
-44:-40
-39:-35
-34:-30
-29:-25
-24:-20
-19:-15
-14:-10
-9:-5
-4:0
1:5
6:10
11:15
16:20
21:25
26:30
31:35
A2 2050
B1 2050
N
95°00E
100°00E
105°00E
110°00E
30°00N
20°00N
10°00N
0°00

Fig. 2. A-B.

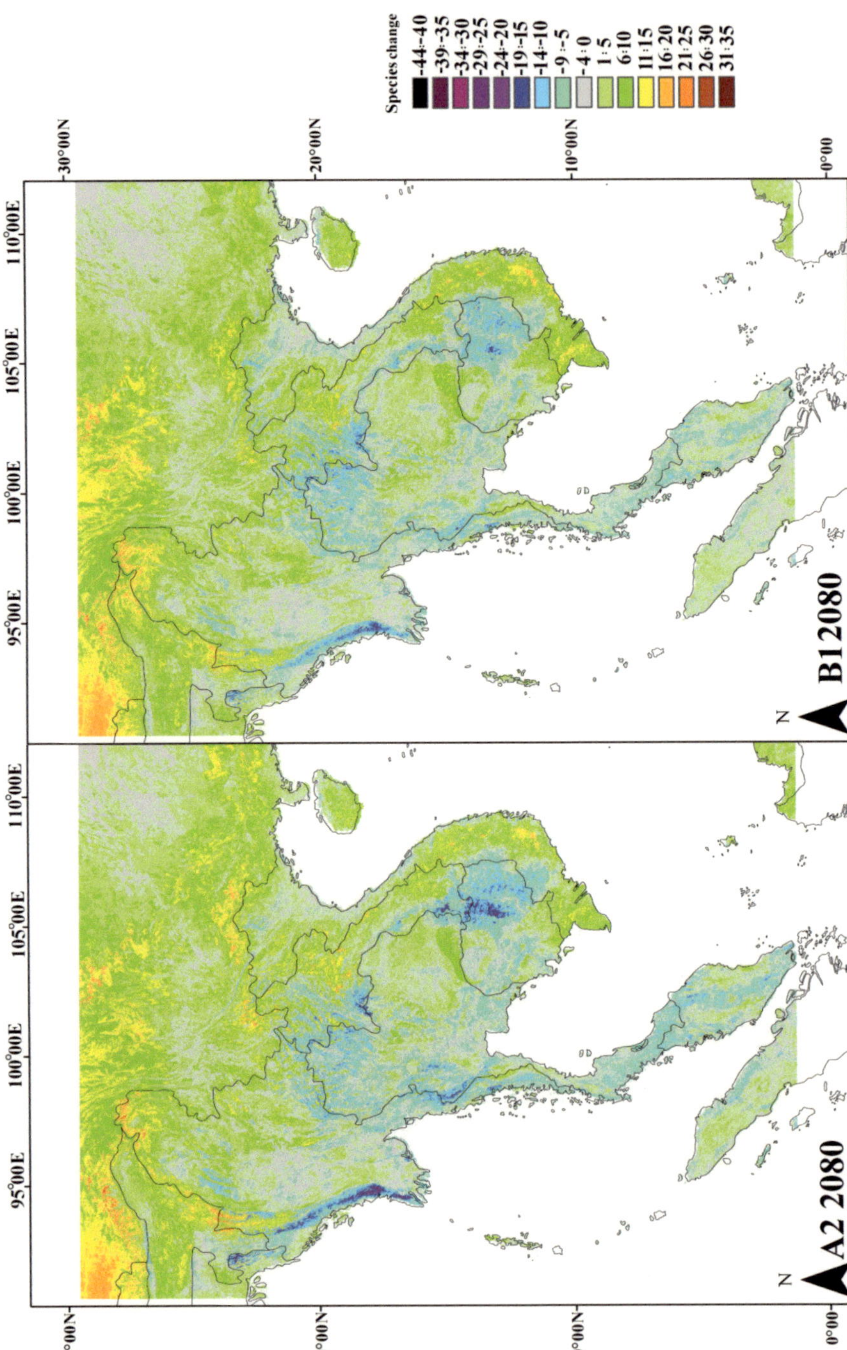

Fig. 2. C-D. These maps display the projected change in the number of bat species under the A2 and B1 climate change scenarios produced by the IPPC. A2 represents the most severe of the climate change scenarios and B1 the mildest. Many regions are projected to lose between five to nineteen species, with some regions projected to lose up to forty-four species.

Predicting species richness by pairing the known distribution of each species with environmental parameters to determine the habitat requirements of each species, (and thus distribution) can help inform and target research and conservation (Box 1). Such models can also aid conservation planning under probable future scenarios, but are targeted to specific questions and can only incorporate some dimensions of ecosystems and must therefore only be interpreted while acknowledging inputs, assumptions and limitations. Models are a powerful tool for predicting the effects of climatic, land-cover and direct anthropogenic change on species richness (if these anthropogenic drivers have been projected). What such models are less good at is incorporating the biotic dimension, the inter-dependence of some species, and temporal interactions such as the flowering of trees and breeding of organisms (L. Hughes, 2000) which can cause resource asynchrony. More complicated ecological interactions and phenomena cannot yet be incorporated in the building of models, but should be included in the interpretation of results; and thus both dimensions of possible ecosystem change can be used to inform and develop appropriate conservation strategies.

Figure 2 shows projections of the effects of climatic change under two potential climate change scenarios (the mildest, and the most severe). Extensive regions are projected to lose between five to nineteen species, with some regions projected to lose up to forty-four species. Though some regions, especially in Northern regions are projected to increase in the number of species, this result should be interpreted with caution for a number of reasons. Firstly these projections only include climatic change, and do not reflect changes in land-cover, and secondly many species are dispersal limited and therefore will not to show the expansions projected here. Also the Northern areas of the projection, which are predicted to gain species here are liable to lose species which were not included in these projections, (which include predominantly tropical dwelling species). These projections are highly conservative, for they only show climate mediated loss of species, this is only one driver of species loss and thus the loss of species in these scenarios is liable to represent a fraction of that when all factors are considered.

4. Mitigating species loss

There are at least three issues that must be addressed if biodiversity is to be most effectively conserved throughout SEA: identification of species and their distributions (Section 3), decreasing the impacts of current threats, and creating ways to allow species to respond to climate change (because halting further climate change is considered impossible, Bowen & Ranger, 2009; Vistor et al., 2009). Each issue requires different actions in order to respond effectively.

As stated previously, accurate species identification requires thorough systematic surveys, taxonomic and acoustic studies, and where possible genetic research. However the scale of this work requires the use of university researchers, students and park rangers. The use of citizen science for survey and monitoring has been advocated by some researchers (Webb et al., 2010). However citizen science is plagued with potential problems in SEA: not only is hunting exceedingly popular, but in many taxonomic groups' cryptic species and the lack of adequate taxonomic knowledge precludes species surveys by non-specialists. However education, and enthusing of the population could allow some citizen science in distinct and recognisable species. School children in some parts of SEA also must complete science projects whilst at high school, and with little training such projects could contribute to this knowledge pool (Sara Bumrungsri pers. comm.). However for successful citizen science to be conducted, people must be educated to the importance of species such as bats (as throughout SEA bats are generally viewed negatively by the public, Kingston et al., 2006). Nature recreation has been implemented in schools, and provides an important means of enthusing the next generations about biodiversity and engendering greater respect for the environment (Pergams & Zaradac, 2008). Education is of paramount importance in the realisation of any level of conservation or mitigation. Without the support and backing of local people no changes to current activities will take place. Projections of species distributions, like those within this study - and subsequent ground truthing (validation and testing) by trained surveyors - can also provide a focus for further research and conservation activity.

Many strategies have attempted to decrease anthropogenic impacts on biodiversity. Recent studies have projected the species richness patterns of bats throughout SEA (Fig 1) (Hughes et al., in review; in prep b), and the regions of high species richness obviously provide a focus for conservation efforts. However within the scientific community there is great dispute as to what criteria should be used to assign conservation priorities. There is debate as to whether regions, species richness, evolutionary uniqueness and richness, numbers of threatened species or specific species should be used in area prioritisation (Corlett, 2009). Under the current circumstances it is not feasible to conserve on a single species basis, because this approach is financially unviable, and it ignores interactions within ecosystems. Furthermore there is currently inadequate knowledge to reliably designate IUCN threat levels for many species throughout SEA, due to the lack of knowledge about the distributions and population sizes of many species, and the presence of cryptic species. However SEA is regarded as an area of both evolutionary, and species richness (Gaston, 1995b; Sechrest, et al., 2002). Most currently species-rich areas are also liable to contain high levels of intraspecific genetic diversity as populations of most species have been predicted to have expanded during the last glacial maximum (LGM), and because current ranges may be restricted compared with ranges occupied during the LGM in tropical areas and may overlie former glacial distributions (Woodruff, 2010). Current species populations within areas of former glacial refugia often contain high genetic diversity in comparison to those in non-refugial areas (Anthony et al., 2007), and genetic heterogeneity and diversity is known to make populations more robust to environmental change and therefore to increase the capacity of such populations to adapt (Aitken et al., 2008). Therefore former refugial areas, many of which fall within current National Parks deserve prioritisation on all grounds. However the current system of National Parks fails to function in many regions (Fuller et al., 2003), and with the high levels of corruption (Global Witness, 2007; EIA/Telepak, 2008) the enforcement of laws such as those governing reserves is difficult.

A variety of schemes and approaches have been developed to try to promote biodiversity conservation and decrease deforestation. The following section evaluates some of these methods in an attempt to formulate a viable method of mitigating biodiversity loss.

Paying for Environmental Services (PES) is one scheme suggested for conservation (Blackman & Woodward, 2010). PES schemes use money generated by environmental service users to pay people who own an area which is (in part) responsible for the service, in order to maintain forest/ biodiversity within the area responsible for that service. For example, forest cover in watersheds may be preserved by using the income generated by hydro-electric dams. Such approaches have potential but must be closely tailored to each site and country, to be economically viable both for those responsible for the maintenance of the area which provides the ecosystem service and those who profit from the service. These schemes have a great number of potential pitfalls which have prevented their success in some areas (Wunder, 2006, 2007). PES-type schemes are obviously unsuitable when the service users earn less than the ecosystem service users, such as the cases of guano miners and durian growers (both of whom have income streams dependent on cave bat populations) and people whom mine karsts (and therefore are responsible for the resource). In situations involving the mining of karsts, determining who should pay for environmental services is difficult, as the income of the miners (who may own the karst) may be higher than those who benefit from bat related services (Wunder, 2006). However when PES-type schemes are well-tailored and targeted to specific areas, they can effectively protect forests, and stipulations can place more emphasis on biodiversity rather than solely forest, as in the case of some "engineered PES" schemes (Wunscher et al., 2006).

Carbon offsets (carbon credits) and the REDD (Reduced Emissions from Deforestation and Forest Degradation) systems also provide a means to fund forest protection (Laurance, 2008b). Afforestation can also be part of such schemes, but in some existing schemes this has included the use of non-native trees. If biodiversity is to be protected it is important that afforestation uses only native species (Corlett, 2009). Afforestation schemes are currently the subject of much debate, however when well applied they have the potential to both decrease rates of biodiversity loss and to mitigate climate change (Canadell & Raupach, 2008). Biodiversity offsets have also been used in some regions (i.e. Uganda), however in some areas (i.e. the USA) the heightened protection of one area has led to greater biodiversity declines elsewhere and thus yielding no net benefit to conservation overall (Ten Kate et al., 2004). Therefore education is necessary alongside offset schemes in order to attempt to prevent greater pressures being deflected elsewhere as a result of conservation within one area (conservation leakage, Gan & McCarl, 2007). Problems also arise when it comes to prioritising areas for conservation based on current risks alone. Hence it is valuable to predict future scenarios based on land cover and climate in assessing conservation priorities (Fig. 2). Although future risk should be part of any assessment criterion, assessment must also analyse other factors, so even if environmental pressure is deflected to other areas as a result of conservation in a particular area: that the most important areas (in terms of biodiversity/uniqueness) are adequately protected (Laumonier et al., 2010). Risk and enforcement can also be projected together and the combined effects predicted to generate the most effective means of minimising deforestation or biodiversity loss within an area (Linkie et al., 2010).

The protection of specific areas still requires funding, as National Parks are currently ineffective in many regions of SEA (Fuller et al., 2003; Steinmetz, et al., 2006). In other countries (i.e. Costa Rica) ecotourism has provided a highly successful means of funding biodiversity protection and educating local people about the value of biodiversity (Jacobson & Robles, 1992; Aylward et al., 1996). Currently although ethnotourism is popular (Zeppel, 2006), ecotourism in mainland SEA is mainly dominated by bird watching tours (Mollmann, 2008). Ecotourism has been shown to work well in parts of Malaysia and Indonesia (Hill et al., 2007, Pearce et al., 2008), and if it were to develop throughout SEA it could provide a viable means of conservation.

Multiple models exist to spatially project species probable distributions used limited spatial data, and in recent years the use of such models has increased dramatically; in 1999-2004 only 74 published studies used species/niche distribution models, however between 2005-2010 this increased to over 850 (Beale and Lennon, in review). Clearly such models represent useful tools for projecting species distributions, and can further allow targeted conservation to either species habitat requirements or the prioritization of areas for research or conservation (Pawar et al., 2007; Sergio et. al., 2007). Recent developments in habitat suitability modelling allow the prediction of a species' potential distribution based on presence-only records (e.g. Hirzel et al., 2002; Phillips et al., 2006). Presence-only modelling is a valuable tool in contemporary conservation biology, and has been applied to a wide range of taxa, from bryophytes (Sergio et. al., 2007) to reptiles (Pawar et al., 2007). Presence-only modelling may be more reliable than presence-absence models for species in which absence records cannot be reliably gathered (i.e. failure to capture a species at a site does not necessarily mean the species is absent- Wintle et al., 2004; MacKenzie, 2005; Elith et al., 2006; Jime´nez-Valverde et al., 2008). One presence-only modelling method that is used widely (Maxent - Phillips et al., 2006) involves maximum entropy modelling and has been used successfully to predict the distributions of bat species in both present day conditions (e.g. Lamb et al., 2008; Rebelo and G. Jones, 2010) and under projected climate change scenarios (Rebelo et al., 2010). Additionally Maxent has been found to be robust to changes in sample size, and still have good predictive ability at low sample sizes, making it the ideal model for the prediction of distributions for rare species (Hernandez et al., 2006; Wisz et al., 2008).

Figures 1 and 2 both use Maxent to project the distributions of 171 bat species for a number of time periods. By pairing known distribution coordinates each species has been recorded at with appropriate environmental variables it is possible to project the probable distribution of each species for any time period for which spatial data exists, and to combine these to calculate species richness (see Hughes et al., In review, for a full account of methods used).

Using projections of future climatic change it is possible to project the probable impacts and develop targeted solutions and effective conservation methods (Prentice et al., 1992; Beerling et al., 1995; Huntley et al., 1995; Sykes et al., 1996; Berry et al., 2001, 2002; Hannah et al., 2002; Midgley et al., 2002). Though improvements in modelling approaches in the future will allow further insights to be generated, such models will take time to be developed and refined. In many areas (such as Southeast Asia) with rapid rates of deforestation, prioritisation of key areas is required to protect areas of high conservation value from deforestation and modelling can facilitate the determination of these priority areas in a region of high conservation importance (Pawar et al., 2007; Sergio et. al., 2007; Gibson et al., 2010).

Box 1. Mapping species distributions using distribution models.

Other countries (e.g. Brazil) with large export markets have also started to produce certified wood for a large proportion of their exports, however few certifications have sufficient biodiversity emphasis (McNeely, 2007). Attempts at certification programs throughout much of SEA have met with little success, as most logged wood is used within the country, and people are not prepared to pay increased prices involved with establishing and maintaining certification (Cashore, et al., 2006; Laurance, 2008b). Until local people value the natural environment, or exports increase, certification will remain an unsuitable scheme for much of SEA. It may be for similar reasons that previous integrated conservation and development projects have met with little success (in terms of impact) despite large-scale investment throughout SEA (Terborgh et al., 2002; McShane & Wells, 2004). Community-based conservation schemes have also been little used outside marine national parks, and their use may be unsuitable for many areas (Gray et al., 2007).

Certification is unlikely to work within SEA, and logging is liable to continue within natural forests (Fredericksen & Putz, 2003). The use of "reduced impact logging" could at least provide a means of providing both humans and biodiversity with a means of existence (Sessions, 2007; Putz et al., 2008). Reduced impact logging would require less human behavioural modification than stopping altogether or certification, and if local people can be educated to perceive it as an efficient way of logging, which preserves ecosystem services then it may provide a means of conservation. However as most logging which takes place in SEA is illegal, enforcement of laws is first essential (EIA/Telepak, 2008). Enforcement is also necessary to restrict hunting, and requires not only education but an enforced system of permits to control it. Logging programmes must also consider that the removal of the most mature trees may have negative consequences for those bat species that roost under bark and in other species which dwell in holes of mature tree (Gibbons & Lindenmayer, 2001; Kunz & Lumsden, 2003; Barclay & Kurta, 2007). As these forest-dwelling species are often the most limited in dispersal abilities, they are liable to suffer most from deforestation (Kingston et al., 2003).

Therefore in the protection of existing highly biodiverse areas, and to prevent an increasingly fragmented landscape further reducing biodiversity (A.C. Hughes et al., in prep b) education and law enforcement are paramount. Well considered funding systems also provide a good opportunity for decreasing biodiversity loss, and ecotourism if well developed could remedy both habitat destruction and overhunting. These are the primary means for protecting areas from anthropogenic direct threats.

4.1 Mitigating the effect of climatic changes on biodiversity

Recently developed models predicted that bat species would both lose areas of suitable habitat in their original range, and would often need to move north if they were to remain in similar niches in response to climatic change (A.C. Hughes et al., in review). However in order to adapt, species must be able to reach suitable habitat. Translocation, and assisted migrations are often put forward as ways of accomplishing this (McLachlan et al., 2007). However many species face the same threats, and so how could species be selected for translocation: by uniqueness, charismatic mega fauna, ecological role or extinction risk? Too many species face the same situation, and too little information exists on many to make translocation a viable solution. Even for species selected as candidates for translocation, (due to IUCN status, or other factors) some species react poorly and show poor survival following translocation (Weinberger et al., 2009). Consequently translocation is not a practicable solution, both due to the number of species that face threat and the variability in the reaction to translocation in particular species, in addition to the financial cost. Human-mediated adaptive strategies should allow species to shift ranges in response to climate change (similar changes have occurred naturally during previous periods of climate change (Hickling et al., 2006; Lenoir et al., 2008)). Movements can be assisted by increasing landscape connectivity, by creating corridors of native forest between existing forest patches particularly in a north-south orientation (Heller & Zavaleta, 2008). These areas must be wide enough so not to act as population sinks, and must contain heterogeneity of both species and genetic variation in order to be viable and sustainable (Lamb et al., 2008, Lamb & Erskine, 2008; Kettle, 2010). Hence there is a need for careful matching of tree species to soil type and area between sites, and corridors should also contain site-appropriate plants including nitrogen-fixing legumes to increase canopy density (Siddique et al., 2008; Suzuki et al., 2009). Afforestation has begun in many countries (UK, Vietnam; McNamara, et al., 2006; McNamara, et al., 2008), and if it is used to connect areas it will give species a higher probability of responding effectively to climatic change, by allowing the species to expand their ranges north as detailed in predictive models (fig.2) and studies in other regions (Malcolm et al., 2006) and therefore not suffer severe reductions in overall range.

5. Conclusion

SEA represents one of the most biodiverse regions on the planet, yet throughout SEA species are at risk due to dynamic interactions between numerous threats, including both direct and indirect drivers of human mediated biodiversity loss. In order to have any chance of preserving a fraction of the current fauna, major changes are needed in human activities, which requires education of people throughout SEA and the minimisation of corruption at all levels. Only if people can gain from the preservation of current biodiversity can it remain, and therefore schemes that use the environment in a sustainable manner present ways for affecting change. Even under fairly minimal impact scenarios modelled, almost all bat species lost original habitat (up to 99%), and many will be unable to reach new suitable areas (A.C. Hughes et al., in review). To allow species to respond to climate change without going extinct will require not only the cessation of destructive activities, but the active intervention of humans to create forested corridors between current forests to allow species an opportunity to reach suitable habitat under changing conditions.

There is no doubt that even with direct conservation action, climatic change and direct environmental change will lead to the loss of species, some as yet undescribed. What cannot yet be quantified is the number of species which will become extinct during the next century, because the number of extinctions is under the direct control of human choices and actions made now. At this point in time humans do have an opportunity to reduce the impacts of destructive human activities and mitigate the effect of climatic change through effective and considered conservation activities, but with further inaction we as a species increase the total number of other species that will become extinct due to our unsustainable human activities.

6. Acknowledgments

I would like to thank my collaborators at Prince of Songkla University, Hat Yai, especially Sara Bumrungsri and Chutamus Satasook, and Paul Bates at the Harrison Institute. I must also thank the British Council for funding for field research which was used in making projections. Additionally I wish to thank those in my laboratories at the University of Bristol and Prince of Songkla University for their support during research.

7. Appendix 1.

Variables included in species distribution models:
Vegetation cover: Globcover-Ionia (http://ionia1.esrin.esa.int/)
Mean annual temperature, minimum and maximum annual temperature, minimum and maximum mouthy precipitation, total annual precipitation, isothermality: www.worldclim.org
Humidity: New et al. 1999 (http://atlas.sage.wisc.edu/)
Elevation: NGDC http://www.ngdc.noaa.gov/mgg/topo/globe.html)
Soil pH :ISRIC-WISE (www.isric.org/)
Distance from waterways and distance from roads: Edited from U.S. Geological Survey (USGS- www.usgs.gov/)
Karsts: Karst portal- School of Environment, University of Auckland, (http://web.env.auckland.ac.nz/our_research/karst/)
Geology :CCOP-Coordinating Committee of Geoscience Programmes in Asia and Southeast Asia (www.ccop.or.th/),Prince of Songkla University's GIS centre, Ministry of Mining in Myanmar.
Human population density: Ciesen (Grump v1: http://sedac.ciesin.columbia.edu/gpw/)
A2 and B1 future climate scenarios: CIAT-GCM (Centro Internacional de Agricultura Tropical-Global Climate Model, - CSIRO-Mk2.0 model: http://ccafs-climate.org/)

8. References

Aitken, S.N.; Yeaman, S.; Holliday, J.A.; Wang, T. & Curtis-McLane, S. (2008). Adaptation, migration or extirpation: climate change outcomes for tree populations, *Evolutionary Applications* Vol.1, No.1, pp.95–111, ISSN 17524563

Anthony, N.M.; Johnson-Bawe, M.; Jeffery, K.; Clifford, S.L.; Abernethy, K.A.; Tutin, C.E.; Lahm, S.A.; White, L.J.T.; Utley, J.F.; Wickings, E.J. & Bruford, M.W. (2007). The role of Pleistocene refugia and rivers in shaping gorilla genetic diversity in central Africa. *Proceeding of the National Academy of Sciences USA,* Vol.104, No.51, pp.20432–20436, ISSN 0027-8424

Aylward, B.; Allen K.; Echeverria J. & Tosi, J. (1996). Sustainable Ecotourism in Costa Rica: The Monteverde Cloud Forest. *Biodiversity and Conservation,* Vol.5, No.3, pp.315-43, ISSN 0960-3115

Bain, R.H.; Lathrop, A.; Murphy, R.W, Orlov, N.L. & Ho, C.T. (2003). Cryptic species of a cascade frog from Southeast Asia: taxonomic revisions and descriptions of six new species. *American Museum Noviates,* Vol.3417, pp.1-60, ISSN 0003-008

Baltzer, J.L.; Davies, S.J.; Bunyavejchewin, S. & Noor, N.S.M. (2008). The role of desiccation tolerance in determining tree species distributions along the Malay-Thai Peninsula. *Functional Ecology,* Vol.22, No. 2, pp.221-231, ISSN 0269-8463

Barclay, R.M.R.; Kurta, A. (2007). Ecology and behaviour of bats roosting in tree cavities and under bark. In: *Conservation and management of bats in forests*, Lacki, M.J.; Kurta, A.; Hayes, J.P. (Eds) pp 17-49. IBSN-10: 0-8018-8499-3, Baltimore, MD Johns Hopkins University Press.

Beale, C. M. & Lennon, J. J. (*In review*). Incorporating uncertainty in predictive species distribution modelling. *Philosophical transactions of the Royal society B.*

Beerling, D. J.; Huntley, B. & Bailey, J. P. (1995). Climate and the distribution of *Fallopia japonica*: use of an introduced species to test the predictive capacity of response surfaces. *Journal of Vegetation Science*, Vol.6, No.6, pp.269–282, ISSN 1100-9233

Berger, L.; Speare, R. & Hyatt, A. D. (1999). Chytrid fungi and amphibian declines: overview, implications and future directions. In: *Declines and disappearances of Australian frogs. Environment Australia*, A. Campbell (Ed.) pp. 23-33. Canberra.

Berry, P. M.; Dawson, T. P.; Harrison, P. A. & Pearson, R. G. (2002). Modelling potential impacts of climate change on the bioclimatic envelope of species in Britain and Ireland. Global Ecology and Biogeography, Vol.11, No.6, pp.453–462, ISSN 1466-822X

Berry, P. M.; Vanhinsberg, D.; Viles, H. A.; Harrison, P. A.; Pearson, R. G.; Fuller, R.; Butt, N. & Miller, F. (2001). Impacts on terrestrial environments. In: Climate change and nature conservation in Britain and Ireland: modelling natural resourse responses to climate change (the MONARCH Project). P. A. Harrison, P. M. Berry and T. P. Dawson (Ed), pp. 43–149. ISBN 978-1-84754-026-3, UKCIP technical report, Oxford. http://www.ukcip.org.uk/model_nat_res/model_nat_res.html

Blackman, A. & Woodward, R. (2010). *User financing in a national payments for environmental services program: Costa Rican hydropower, resources for the future,* Washington, DC RFFDP09-04-RE. http://ideas.repec.org/a/eee/ecolec/v69y2010i8p1626-1638.html

Blehert, D.S.; Hicks, A.C.; Behr, M.; Meteyer, C.U.; Berlowski-Zier, B.M.; Buckles, E.L.; Coleman, J.T.H.; Darling, S.R.; Gargas, A.; Niver, R.; Okoniewski, J.C.; Rudd, R.J. & Stone W.B. (2009). Bat white-nose syndrome: an emerging fungal pathogen? *Science*, Vol.323, No.5911, pp.227. ISSN 1095-9203

Bowen, A. & Ranger, N. (2009). Mitigating climate change through reductions in greenhouse gas emissions: the science and economics of future paths for global annual emissions, *Grantham Research Institute on Climate Change and the Environment.* London.
http://personal.lse.ac.uk/RANGERN/Bowen-Ranger%20policy%20brief.pdf

Boyles, J.G. & Willis, C.K.R. (2010). Could localized warm areas in cold caves reduce mortality of hibernating bats affected by white-nose syndrome? *Frontiers in Ecology and Environment*, Vol.8, No. 2, pp.92-98, doi:10.1890/080187

Brodie, J.F.; Helmy, O.E.; Brockelman, W.Y. & Maron, J.L. (2009). Bushmeat poaching reduces the seed dispersal and population growth rate of a mammal-dispersed tree. *Ecological Applications,* Vol.19, No.4, pp.854–863, ISSN 1051-0761

Brook, B.W.; Bradshaw. C.J.A.; Koh, L.P. & Sodhi, N.S. (2006). Momentum drives the crash: mass extinction in the tropics. *Biotropica*, Vol.38, No.3, pp.302-305, ISSN 0006-3606

Bumrungsri, S.; Harrison, D.L.; Satasook, C.; Prajukjitr, A.; Thong-Aree, S. & Bates, P.J.J (2006). A review of bat research in Thailand with eight new species records for the country. *Acta Chiropterologica*, Vol.8, No.2, pp.325-359, ISSN 1733-5329

Canadell, J.G. & Raupach, M.R. (2008). Managing forests for climate change mitigation. *Science*, 320, No.5882, pp.1456–1457, ISSN 1095-920

Carignan, V. & Villard, M. (2002) Selecting indicator species to monitor ecological integrity: a review. *Environmental Monitoring and Assessment*, Vol.78, No.1, pp.45-61, ISSN 0167-6369

Cashore, B.; Gale, F.; Meidinger, E. & Newsom, D. (2006). Confronting Sustainability: Forest Certification in Developing and Transitioning Countries, ISBN 10: 0970788258, Yale FandES Publication Series, New Haven.

Center for International Earth Science Information Network (CIESIN) (2002). Country-level Population and Downscaled Projections based on the B2 Scenario, 1990-2100, Palisades, NY: CIESIN, Columbia University. http://www.ciesin.columbia.edu/datasets/downscaled

Clements, R.; Sodhi, N.S.; Ng, P.K.L. & Schilthuizen, M. (2006). Limestone karsts of Southeast Asia: imperiled arks of biodiversity. *Bioscience*, Vol.56, No.9, pp.733-742, ISSN 0006-3568

Corlett, R.T. (2008). Frugivory and seed dispersal by vertebrates in the Oriental (Indomalayan) region. *Biological Reviews*, Vol.73, No.4, pp.413-448, ISSN 1464-7931

Corlett, R. T. (2009). The ecology of tropical East Asia. ISBN-10: 019953246X, Oxford University Press, New York.

Cox, C. B. & Moore, P. D. (2010). *Biogeography: An ecological and evolutionary approach*, 8th Edn. ISBN-10: 0470637943, Oxford: Blackwell.

Day, M.J. & Urich, P.B. (2000). An assessment of protected karst landscapes in Southeast Asia. *Cave and Karst Science*, Vol.27, No. 2, pp.61-70, ISSN 1356191X

De Bruyn, M.E.; Nugroho, M.M. Hossain, J.C.; Wilson, J.C. & Mather, P.B. (2004). Phylogeographic evidence for the existence of an ancient biogeographic barrier: the Isthmus of Kra seaway. *Heredity*, Vol.94, pp.370-378, ISSN 1365-2540

Douangboubpha, B.; Bumrungsri, S.; Satasook, C.; Hammond, N. & Bates, P.J.J. (2010). A taxonomic review of *Hipposideros bicolor* and *H. pomona* (Chiroptera: Hipposideridae) in Thailand. *Acta Chiropterologica*, Vol.12, No.1, pp.415-438, ISSN 1733-5329

Duckworth, J.W. Hedges, S. (1998). Bird records from Cambodia in 1997, including records of sixteen species new for the country. *Forktail*, Vol.14, pp.29-36.

EIA/Telapak. (2008). Borderlines. Vietnam's Booming Furniture Industry and Timber Smuggling in the Mekong Region. Environmental Investigation Agency and Telapak, ISBN 0-9540768-6-9, London.

EIA. (2004). The tiger skin trail. Environmental Investigation Agency, London, UK. http://www.eia-international.org/files/reports85-1.pdf

Elith, J.; Graham, C. H.; Anderson, R. P.; Dudík, M.; Ferrier, S.; Guisan, A.; Hijmans, R. J.; Huettman, F.; Leathwick, J. R.; Lehmann, A.; Li, J.; Lohmann, L. G.; Loiselle, B. A.; Manion, G.; Moritz, C.; Nakamura, M.; Nakazawa, Y.; Overton, J. M.; Peterson, A. T.; Phillips, S. J.; Richardson, K. S.; Scachetti Pereira, R.; Schapire, R. E.; Soberón, J.; Williams, S.; Wisz, M. S. & Zimmermann, N. E. (2006). Novel methods improve prediction of species' distributions from occurrence data. *Ecography*, Vol.29, No.2, pp.129–151, ISSN 0906-7590

Ellis, R. (2005). Tiger bone and rhino horn: The destruction of wildlife for traditional Chinese medicine. ISBN-10: 1559635320, Island Press, Washington, DC.

Epstein, J.H.; Olival, K.J.; Pulliam, J.R.C.; Smith, C.; Westrum, J.; Hughes, T.; Dobson, A.P.; Zubaid, A.; Rahman, S.A.; Basir, M.M.; Field, H.E. & Daszak, P. (2009). *Pteropus vampyrus*, a hunted migratory species with a multinational home-range and a need for regional management. *Journal of Applied Ecology*, Vol.46, No.5, pp.991-1002, ISSN 0021-8901

Francis, C.M.; Borisenko, A.V.; Ivanova, N.V.; Judith, L.; Eger, E.R.; Burton, J.L.; Lim, K.; Guillén-Servent, A.; Sergei V. Kruskop, S.C.; Mackie, I. & Hebert, P. D.N. (2010). The Role of DNA Barcodes in Understanding and Conservation of Mammal Diversity in Southeast Asia. *PLoS ONE*, 9: e12575. doi:10.1371/journal.pone.0012575

Fredericksen, T.S. & Putz, F.E.; (2003). Silvicultural intensification for tropical forest conservation. *Biodiversity and Conservation*, Vol.12, No.7, pp.1445–1453, ISSN 0960-3115

Fujita, M.S. & Tuttle, M.D. (1991). Flying foxes (Chiroptera: Pteropodidae): threatened animals of key ecological and economic importance. *Conservation Biology*, Vol.4, No.6, pp.455-463, ISSN 0888-8892

Fuller, D.O.; Jessup, T.C.; Salim, A. (2003). Loss of forest cover in Kalimantan, Indonesia, since the 1997–1998 El Niño. *Conservation Biology*, Vol.18, No.1, pp.249–254, ISSN 0888-8892

Gan, J. & McCarl, B.A. (2007). Measuring transnational leakage of forest conservation. *Ecological Economics*, Vol.64, No.2, pp.423–432, ISSN 0921-8009

Gaffin, S.R.; Xing, X. & Yetman, G. (2003). Downscaling and geo-spatial gridding of socio-economic projections from the IPCC Special Report on Emissions Scenarios (SRES), *Global Environmental Change Part A*, Vol.14, No.2, pp.105-123, doi:10.1016/j.physletb.2003.10.071

Gaston, K.J.; Williams, P.H.; Eggleton. P.J. & Humphries, C.J. (1995a). Large scale patterns of biodiversity: spatial variation in family richness. *Proceedings of the Royal Society. London* B, Vol.260, No.1358, pp.149-154, ISSN 09628452

Gaston, K.J. & Blackburn, T.M. (1995b). Mapping biodiversity using surrogates for species richness: macro-scales and New World birds. *Proceedings of the Royal Society of London B*, Vol.262, No.1365, pp.335-341, ISSN 0962-8452

Giam, X.; Ng, T.H.; Yap, V.B.; Tan, H.T.W. (2010). The extent of undiscovered species in Southeast Asia. *Biological Conservation*, Vol.19, No.4, pp.943-954, ISSN 1572-9710

Global Witness (2007). Cambodia's family trees: Illegal logging and the stripping of public assets by Cambodia's elite, Washington DC.

Gibbons, P. & Lindenmayer, D. (2001). Tree hollows and wildlife conservation in Australia. pp. 240, ISBN: 9780643090033 Canberra: CSIRO.

Gibson. L.; McNeill, A.; De Tores, P.; Waynec, A. & Yates, C. (2010). Will future climate change threaten a range restricted endemic species, the quokka (*Setonix brachyurus*), in south west Australia? *Biological Conservation*, Vol.143, No.11, pp.2453-2461, ISSN 0006-3207

Gray, T.N.E.; Chamnan, H.; Borey, B.; Collar, J. & Dolman, P.M. (2007). Habitat preferences of a globally threatened bustard provide support for community-based conservation in Cambodia. *Biological Conservation*, Vol.138, No3-4, pp.341–350, ISSN 0006-3207

Hall, R. (2002). Cenozoic geological and plate tectonic evolution of SE Asia and the SW Pacific: computer-based reconstructions and animations. *Journal of Asian Earth Sciences*, Vol.20, No.4, pp.353-434, ISSN 1367-9120

Hannah, L.; Midgley, G. F.; Lovejoy, T.; Bond, W. J.; Bush, M.; Lovett, J. C.; Scott, D. & Woodward, F. I. (2002). Conservation of biodiversity in a changing climate. *Conservation Biology*, Vol.16, No.1, pp.264–268, ISSN 0888-8892

Hansen, M.C. & DeFries, R. (2004). Detecting long-term global forest change using continuous fields of tree-cover maps from 8-km advanced very high resolution radiometer (AVHRR) data for the years 1982-99. *Ecosystems*, Vol.7, No.7, pp.695-716, ISSN 1432-9840

Heller, N.E. & Zavaleta, E.S. (2009). Biodiversity management in the face of climate change: a review of 22 years of recommendations. *Biological Conservation*, Vol.142, No.1, pp.14-32, ISSN 0006-3207

Hernandez, P. A.; Graham, C. H.; Master, L. L. & Albert, D. L. (2006). The effect of sample size and species characteristics on performance of different species distribution modelling methods. *Ecography*, Vol.29, No.5, pp. 773–785, ISSN 0906-7590

Hickling, R.; Roy, D.B.; Hill, J.K.; Fox, R. & Thomas, C.D. (2006). The distributions of a wide range of taxonomic groups are expanding polewards. *Global Change Biology*, Vol.12, No.3, pp.450–455, ISSN I1354-1013

Hill, J.; Woodland, W. & Gough, G. (2007). Can visitor satisfaction and knowledge about tropical rainforests be enhanced through biodiversity interpretation, and does this promote a positive attitude towards ecosystem conservation? *Journal of Ecotourism*, Vol.6, No.1, pp.75–85, ISSN 1472-4049

Hirzel, A. H.; Hausser, J.; Chessel, D. & Perrin, N. (2002). Ecological-niche factor analysis: how to compute habitat suitability maps without absence data. *Ecology*, Vol.83, No.7, pp2027-2036, ISSN: 00129658

Hodgkison, R.; Balding, S.T.; Zubaid, A. & Kunz, T.H. (2004). Temporal variation in the relative abundance of fruit bats (Megachiroptera: Pteropodidae) in relation to the availability of food in a lowland Malaysian rain forest. *Biotropica*, Vol.36, No.4, pp.522-533, ISSN 0006-3606

Hughes, A.C.; Satasook, C.; Bates, P.J.J.; Bumrungsri, S. & Jones, G. (2011). Explaining the causes of the zoogeographic divide at the Isthmus of Kra: using bats as a case study. *Journal of Biogeography*, doi/10.1111/j.1365-2699.2011.02568.x/pdf

Hughes, A.C.; Satasook, C.; Bates, P.J.J.; Soisook, P.; Sritongchuay, T.; Jones, G. & Bumrungsri, S. (2010). Echolocation call analysis and presence-only modelling as conservation monitoring tools for rhinolophoid bats in Thailand. *Acta Chiropterologica*, Vol.12, No.2, pp.311-327, ISSN 1733-5329

Hughes, A.C.; Satasook, C.; Bates, P.J.J.; Bumrungsri, S. & Jones, G. (*in press*). Using echolocation call structure to identify Thai bat species: Vespertilionidae, Emballonuridae, Nycteridae and Megadermatidae. *Acta Chiropterologica,*

Hughes, A.C.; Satasook, C.; Bates, P.J.J.; Bumrungsri, S. & Jones, G. (*In review*). The projected effects of climatic and vegetation changes on the distribution and diversity of Southeast Asian bats.

Hughes, A.C.; Satasook, C.; Bates, P.J.J.; Soisook, P. & Bumrungsri, S. (*In prep a*). Cryptic clues to a complex problem.

Hughes, A.C.; Satasook, C.; Bates, P.J.J.; Bumrungsri, S. & Jones, G. (*In prep b*). Southeast Asian bat biodiversity, and an analysis of the relationship between anthropogenic threat factors.

Hughes, J.B.; Round, P.D. & Woodruff, D.S. (2003). The Sundaland-Asian faunal transition at the Isthmus of Kra: an analysis of resident forest bird species distributions. *Journal of Biogeography,* Vol.30, No.4, pp.569-580, ISSN 0305-0270

Hughes, L. (2000). Biological consequences of global warming: is the signal already apparent? *Trends in Ecology and Evolution,* Vol.15, No.2, pp.56–61, ISSN 0169-5347

Huntley, B.; Berry, P. M.; Cramer, W. & Mcdonald, A. P. (1995). Modelling present and potential future ranges of some European higher plants using climate response surfaces. *Journal of Biogeography,* Vol.22, No.6, pp.967–1001, ISSN 0305-0270

Jacobson, S.K. & Robles, R. (1992). Ecotourism, sustainable development, and conservation education: development of a tour guide training program in Tortuguero, Costa Rica. *Environmental Management,* Vol.16, No.6, pp. 701-713, ISSN: 0364152X

Jime´nez-Valverde, A.; Lobo, J. M. & Hortal, J. (2008). Not as good as they seem: the importance of concepts in species distribution modelling. *Diversity and Distributions,* Vol.14, No.6, pp.885–890 ISSN 1366-9516

Jones, G.; Jacobs, D.S.; Kunz, T.H.; Willig, M.R. & Racey, P.A. (2009). Carpe noctem: the importance of bats as bioindicators. *Endangered Species Research,* Vol.8, No.1-2, pp.93-115, doi: 10.3354/esr0018

Jones, G.; Van Parijs, S.M. (1993). Bimodal echolocation in pipistrelle bats: are cryptic species present? *Proceedings of the Royal Society of London, B. Biological Sciences,* Vol.251, No.1331, pp.119-125, ISSN 0962-8452

Jones, K. E.; Russ, J. A.; Bilhari, Z.; Catto, C.; Csosz, I.; Gorbachev, A.; Gyorfi, P.; Hughes, A.; Ivashkiv, I.; Koryagina, N.; Kurali, A.; Langton, S.; Maltby, A.; Margina, G.; Pandourski, I.; Parsons, S.; Prokofev, I.; Szodoray-Paradi, A.; Szodoray-Paradi, F.; Taras-Bashta, A.; Tilova, E.; Walters, C.; Weatherill, A. & Zavarzin, O. (2011). Monitoring ultrasonic biodiversity: using bats as biodiversity indicators. *Biodiversity monitoring and conservation: bridging the gaps between global commitment and local action,* B. P. Collen, N. Pettorelli, S. M. Durant, L. Krueger, J. Baillie (Eds). ISBN-10: 1444332929, Blackwell Press, London

Kettle, C.J. (2010). Ecological considerations for using dipterocarps for restoration of lowland rainforest in Southeast Asia. *Biodiversity and Conservation,* Vol.19, No.4, pp.1137-1151, ISSN 1572-9710

Kingston, T. (2010). Research priorities for bat conservation in Southeast Asia: a consensus approach. *Biodiversity and conservation,* Vol.19, No.2, pp.471-484, ISSN 1572-9710

Kingston, T.; Francis, C.M.; Zubaid, A. & Kunz, T.H. (2003). Species richness in an insectivorous bat assemblage from Malaysia. *Journal of Tropical Ecology,* Vol.19, No.1, pp.67-79, ISSN 0266-4674

Kingston, T.; Juliana, S.; Rakhmad, S. K, Fletcher, C. D.; Benton-Browne, A.; Struebig, M.; Wood, A.; Murray, S. W.; Kunz, T. H. & Zubaid, A. (2006). The Malaysian Bat Conservation Research Unit: research, capacity building and education in an Old World hotspot. 41-60 In: *Proceedings of the National Seminar On Protected Areas,* Othman, S. H. Yatim, S. Elaguipillay, Sh. Md. Nor, N. A. Sharul and A. M. Sah. (Eds) Department of Wildlife and National Parks.

Kotiaho, J.S.; Kaitala, V.; Kolmonen, A. & Paivinen, J. (2005). Predicting the risk of extinction from shared ecological characteristics. *Proceedings of the National Academy of Sciences USA*, Vol.102, No.6, pp.1963–1967, ISSN 0027-8424

Kunz, T.H. & Lumsden, L.F. (2003). Ecology of cavity and foliage roosting bats. In: *Bat ecology,* T. H. Kunz & M. B. Fenton (Eds.) pp. 3-90. Chicago, IL University of Chicago Press.

Lamb, D. & Erskine, P. (2008). Forest restoration on a landscape scale. In: *Living in a dynamic forest environment,* N.E. Stork and S.M. Turton (Eds.) pp.469-484, ISBN 978-1-4051-5643-1, Blackwell, Malden.

Lamb, J. M.; Ralph, T.M.C.; Goodman, S. M.; Bogdanowicz, B, Fahr, J.; Gajewska, M.; Bates, P. J. J.; Eger, J.; Benda, P. & Taylor, P. J. (2008). Phylogeography and predicted distribution of African-Arabian and Malagasy populations of giant mastiff bats, *Otomops* spp. (Chiroptera: Molossidae). *Acta Chiropterologica*, Vol.10, No.1, pp21-40, ISSN 1508-1109

Laumonier, Y.; Uryu, Y.; Stüwe, M.; Budiman, A.; Setiabudi, B. & Hadian, O. (2010). Eco-floristic sectors and deforestation threats in Sumatra: identifying new conservation area network priorities for ecosystem-based land use planning. *Biodiversity and Conservation*, Vol.19, No.4, pp.1153-1174, ISSN 1572-9710

Laurance, W.F. (2007). Have we overstated the tropical biodiversity crisis? *Trends in Ecology and Evolution*, Vol.22, No.2, pp.65-70, ISSN 0169-5347

Laurance, W.F. (2008a). The real cost of minerals. *The New Scientist,* Vol.199, pp.16. http://hdl.handle.net/10088/12008

Laurance, W.F. (2008b). Can carbon trading save vanishing forests? *BioScience*, Vol.58, No.8, pp.286–287, ISSN 0006-3568

Lepers, E.; Lambin, E.F.; Janetos, A.C.; DeFries, R.; Achard, F.; Ramankutty, N. & Scholes, R. J. (2005). A synthesis of information on rapid land-cover change for the period 1981-2000. *BioScience*, Vol.55, No.2, pp.115-124, ISSN 0006-3568

Lenoir, J.; Gegout, J.C.; Marquet, P.A.; De Ruffray, P. & Brisse, H. (2008). A significant upward shift in plant species optimum elevation during the 20th century. *Science*, Vol.320, No.5884, pp.1768–1771, ISSN 0036-8075

Linkie, M.; Rood, E. & Smith, R.J. (2010). Modelling the effectiveness of enforcement strategies for avoiding tropical deforestation in Kerinci Seblat National Park, Sumatra. *Biodiversity and Conservation*, Vol.19, No.4, pp.973-984, ISSN 1572-9710

Mace, G.M. & Baillie, J.E.M. (2007). The 2010 biodiversity indicators: challenges for science and policy. *Conservation Biology,* Vol.21, No.6, pp.1406-1413, doi/10.1111/j.1523-1739.2007.00830.x/full

MacKenzie, D. I. (2005). Was it there? Dealing with imperfect detection for species presence/absence data. *Australia and New Zealand Journal of Statistics*, Vol.47, No.1, pp.65–74, ISSN 13691473

Macswiney, G. M.; Clarke, C. F. & Racey, P. A. (2008). What you see is not what you get: the role of ultrasonic detectors in maximising inventory completeness in Neotropical forests. *Journal of Applied Ecology*, Vol.45, No.5, pp.1364-1371, ISSN: 00218901

Malcolm, J.R.; Liu, C.; Neilson, R.P.; Hansen. L. & Hannah, L. (2006). Global warming and extinctions of endemic species from biodiversity hotspots. *Conservation Biology,* Vol.20, No.2, pp.538–548, ISSN 0888-8892

McLachlan, J.S.; Hellmann, J. & Schwartz, M. (2007). A framework for debate of assisted migration in an era of climate change. *Conservation Biology*, Vol.21, No.2, pp.297–302, ISSN 0888-8892

McMahon, G.; Subdibjo, E.R.; Eden, J.; Bouzaher, A.; Dore, G. & Kunanayagan, R. (2000). Mining and the environment in Indonesia: Long-term trends and repercussions of the Asian economic crisis (EASES Discussion Paper Series). Washington, DC: World Bank, East Asia Environment and Social Development Unit, www.natural-resources.org/minerals/cd/docs/twb/mining_indonesia.pdf

McNamara, S.; Duong Viet Tinh, Erskine, P. D.; Lamb, D.; Yates, D. & Brown, S. (2006). Rehabilitating degraded forest land in central Vietnam with mixed native species plantings. *Forest Ecology and Management*, Vol.233, No.2, pp.358-365, DOI: 10.1111/j.1749-4877.2007.00048.x

McNamara, N.P.; Black, H.I.J, Piearce, T.G.; Reay, D.S. & Ineson, P. (2008). The influence of afforestation and tree species on soil methane fluxes from shallow organic soils at the UK gisburn forest experiment. *Soil Use Management*, Vol.24, No.1, pp.1–7, ISSN 0266-0032

McNeely, J.A. (2007). A zoological perspective on payments for ecosystem services. *Integrative Zoology*, Vol.2, No.2, pp.68–78, DOI: 10.1111/j.1749-4877.2007.00048.x

McShane, T.O. & Wells, M.P. (2004). *Getting biodiversity projects to work: towards more effective conservation and development.* New York: Columbia University Press.

Meyer, C. F. J.; Aguiar, M. W. S.; Aguirre, L. F.; Baumgarten, J.; Clarke, F. M.; Cosson, J. F.; Villegas, S. E.; Fahr, J.; Faria, D.; Furey, N.; Henry, M.; Hodgkison, R.; Jenkins, R. K. B.; Jung, K. G.; Kingston, T.; Kunz, T. H.; M. MacSwiney G, M. C.; Moya, I.; Pons, J-M.; Racey, P. A.; Rex, K.; Sampaio, E. M.; Stoner, K. E.; Voigt, C. C.; von Staden, D.; Weise, C. D. & Kalko, E. K. V. (2010). Long-term monitoring of tropical bats for anthropogenic impact assessment: Gauging the statistical power to detect population change, *Biological Conservation*, Vol.143, No.11, pp.2797-2807, ISSN 0006-3207

Mickleburgh, S.; Waylen, K. & Racey, P.A. (2009). Bats as bushmeat - a global review, *Oryx*, Vol.43, No.2, pp.217–234, DOI: 10.1017/S0030605308000938

Midgley, G. F.; Hannah, L.; Millar, D.; Rutherford, M. C. & Powerie, L. W. (2002). Assessing the vulnerability of species richness to anthropogenic climate change in a biodiversity hotspot. *Global Ecology and Biogeography*, Vol.11, No.6, pp.445–451, ISSN 1466-822X

Mollmann, S. (2008). Birders flock east. *The Wall Street Journal,* September 12, W1. http://online.wsj.com/article/SB122115338586124101.html

Mohd-Azlan, J.; Zubaid, A. & Kunz, T.H. (2001). Distribution, relative abundance, and conservation status of the large flying fox, *Pteropus vampyrus*, in peninsular Malaysia: a preliminary assessment. *Acta Chiropterologica*, Vol.3, No.2, pp.149-162, ISSN 1508-1109

Morley, R.J. & Flenley, J.R. (1987). Late Cainozoic Vegetational and Environmental Changes in the Malay Archipelago', In: *Biogeographical Evolution in the Malay Archipelago, Biogeography,* T. C. Whitmore (Ed.), pp.50-59, ISBN-10 0198541856, Oxford Monographs.

Myers, N. (1987) The extinction spasm impending: synergisms at work. *Journal of Conservation Biology*, Vol.1, No.1, pp.14-21, ISSN 0888-8892

Myers, N. (1988). Mass extinction-profound problem, splendid opportunity. *Oryx,* Vol.22, pp.205-15, DOI: 10.1017/S003060530002233X

Myers, N.; Mittermeier, R. A.; Mittermeier, C. G.; Da Fonseca, G. A. B. & Kent, J. (2000). Biodiversity hotspots for conservation priorities. *Nature,* Vol.403, No.6772, pp. 853-858, ISSN 0028-0836

Nijman, V. (2010). An overview of international wildlife trade from Southeast Asia. *Biodiversity and Conservation,* Vol.19, No.4, pp.1101-1114, ISSN 1572-9710

Olson, D.M.; Dinerstein, E.; Wikramanayake, E.D.; Burgess, N.D.; Powell, G.V.N.; Underwood, E.C. D'Amico, J.A.; Itoua, I.; Strand, H.E.; Morrison, J.C.; Loucks, C.J.; Allnutt, T.F.; Ricketss, T.H.; Kura, Y.L.J.F.; Wettengel, W.W.; Hedao, P. & Kassem, K.R. (2001). Terrestrial ecoregions of the World: a new map of life on Earth. *BioScience,* Vol.51, No.11, pp.933-938, ISSN 0006-3568

Pawar, S.; Koo, M. S.; Kelley, C.; Firoz Ahmed, M.; Chaudhuri, S. & Sarkar, S. (2007). Conservation assessment and prioritization of areas in Northeast India: priorities for amphibians and reptiles. *Biological Conservation,* Vol.136, No.3, pp.346-361, ISSN 0006-3207

Pearce, P.L. (2008). The nature of rainforest tourism: insights from a tourism social science research programme.In: *Living in a dynamic forest environment,* N.E. Stork and S.M. Turton (Eds.) pp.94-104. ISBN 978-1-4051-5643-1, Blackwell, Malden.

Pearman, P. B.; Guisan, A. & Zimmermann, N.E. (2010). Impacts of climate change on Swiss biodiversity: an indicator taxa approach. *Biological Conservation,* Vol.144, No.2, pp.866-875, doi:10.1016/j.biocon.2010.11.020

Peh, K.S.H. (2010). Invasive species in Southeast Asia: the knowledge so far. *Biodiversity and Conservation,* Vol.19, No.4, pp.1083-1099, ISSN 1572-9710

Pergams, O.R.W. & Zaradic, P.A. (2008). Evidence for a fundamental and pervasive shift away from nature-based recreation. *Proceedings of the National Academy of Sciences,* Vol.105, No.7, pp.2295–2300, ISSN 0027-8424

Phillips, S.J.; Anderson, R.P. & Schapire, R.E. (2006). Maximum entropy modelling of species geographic distributions. *Ecological Modelling,* Vol.190, No3-4, pp.231-259, ISSN 0304-3800

Pounds, J.A.; Bustamente, M.R.; Coloma, L.A.; Consuegra, J.A.; Fogden, M.P.L.; Foster, P. N.; La Marca, E.; Masters, K.L.; Merino-Viteri A.; Puschendorf, R.; Santiago R.R.; Sánchez-Azofeifa, G.A.; Still, C.J. & Young, B.E. (2006). Widespread amphibian extinctions from epidemic disease driven by global warming. *Nature,* Vol.439, No.7073, pp.161–67, ISSN 0028-0836

Prentice, I. C.; Cramer, W.; Harrison, S. P.; Leemans, R.; Monserud, R. A. & Solomon, A. M. (1992). A global biome model based on plant physiology and dominance, soil properties and climate. *Journal of Biogeography,* Vol.19, No.2, pp.117–134, ISSN: 03050270

Putz, F.E.; Sist P.; Fredericksen T.S. & Dykstra D. (2008). Reduced-impact logging: challenges and opportunities. *Forest Ecology and Management,* Vol.256, No.7, pp.1427–1433, ISSN 0378-1127

Rebelo, H. & Jones, G. (2010). Ground validation of presence-only modelling with rare species: a case study on barbastelles *Barbastella barbastellus* (Chiroptera: Vespertilionidae). *Journal of Applied Ecology,* Vol.47, No.2, pp.410-420, ISSN 0021-8901

Rebelo, H.; Tarroso, P, Jones, G. (2010). Predicted impact of climate change on European bats in relation to their biogeographic patterns. *Global Change Biology*, Vol.16, No.2, pp.561-576, ISSN 1354-1013

Redford, K.H. (1992). The empty forest. *BioScience*, Vol.42, No.6, pp.412–422, ISSN 0006-3568

Riley, S.P.D.; Busteed, G.T.; Kats, L.B.; Vandergon, T.L.; Lee, L.F.S.; Dagit, R.G.; Kerby, J.L.; Fisher, R.N. & Sauvajot, R.M. (2005). Effects of urbanization on the distribution and abundance of amphibians and invasive species in southern California streams. *Conservation Biology*, Vol.19, No.6, pp.1894–1907, ISSN 0888-8892

Salazar, L.F.; Nobre, C.A. & Oyama, M.D. (2007). Climate change consequences on the biome distribution in tropical South America. *Geophysical Research Letters*, Vol.34, No.9, pp.1–6, ISSN 0094-8276

Sechrest, W.; Brooks, T.M.; da Fonseca, G.A.B.; Konstant, W.R.; Mittermeier, R.A.; Purvis, A.; Rylands, A.B. & Gittleman, J.L. (2002). Hotspots and the conservation of evolutionary history. *Proceedings of the National Academy of Sciences USA*, Vol.99, No.4, pp.2067–2071, ISSN 0027-8424

Sérgio, C.; Figueira, R.; Draper, D.; Menezes, R. & Sousa, A.J. (2007). Modelling bryophyte distribution based on ecological information for extent of occurrence assessment. *Biological Conservation*, Vol.135, No.3, pp.341-351, ISSN 0006-3207

Sessions, J. (2007). Forest Road Operations in the Tropics. Tropical Forestry Series, Springer-Verlag, pp. 168, ISBN-10 3642079776, Berlin.

Seto, K.; Sanchez-Rodrıguez, R. & Fragkias, M. (2010). The new geography of contemporary urbanization and the environment. *Annual review of environment and resources*, Vol.35, pp.167–94, ISSN 1543-5938

Siddique, I.; Engel, V.L.; Parrotta, J.A.; Lamb, D.; Nardoto, G.B.; Ometto, J.; Martinelli, L.A. & Schmidt, S. (2008). Dominance of legume trees alters nutrient relations in mixed species forest restoration plantings within seven years. *Biogeochemistry*, Vol.88, No.1, pp.89–101, ISSN 0168-2563

Simmons, N. (2005). Order Chiroptera, In: *Mammal species of the World: a taxonomic and geographic reference*, Wilson and D. M. Reeder(Eds) pp 312-529. ISBN-10 0801882214, John Hopkins University Press, Baltimore.

Simmons, N. B. & Wetterer, A.L (2011) Estimating diversity: how many bat species are there? The Second International South-East Asian Bat Conference Bogor, 6-9 June 2011.

Sodhi, N.S.; Posa, M.R.; Lee, T.M.; Bickford, D.; Koh, L.P. & Brook, B.W. (2010). The state and conservation of Southeast Asian biodiversity. *Biodiversity and Conservation*, Vol.19, No.2, pp.317-328, ISSN 1572-9710

Soisook, P.; Bumrungsri, S.; Satasook, C.; Thong, V.D.; Bu, S.S.H.; Harrison, D.L. & Bates, P.J.J (2008). A taxonomic review of *Rhinolophus stheno* and *R. malayanus* (Chiroptera: Rhinolophidae) from continental Southeast Asia: an evaluation of echolocation call frequency in discriminating between cryptic species. *Acta Chiropterologica*, Vol.10, No.2, pp.221-242, ISSN 1508-1109

Soisook. P.; Niyomwan. P.; Srikrachang. M.; Srithongchuay, F. & Bates, P.J.J. (2010). Discovery of *Rhinolophus beddomei* (Chiroptera: Rhinolophidae) from Thailand with a brief comparison to other related taxa. *Tropical Natural History*, Vol.10, No.1, pp.67-79.

Stibig, H.J.; Belward, A.S.; Roy, P.S.; Rosalinea-Wasrin, U.; Agrawal, S.; Joshi, P.K.; Hildanus Beuchle, S.; Fritz, S.; Mubareka, S. & Giri, C. (2007). A land-cover map for South and Southeast Asia derived from SPOT-VEGETATION data. *Journal of Biogeography*, Vol.34, No.4, pp.625-637, ISSN 0305-0270

Steinmetz, R.; Chutpong, W. & Seuaturien, N. (2006). Collaborating to conserve large mammals in Southeast Asia. *Conservation Biology*, Vol.20, No.5, pp.1391–1401, ISSN 0888-8892

Stralberg, D.; Jongsomjit, D.; Howell, C.A.; Snyder, M.A.; Alexander, J.D.; Wiens, J.A.& Root, T.L. (2009). Re-shuffling of species with climate disruption: a no-analog future for California birds? PLoS One, Vol.4, No.9, e6825, doi:10.1371/journal.pone.0006825.

Struebig, M.; Kingston, T.; Zubaid, A.; Adnan, A. & Rossiter, S. (2008). Conservation value of forest fragments to Palaeotropical bats. *Biological Conservation*, Vol.141, No.8, pp.2112-2126, ISSN 0006-3207

Suzuki, R.; Numata, S.; Okuda, T.; Supardi, M.N.; Kachi, N. (2009). Growth strategies differentiate the spatial patterns of 11 dipterocarp species coexisting in a Malaysian tropical rain forest. *Journal of Plant Research*, Vol.122, No.1, pp.81–93, ISSN 1618-0860

Sykes, M. T.; Prentice, I. C. & Cramer, W. (1996). A bioclimatic model for the potential distributions of north European tree species under present and future climates. *Journal of Biogeography*, Vol.23, No.2, pp.203–233, ISSN 0305-0270

Taylor, D.; Saksena, P.; Sanderson, P.G. & Kucera, K. (1999). Environmental change and rain forests on the Sunda Shelf: drought, fire and the biological cooling of biodiversity hotspots. *Biodiversity and Conservation*, Vol.8, No.8, pp.1159–1177, ISSN 0960-3115

Ten Kate, K.; Bishop, J.; Bayron, R. (2004). *Biodiversity Offsets: Views, Experience and the Business Case. International Union for Conservation of Nature and Natural Resources and Insight Investment, London.* http://www.eldis.org/static/DOC16610.htm

Terborgh, J.; Van Schaik, C.P.; Davenport, L. & Rao, M. (2002). *Making parks work: strategies for preserving tropical nature*. Island Press, Washington , D.C.

Tungittiplakorn, W. & Dearden, P. (2002). Biodiversity conservation and cash crop development in Northern Thailand. *Biodiversity and Conservation*, Vol.11, No.11, pp.2007–2025, DOI: 10.1023/A:1020812316749

United Nations Population Division (2008). World Urbanization Prospects: The 2007 Revision Population Database (http://esa.un.org/unup/)

Vistor, D. G.; Morgan, M. G.; Apt, J., Steinbruner, J. & Ricke, K. (2009) The geoengineering option: a last resort against Global warming? *Foreign Affairs*, March/April http://iis-db.stanford.edu/pubs/22456/The_Geoengineering_Option.pdf

Webb, C.O.; Slik, J.W.F. & Triono, T. (2010). Biodiversity inventory and informatics in Southeast Asia. *Biodiversity and Conservation*, Vol.19, No.4, pp.955-972, ISSN 1572-9710

Whitmore, T.C. (1981). Palaeoclimate and vegetation history.In: *Wallace's Line and Plate Tectonics*, T.C. Whitmore (Ed.), pp.36-42. ISBN-10: 0198545452, Clarendon Press, Oxford.

Wintle, B. A.; McCarthy, M. A.; Parris, K. M. & Burgman, M. A. (2004). Precision and bias of methods for estimating point survey detection probabilities. *Ecological Applications*, Vol.14, No.3, pp.703–712, http://www.jstor.org/stable/4493574

Wisz, M. S.; Hijmans, R. J.; Li, J.; Peterson, A. T.; Graham, C. H. & Guisan, A. (2008). Effects of sample size on the performance of species distribution models. *Diversity and Distributions*, Vol.14, No.5, pp.763–773, ISSN 1366-9516

Witmer, F. (2005). Simulating future global deforestation using geo-graphically explicit models, IIASA *Interim Report* IR-05-010

Woodruff, D.S. (2003). Neogene marine transgressions, paleogeography and biogeographic transitions on the Thai-Malay Peninsula. *Journal of Biogeography*, Vol.30, No.4, pp.551-567, ISSN 0305-0270

Woodruff, D.S (2010). Biogeography and conservation in Southeast Asia: how 2.7 million years of repeated environmental fluctuations affect today's patterns and the future of the remaining refugial-phase biodiversity. *Biodiversity and Conservation*, Vol.19, No.4, pp.919-941, ISSN 1572-9710

Wright, S.J.; Stoner, K.E.; Beckman, N.; Corlett, R.T.; Dirzo, R.; Muller-Landau, H.C.; Nunez-Iturri, G.; Peres C.A. & Wang, B.C. (2007). The plight of large animals in tropical forests and the consequences for plant regeneration. *Biotropica*, Vol.39, No.3, pp.289–291, DOI: 10.1111/j.1744-7429.2007.00293.

Wunder, S. (2006). Are direct payments for environmental services spelling doom for sustainable forest management in the tropics? *Ecology and Society*, Vol.11, No.2, pp.25, http://www.ecologyandsociety.org/vol11/iss2/art23/

Wunder, S. (2007). The efficiency of payments for environmental services in tropical conservation. *Conservation Biology*, Vol.21, No.1, pp.48–58, ISSN 0888-8892

Wunscher, T.; Engel S. & Wunder, S. (2006). Payments for environmental services in Costa Rica: increasing efficiency through spatial differentiation. *Quarterly Journal of International Agriculture*, Vol.45, No.4, pp.317-335, http://131.220.109.9/module/register/media/a9b0_Wuenscher%20Engel%20Wunder%202006.pdf

Zeppel, H. (2006). Indigenous ecotourism: Sustainable development and management, pp.272, ISBN-10: 1845931246, CABI, Wallingford, UK.

2

Protein Limitation Explains Variation in Primate Colour Vision Phenotypes: A Unified Model for the Evolution of Primate Trichromatic Vision

Kim Valenta[1] and Amanda D. Melin[2]
[1]*University of Toronto, Department of Anthropology, Toronto, Ontario,*
[2]*Dartmouth College, Department of Anthropology, Hanover, New Hampshire,*
[1]*Canada*
[2]*USA*

1. Introduction

Primate colour vision has intrigued scientists for many decades and will likely continue to do so for the foreseeable future. Primates are the most visually adapted order of mammals and a considerable proportion of their large brain size is devoted to processing visual information (e.g. Barton, 2006). Most eutherian mammals have dichromatic (two-colour) vision, and chromatic distinctions are based on discriminating relatively shorter from relatively longer wavelengths within the visual spectrum (~400-700nm). These distinctions are made by neural comparison of cone cells possessing short (S) wavelength-sensitive photopigments, which are maximally sensitive to bluish light, and long (L) wavelength-sensitive pigments, which are maximally sensitive to greenish light. These photopigments are encoded by an autosomal S opsin gene and an X-chromosomal L opsin gene respectively. Primates have an additional colour channel enabling trichromatic vision via a duplication and divergence of the L opsin gene, resulting in long and middle (L-M) wavelength-sensitive photopigments (reviewed in Hunt et al., 2009; Regan et al., 2001). This arrangement permits enhanced discrimination of light and perception of different shades of green, yellow, orange and red.

Old World monkeys, apes and humans are routinely trichromatic, having two loci for L-M opsin genes on each X-chromosome due to a gene duplication event (Jacobs, 2008), in addition to the autosomal S opsin locus. Alternatively, New World monkeys and some lemurs exhibit polymorphic trichromacy (Jacobs, 2007; Tan & Li, 1999; Tan et al., 2005; Veilleux & Bolnick, 2009). This alternate path to trichromacy results from a genetic polymorphism at a single locus of the X-chromosomal L-M opsin gene. Females with two different L-M opsin genes, combined with the common autosomal S opsin gene, possess trichromacy (Mollon et al., 1984). However, males, being hemizygotes, can inherit only one L-M opsin gene and are always dichromatic; homozygous females are also dichromatic. Variation in primate trichromacy and the selective pressures that led to trichromacy are under considerable debate (Caine et al., 2010).

In order to evaluate routine and polymorphic trichromacy in primates within an evolutionary framework, it is necessary to review the behavioural, experimental, genetic, biogeographical and ecological evidence pertaining to primate trichromacy. We focus this review on the evidence for and against foraging hypotheses because the vast majority of research to date evaluates and supports the utility of trichromacy for finding and selecting food. We begin with an overview of the fruit foraging and the young leaf hypotheses. We additionally introduce a new hypothesis - that both routine and polymorphic trichromacy confer a selective advantage to primates in the detection and selection of proteinaceous foods (either leaves or fruit), and present evidence for *Ficus* (Moraceae) as a protein source for neotropical primates. To explain the foundation for this new hypothesis, we review primate foraging patterns on *Ficus*, and discuss the abundance and density of *Ficus* in African and neotropical forests.

2. The frugivory hypothesis

The oldest hypothesis for the evolution of primate colour vision is that trichromacy is an adaptation to frugivory (Allen, 1879; Polyak, 1957). This hypothesis builds on the observation that most primates rely heavily on dietary fruit (Chapman & Onderdonk, 1998; Fleagle, 1999). Given that many tropical plants produce red, orange or yellowish ("colourful") fruits and that trichromatic colour vision enhances primates' abilities to detect these colours amidst green foliage and unripe fruits, trichromatic colour vision should facilitate the detection and selection of edible fruits (Regan et al., 2001). Primates in turn may provide high-quality seed dispersal services to these fruiting plants (Garber & Lambert, 1998). The ability to distinguish between ripe and unripe fruits, fruits from leaves, and edible from inedible species is a complex task that could be facilitated by trichromacy (Smith et al., 2003), and there is compelling experimental and behavioural evidence for trichromatic advantages in these tasks.

The frugivory hypothesis was originally proposed and tested for catarrhine primates (Sumner & Mollon, 2000b) because until relatively recently platyrrhines (Mollon et al., 1984) and strepsirrhines (Tan & Li, 1999) were not known to possess trichromacy. The frugivory hypothesis has since been extended to platyrrhines and tested behaviourally (Caine et al., 2003). Theoretical studies have also assessed the conspicuity of dietary fruits to trichromatic versus dichromatic platyrrhines (Riba-Hernández et al., 2005; Stoner et al., 2005). Research evaluating the frugivory hypothesis is reviewed briefly below.

Tropical fruits exhibiting a "primate dispersal syndrome" (medium-sized red, orange or yellow fruits with succulent pulp, large seeds, tough exocarps; (Gautier-Hion et al., 1985)) have been found to occupy a narrow region of colour space that is detectable to trichromatic platyrrhines (Regan et al., 2001). Furthermore, the reflectance spectra of most primate-consumed fruits in Uganda were found to show chromatic changes as they ripen, and trichromatic catarrhines should be well adapted to discriminate fruit ripeness (Sumner and Mollon, 2000b). In the New World, the peak spectral sensitivities of trichromatic platyrrhine cone pigments were found to be well suited to the detection of fruits against a background of leaves under photopic conditions (Regan et al., 2001). The spectral tuning of the L-M cone pigments in the trichromatic platyrrhine *Alouatta seniculus* (Linnaeus, 1766) was found to be optimal for detecting ripe fruits against a background of leaf "noise" (Regan et al., 1998). Among six phenotypes in one white-faced capuchin

(*Cebus capucinus*, Linnaeus, 1758) population, monkeys possessing the phenotype with the most spectrally separated L-M opsin alleles showed the highest acceptance index for red fruits (Melin et al., 2009). For this same population of capuchin monkeys, colourful fruits comprise 60% of the diet and are preferred to cryptically coloured fruits (Melin, 2011).

Trichromatic vision may also afford a selective advantage by allowing trichromats to forage on fruit under a greater range of light conditions than dichromats (Yamashita et al., 2005). In two species of *Saguinus* (Hoffmannsegg, 1807), captive trichromatic individuals were more efficient than dichromats at selecting ripe fruits, both in isolation and against a background of leaves (Smith et al., 2003). In a study linking nutritional value to fruit colouration, Riba-Hernandez et al. (2005) have provided additional support for this hypothesis. Arguing that the principle reward of ripe fruits is sugar, these authors demonstrated a positive correlation between fruit glucose content and the red-green colour channel of spider monkeys (*Ateles geoffreyi*, Kuhl, 1820).

In summary, the evidence for a ripe fruit foraging advantage for trichromats is compelling. Behavioural evidence for ripe fruit foraging advantages in wild and captive trichromats, the trichromat-accessible colour space of tropical fruits, the dietary importance of colourful fruits, and a positive correlation between nutritional value and fruit colouration all indicate that a ripe fruit foraging advantage may indeed result in the evolution and persistence of trichromacy. However, the frugivory hypothesis is problematic on several accounts. In the sections below, we review problems with the frugivory hypothesis and alternate hypotheses regarding the evolution of trichromacy.

2.1 Problems with the frugivory hypothesis

There is compelling empirical evidence that trichromacy is not necessary for detection of ripe fruits. Dominy and Lucas (2001) found that four routinely trichromatic catarrhine species in Uganda did not choose fruit on the basis of chromaticity, and that consumed fruits did not differ in chromaticity from unconsumed fruits. Comparing routinely trichromatic *Alouatta palliata* (Gray, 1849) fruit choices with polymorphically trichromatic *Ateles geoffroyi* in Costa Rica, Stoner et al. (2005) found that, contrary to expectations of a fruit foraging advantage for trichromats, the polymorphic *Ateles* consumed reddish fruits more often while routinely trichromatic *Alouatta* concentrated primarily on green fruits. Further studies of wild primates have also demonstrated that trichromats do not feed on brightly coloured fruits at faster rates than dichromats (Hiramatsu et al., 2008; Vogel et al., 2007). Additionally many fruits that are consumed and dispersed by primates are dull, and green or brown (Janson, 1983; Link & Stevenson, 2004), colours that can be detected through the blue-yellow colour channel of the dichromat (Dominy & Lucas, 2001). Intriguingly, the only genus of routinely trichromatic platyrrhine, *Alouatta*, is not a frugivore, but a folivore (Araújo et al., 2008; Jacobs et al., 1996).

3. The young leaf hypothesis

Many frugivorous primates, and specifically catarrhines, rely on leaves as fallback foods during periods of habitat-wide fruit scarcity (Lucas et al., 2003). Building on this observation, Dominy and Lucas (2001) argued that it is the ability to distinguish young leaves from mature leaves, rather than the ability to distinguish ripe fruits, that provided the selective force behind routine trichromacy. Measuring the luminance of young leaves and fruit consumed by catarrhines in Kibale, Uganda they found that while fruits could be reliably distinguished by the yellow-blue colour channels alone, the detection of young, proteinaceous leaves required the red-green colour channel of the trichromat. They additionally demonstrate that red-greenness of young leaves is positively correlated with the ratio of protein content to toughness. These observations formed the basis for the hypothesis that the evolution of both the routine trichromacy of catarrhines and the polymorphic trichromacy of platyrrhines results from the ability to detect fallback foods (Dominy et al., 2003).

Dominy et al. (2003) hypothesize that the evolution of routine trichromacy in catarrhines is the result of selection for colour vision that allows for the detection of fallback foods that sustain catarrhine populations during periods of low overall fruit availability[1]. In the case of catarrhines, trichromacy thus evolved as a means of detecting young, red leaves as fallback foods. Fallback foods have been operationally defined as "items assumed to be of relatively poor nutritional quality and high abundance, eaten particularly during periods when preferred foods are scarce" (Marshall & Wrangham, 2007, p. 1220). Dominy et al. (2003) posit that climatic cooling at the end of the Eocene resulted in the local extinction or decimation of figs and palms which led catarrhines to "fall back" on leaf resources in periods of low fruit availability. In the neotropics, where figs and palms remain abundant, platyrrhines have evolved mixed capabilities for the detection of these cryptic fallback resources.

Young leaves in the Old World are often red (Dominy et al., 2002). An estimated 50-62% of Old World species display red young leaf flush, compared to only 18-36% of New World species (Dominy & Lucas, 2001). The redness of young leaves results from delayed greening wherein plants postpone chloroplast function until full leaf expansion (Dominy & Lucas, 2001), coupled with the presence of anthocyanin pigment (Lee et al. 1987) which has several potential selective advantages to plants. Advantages to plants include fungicidal properties, photoprotection against UV damage of new leaves, the prevention of photoinhibition, and crypsis to protect against dichromatic herbivores (Coley & Aide, 1989; Gould et al., 1995; Stone, 1979). In a review of the evidence for and against these hypotheses, Dominy et al. (2003) conclude that crypsis is the most plausible, and that young leaf reddening is thus a plant strategy to reduce new leaf damage by herbivores. Providing evidence from the colour space occupied by young leaves, Dominy et al. (2002) conclude that to a dichromatic herbivore a young, red leaf would appear dark, dead, and not worth consuming. In Old World forests where young leaf flush coincides with low fruit availability, the ability to detect these important fallback foods would provide a clear advantage favoring trichromacy (Dominy & Lucas, 2001).

---

[1] Though Dominy et al. (2003) use the term "keystone resources" to refer to young leaves and figs, they define these as "resources (that) sustain frugivore populations during crucial periods" and that "consistently provide food during community-wide periods of fruit dearth" (p. 27). Keystone resources are more commonly defined as preferred resources (Terborgh, 1986). Dominy et al.'s (2003) definition is more commonly applied to foods that are considered "fallback foods" (Marshall & Wrangham, 2007) and as such we use that term here.

4. The protein limitation hypothesis

Trichromacy is useful for visually detecting long-wavelength colours of food targets against a green mature leafy background, whether the food targets are young red leaves or ripe fruits. The debate regarding whether frugivory or folivory has favored trichromacy continues. Both sides are supported and perhaps both the frugivory and the folivory hypotheses are partially correct. Here we present a new hypothesis that builds on both the frugivory and the young leaf hypotheses. We suggest that trichromacy is adaptive for finding a limiting resource critical to primates - protein. We further suggest that selective pressures may vary between catarrhine and platyrrhine primates based on food availability and body size constraints. Specifically, while protein sources for Old World primates are predominantly red leaves, among New World primates, they are red figs. Below we outline the basis for our unifying protein limitation hypothesis by reviewing the evidence that protein is a crucial limiting resource and the evidence that red figs provide a consistent and favored source of protein for New World primates and that trichromats have a foraging advantage on these fruits.

4.1 Protein as a limiting resource to frugivorous primates

Liebig's Law of the Minimum states that the functioning of a given organism is controlled (or limited) by an environmental factor or combination of factors present in the least favorable amount (Taylor, 1934). Since its definition, limiting resources have been shown to have profound effects on a wide array of species (Interlandi & Kilham, 2001). For primates that rely on ripe fruit as a dietary staple, that limiting factor is protein (Ganzhorn et al., 2009). Fruits are not considered to be sufficiently high in protein to meet the nutritional needs of primates (Kay, 1984; Milton, 1979; Oftedal, 1991), without supplementation from either leaves (in large bodied primates) or insects (in small bodied primates) (Fleagle, 1999). Obligate frugivores should feed on proteinaceous fruits whenever available, whereas facultative frugivores can supplement fruits with insects, seeds and leaves (Kunz & Diaz, 1995; Snow, 1981). Models of the evolution of primate diversity have identified the importance of the constraints of protein availability (Milton, 1979). The biomass of folivorous primates has been linked to nitrogen: fiber ratios of leaves in forests indicating a profound effect of protein availability on folivores (Chapman et al., 2004; Ganzhorn, 1992). Likewise in frugivores, protein as well as mineral content have been identified as factors influencing dietary selectivity (Barclay, 1995; Felton et al., 2009; Kunz & Diaz, 1995; O'Brien et al., 1998; Thomas, 1984; Wendeln et al., 2000). Further evidence for the importance of protein comes from nutritional analyses of fruits consumed by spider monkeys (*Ateles chamek*, Humboldt, 1812). Spider monkeys were found to regulate their daily protein intake much more tightly than either carbohydrates or fats, and their protein intake did not vary across seasons regardless of fluctuations in food availability (Felton et al., 2009).

Taken together, there is compelling evidence that protein is a limiting resource for primates. We suggest that acquisition of protein can provide a unified explanation of the adaptiveness of trichromacy. Protein acquisition by Old World monkeys and apes is facilitated by trichromacy via improved search efficiency for young, proteinaceous red leaves (after Dominy & Lucas, 2001). However, platyrrhines generally consume far fewer leaves than catarrhines (Dominy and Lucas, 2001). Given the general lack of leaf consumption by frugivorous platyrrhines and the abundance of proteinaceous figs in the neotropics, *Ficus* plants are a strong candidate for helping to meet the protein requirements of platyrrhines. For the smaller bodied, less folivorous neotropical primates, we suggest that improved foraging efficiency for red, proteinaceous figs favors trichromacy. In New World tropical forests primates prefer feeding on figs over other ripe fruits (Felton et al., 2009; Melin et al., 2009; Parr et al., In Press) and polymorphic trichromacy is the rule for all but one monochromatic nocturnal genus (*Aotus*, Gray, 1870), and one routinely trichromatic folivorous genus (*Alouatta*, Gray, 1849) (Jacobs, 2008). Following previous research, we submit that this may be maintained by heterozygote advantage (Hedrick, 2007) to trichromatic females via an increased ability to detect ripe fruits (Osorio et al., 2004; Osorio & Vorobyev, 1996; Smith et al., 2003; Sumner & Mollon, 2000a; Surridge et al., 2003). However, we emphasize that the primary advantage lies in the detection of conspicuous figs, which differ from other fruits in their high protein content. Our hypothesis differs from previous ideas concerning fruit foraging, which suppose fruit sugars are the primary reward. We suggest that protein, relatively abundant in figs, is the primary element favoring trichromacy.

In a comprehensive review of paleogeographic and biogeographic evidence for the dearth of *Ficus* plants in the Old World, Dominy et al. (2003) posit that climatic cooling at the end of the Eocene resulted in the local extinction or decimation of cryptically coloured (greenish-brownish), relatively nutritionally poor fruits of figs and palms in the Old World. They suggest that where figs and palms remain abundant in the neotropics, platyrrhine polymorphism (as opposed to routine trichromacy) is adaptive because dichromacy is suitable for the detection of these cryptic fallback resources. The hypothesis is compelling, elegant and has received empirical support. Additionally, recent studies have corroborated the importance of fallback foods in the evolution of primate traits (Lambert et al., 2004). However, we suggest that two components require a closer look: 1) the role of figs in the foraging decisions of platyrrhines, and 2) the nutritional value of figs as a key source of protein. Here, we suggest that properly identifying the role of *Ficus* species allows for an ecological refinement of the young leaf hypothesis. Rather than figs as fallback resources, as they may be for Old World primates such as orangutans (Marshall et al., 2009) we propose that at least some species (i.e those producing red fruits) are preferred and limiting resources to neotropical primates.

In the sections below, we present evidence for the nutritional value of fruits of the genus *Ficus*, behavioural evidence of primate feeding preference for *Ficus*, and comparative abundance data for *Ficus* in Africa and the neotropics. We additionally review behavioural and molecular evidence for the adaptive value of primate dichromacy and trichromacy, and the balancing selection that might serve to maintain polymorphic trichromacy.

4.2 Ficus as protein

There are approximately 750 extant species of the genus *Ficus* which together constitute one of the most distinctive and peculiar genera of tropical flora in the world (Shanahan et al., 2001). The unique complex obligatory mutualism between *Ficus* species and pollinating agaonid fig wasps results in a universal pattern of asynchronous and unpredictable mast fruiting events (Janzen, 1979). Fruits of the genus *Ficus* are not in fact strictly fruits, but syconia, or inflorescences. Syconia are clusters of multiple flowers used in pollination, and bundled together in a single package with thousands of miniscule seeds (Janzen, 1979). These inflorescences are entered by one or more species-specific pollinating agaonid fig wasps through an ostiole in the pericarp. Wasps pollinate the florets inside the fruit as well as laying a single egg in each of the ovaries. Wasp larvae live inside the seed coat and eat developing seeds. A month or so later, wingless male wasps emerge from seed coats and mate with females through holes they have cut in the sides of ovaries. Developed female wasps are dependent upon males for this service, and where males fail to cut holes in the sides of *Ficus* ovaries, females remain trapped in their larval home. In addition to these mutualists, numerous species of parasitic Hymenoptera have been discovered. These species oviposit through the wall of the unripe fig without providing pollination services (Janzen, 1979).

The mutualism between figs and wasps, and its interruption by parasitic ovipositing Hymenoptera render the fig syconia (which we shall hereafter refer to as fruit) a bundle of protein-rich insects and larvae. Not only can grown female wasps become trapped inside of ovaries, but at some times of year parasitic larvae become so numerous that hundreds of fig fruits are eaten hollow and found completely filled with larvae (Janzen, 1979). In addition to wasp larvae, fig fruits can contain several other arthropod species, including mites, dipterans and nematodes (Frank, 1989). One analysis of a neotropical *Ficus* species found that animal matter from insects contributed 2.9% to fig dry matter, increasing protein content by 15.3% (Urquiza-Haas et al., 2008).

In addition to the presence of protein-rich insects, there is evidence of unique nutritional qualities of the actual fig fruits themselves. Many authors have referred to figs as low quality foods (Bronstein & Hoffman, 1987; Herbst, 1986; Jordano, 1983; Lambert, 1989). However, overall, nutritional analyses of tropical figs indicate that fig fruits contain moderate-to-high levels of protein relative to other fruits and relative to frugivore protein requirements (Conklin & Wrangham, 1994; Dierenfeld et al., 2002; Ganzhorn, 1988; Ganzhorn et al., 2009; Goodman et al., 1997; Kendrick et al., 2009; Molloy & Hart, 2002; O'Brien et al., 1998; Rogers et al., 1990; Wendeln et al., 2000). In one study, African fig fruits - without supplementation - were found to provide an acceptable baseline level of protein (Conklin & Wrangham, 1994). Neotropical figs may have nutritional value adequate to sustain at least some frugivores without additional food (Wendeln et al., 2000). A study of two neotropical *Ficus* species found they contained 4.5% - 6.1% (dry weight) protein, which is 2-3 times higher than that of a sympatric neotropical angiosperm, *Spondias mombin* (Anacardiaceae), and are more nutritionally balanced with higher protein:TNC and lipid ratios than sympatric species (Hladik et al., 1971). Evidence for the nutritionally balanced nature of figs is further indicated by the fact that some species of bat subsist entirely on certain *Ficus* species (Janzen, 1979; Wendeln et al., 2000) and capuchin monkeys decrease time spent foraging for insects, their primary protein source, on days they visit figs (Parr et al., In Press). Comparison of protein content (as proxied by percent nitrogen) of New World and Malagasy figs indicates that neotropical fruits contain enough protein to satisfy the needs of primates (7-11% protein, 1.1- 1.8% nitrogen), while those in Madagascar do not (Ganzhorn et al., 2009). Figs, then, might be well identified as important sources of protein for New World frugivorous primates. In order to determine the possible extent of this importance, it is necessary to review not only the nutritional value of figs, but also primate feeding patterns.

4.3 Who eats figs?

The concise answer to the question of who eats figs was offered by Daniel Janzen (1979): "Everybody." Indeed, figs are notable for the number of bird and mammal species that consume their fruits and they comprise a part of the diet for more species of animal than any other genus of wild tropical perennial fruit (Janzen, 1979). In this section, we focus on the role of figs in the diet and feeding ecology of platyrrhines.

A review of the literature on neotropical primates reveals that all diurnal frugivorous genera with a body mass exceeding Kay's 500g threshold (Fleagle, 1999) include figs in their diet to varying extents, with the exception of uakari monkeys of the genus *Cacajao* (Mivart, 1865) for which dietary information is scant (Boubli, 1999). Tamarins (*Saguinus* species) and marmosets (*Callithrix kuhlii*, Coimbra-Filho, 1985) have been found to include a variety of fig species in their diet across sites (Knogge et al., 2003; Raboy et al., 2008; Terborgh, 1983). Three of the four *Ficus* species present at a Brazilian site are included in the 55 "extremely valuable" species consumed by golden-lion tamarins (*Leontopithecus chrysomelas*, Kuhl, 1820) (Oliveira et al., 2009). The seventh most frequently consumed fruit of black titi monkeys (*Callicebus torquatus lugens*, Hoffmannsegg, 1807) in Columbia was the single species *Ficus mathewsii* (Palacios et al., 1997). At five different sites in South America (Yasuni, Caparu, Urucu, Tinigua and Pacaya) figs were one of the top seven genera consumed by woolly monkeys (*Lagothrix lagotricha*, Humboldt, 1812) at each site (di Fiore, 2004). In Cocha Cashu, Peru, brown capuchin monkeys (*Cebus apella*, Linnaeus, 1758) were found to eat figs whenever they were available, and South American squirrel monkeys (*Saimiri sciurius*, Linnaeus, 1758) and white-fronted capuchin monkeys (*Cebus albifrons*, Humboldt, 1812) concentrated feasting on figs whenever they were available (Terborgh, 1983). White-faced capuchins (*Cebus capucinus*) in Santa Rosa, Costa Rica prefer figs over all other ripe fruits, and concentrate their foraging efforts on *Ficus* whenever available, regardless of the presence and abundance of other ripe fruits in their habitat (Melin et al., 2009). Figs comprised nearly one-third of capuchin annual fruit foraging effort which represents over-selection relative to fig abundance (Melin et al., 2009). For black-faced spider monkeys (*Ateles chamek*) in the Guarayos Forest Reserve in Bolivia, the preference for figs relative to all other ripe fruits is striking, and when combined with data on habitat-wide fruit availability indicates that "spider monkeys consumed a diverse array of ripe fruits to overcome periods of fig scarcity rather than vice versa" (Felton et al., 2009). Even species that are considered to be folivorous show a preference for *Ficus* fruits. One study of Mexican howler monkeys (*Alouatta palliata mexicana*, Merriam, 1902) found that six *Ficus* species were amongst the eight most important species consumed at one site in Veracruz, Mexico (Serio-Silva et al., 2002). Diurnal frugivorous and folivorous neotropical primates thus all eat figs to a greater or lesser extent, with some species showing an unequivocal preference for them even during periods of high habitat wide fruit availability. The feeding ecology of platyrrhines, coupled with preliminary evidence for the protein content of neotropical figs indicate that rather than viewing figs as a fallback foods in times of habitat-wide fruit scarcity (Dominy et al., 2003; Shanahan et al., 2001) it may be more appropriate to view them as important, preferred and limiting resources.

4.4 The relative abundance of Ficus species in Africa and the Neotropics

In order for figs to be a more important source of protein for New World monkeys vis-à-vis catarrhines, it is necessary to demonstrate that figs are abundantly present in the neotropics relative to Old World forests. Fig abundance is especially salient given its asynchronous and unpredictable fruiting phenology (Janzen, 1979). We reviewed a worldwide database of plant species abundance maintained by the Center for Tropical Forest Science of the Smithsonian Tropical Research Institute (STRI) as well as our own data and published species abundance data to test the prediction that *Ficus* is more abundant in the neotropics than in Africa. Using the online database maintained by STRI, we counted the number of *Ficus* species and the number of individuals of each species recorded during the most recent census of each forest plot in Africa and in the neotropics for which data are currently available (N=3 African plots and 5 neotropical plots). A search of tree abundance literature uncovered three additional datasets containing information on *Ficus* abundance: one from Budongo Forest Reserve in Uganda (Tweheyo & Babweteera, 2007), one from Agaltepec Island in Mexico (Serio-Silva et al., 2002), and one from Santa Rosa National Park, Costa Rica (Parr et al., In Press).

In all plots censused by STRI all *Ficus* individuals >10cm diameter at breast height (DBH) were recorded. In the Budongo Forest Reserve in Uganda all *Ficus* trees >4cm DBH were recorded, and as such abundance data are overestimated. In the Agaltepec Island plot all *Ficus* trees >30cm DBH were recorded, and as such abundance data may be an underestimate. Parr et al. (In Press) recorded all *Ficus* trees >10cm DBH encountered in a transect survey of Sector Santa Rosa, Costa Rica.

Fig species are both more diverse and abundant in the New World relative to the Old World (Table 1). In the African plots *Ficus* species diversity ranges from two to four species per plot (N=70 trees, 114 hectares, 4 plots), with a mean number of 0.73 *Ficus* trees per hectare (range = 0.06 - 2.45). In the neotropical plots *Ficus* species diversity ranges from three to 12 species per plot (N = 458 trees, 145.96 hectares, 7 plots), with a mean number of 3.89 *Ficus* trees per hectare (range = 0.5 - 11.56). While the sample size is inappropriate for a statistical analysis of *Ficus* abundance data, the pattern in the data is clear. Figs are both more speciose and abundant in the neotropical plots relative to the African plots. This pattern holds despite one of the four African plot values representing an abundance overestimate (Budongo Forest Reserve, Uganda), and one of the six neotropical plot abundance values representing an underestimate (Agaltepec Island, Mexico). Removing the overestimated abundance data from Budongo Forest Reserve, the *Ficus* abundance in the remaining three African plots never meets or exceeds the abundance of the most *Ficus*-poor neotropical plot (Sherman, Panama, 0.5 individuals per hectare). This dataset supports the prediction of relatively high abundance of neotropical figs relative to African figs that has been noted elsewhere (Gautier-Hion & Michaloud, 1989; Shanahan et al., 2001).

4.5 Fig foraging

Colour vision polymorphism has persisted in primates for up to 14 million years (Surridge & Mundy, 2002). In order for polymorphic trichromacy in platyrrhines to be considered as an adaptive strategy it must be demonstrated that there is some advantage to the maintenance of different colour vision phenotypes within a population (Melin et al., 2008). There exists ample evidence for a trichromat advantage during ripe fruit foraging (Melin et al., 2009; Regan et al., 1998; Riba-Hernández et al., 2005; Smith et al., 2003; Sumner & Mollon, 2000b; Yamashita et al., 2005). To our knowledge, the only published study on fig foraging by wild platyrrhines (Melin et al., 2009) found that trichromatic white-faced capuchin monkeys (*Cebus capucinus*) demonstrate a foraging advantage when feeding on ripe figs. Conspicuous figs were more common than cryptic figs at the research site, and were fed on much more frequently. Trichromatic monkeys had a higher acceptance index of conspicuous figs than did dichromats, suggesting an ability to select riper figs based on visual cues. Dichromats, on the other hand, used longer and more diverse foraging sequences (e.g. frequent sniffing) when assessing figs. While the variation in foraging behaviour did not result in a net variation of fig intake rate, the authors suggest that it is possible that the improved discrimination ability of trichromats may lead to higher feeding rates under conditions where ripe fruit is less available and quickly depleted. Additionally, trichromatic monkeys seemed better able to select the ripest figs, while dichromats consumed mid-ripe (yellowish) figs more often (A.D Melin, pers. obs).

Site	Plot Size (hectares)	Census Date	N species	N Individuals	N Individuals per hectare	Source
Korup, Cameroon	50	1998	N/A	3	0.06	STRI
Edoro, Democratic Republic of Congo	20	2000	2	2	0.10	STRI
Lenda, Democratic Republic of Congo	20	2000	4	6	0.30	STRI
Budongo Forest Reserve, Uganda	24.13	2006	4*	59	2.45	Tweheyo and Babweteera 2007
La Planada, Columbia	25	2003	7	289	11.56	STRI
Yasuni, Ecuador	50	2003	12	30	0.60	STRI
BCI, Panama	50	2005	8	42	0.84	STRI
Cocoli, Panama	4	1998	4	9	2.25	STRI
Sherman, Panama	6	1999	3	3	0.50	STRI
Sector Santa Rosa, Costa Rica	2.66	2010	8*	4	1.5**	Parr et al, In Press
Agaltepec Island, Mexico	8.3	1996	6	81	9.76	Serio-Silva et al. 2002

* 8 species of *Ficus* are known in Sector Santa Rosa (Melin et al., 2009) however only 2 species have been encountered to date in transects (Parr et al., In Press).
** If only those fig trees with a DBH of 10cm or greater are included, then the number of individuals drops to three, and density drops to 1.13 (N.Parr, pers. comm.).

Table 1. Number of species and individuals of the genus *Ficus* in African and neotropical forests.

5. Conclusion

While most hypotheses for the variation in colour vision capabilities of catarrhines and platyrrhines posit the importance of either the ability to detect ripe fruit (Regan et al., 2001) or young leaves (Dominy et al., 2003) there is evidence that this variation stems from visual adaptive strategies aimed at the detection of proteinaceous foods. In continental Africa where frugivorous primates rely on young, reddish leaves as protein sources (Dominy et al., 2003), routine trichromacy is the rule. In New World tropical forests efficient foraging for conspicuously coloured proteinaceous figs may favor trichromats and maintain colour vision polymorphism via heterozygote advantage. It is also worthwhile to mention that the selection pressures acting on trichromacy may be less in the New World because insects represent a protein source to many neotropical monkeys and dichromacy is useful in the detection of surface-dwelling insects (Melin et al., 2007; Melin et al., 2010) via the enhanced ability of dichromats to break camouflage (Caine et al., 2010; Morgan et al., 1992).

Rather than viewing figs as fallback resources (Dominy et al., 2003; Shanahan et al., 2001) we propose that they are preferred and limiting resources. Additionally, given the timing of Old World phenophases, where young leaves are most widely available during periods of overall fruit scarcity (Dominy et al., 2003), it is not possible to clearly identify young leaves as fallback foods, since the definition of a fallback food requires spatial and temporal ubiquity (Marshall & Wrangham, 2007). The evolution and maintenance of polymorphic trichromacy is a response to variation in the detectability not of fallback foods, but of the limiting resources of frugivorous primates: proteinaceous foods.

Our analysis has been restricted to continental Africa and the neotropics due to the availability of data from these continents. However, the recent discovery of polymorphic trichromacy amongst some Malagasy strepsirrhines (Tan & Li, 1999; Veilleux & Bolnick, 2009) will provide an interesting comparative dataset against which to test hypotheses. Under the protein limitation hypotheses, we predict that balancing selection will favor polymorphic trichromacy via a heterozygote advantage in frugivorous strepsirrhines whose protein source is conspicuous. Given current evidence for the low abundance of *Ficus* in Malagasy forests and the recent discovery of a paucity of available protein in Malagasy *Ficus* fruits (Ganzhorn et al., 2009), it seems likely that Malagasy protein resources and resulting effects on strepsirrhine colour vision capabilities will be found elsewhere.

Many hypotheses regarding primate morphological adaptations hinge on the importance of fallback foods (Dominy & Lucas, 2001; Marshall & Wrangham, 2007), and foods that are dominant in primate diets (Osorio & Vorobyev, 1996; Regan et al., 2001; Smith et al., 2003). This represents an unusual departure from studies of primate behaviour which tend to focus on the centrality of limiting resources (e.g. Sterck et al., 1997; van Schaik, 1989; Wrangham, 1980). Studies of primate foraging adaptations might benefit from the additional consideration of morphological and sensory adaptations to foraging on limiting resources.

6. Acknowledgements

We would like to thank the staff of the Área de Conservación Guanacaste, especially Roger Blanco, Maria Marta Chavarria and the administration of the Sector Santa Rosa, for help and for permission to conduct research in the park. We also thank the STRI Center for Tropical Forest Science: all *Ficus* data denoted as being from STRI came from their website: www.ctfs.si.edu. The authors gratefully acknowledge Adrián Guadamuz, John Addicott, Mike Lemmon, Barbara Kowalsic and Nigel Parr for contributions to the *Ficus* project, as well as Linda Fedigan, Carl Toborowsky, Adrienne Blauel, Mackenzie Bergstrom, Brandon Klug, Laura Weckman, Fernando Campos and Valerie Schoof for their assistance. We thank

E.C. Kirk and N.J. Dominy for helpful comments on the manuscript. For grant support we thank the Leakey Foundation, the Wenner-Gren Foundation, Alberta Ingenuity Fund, NSERC, Animal Behaviour Society (ADM), NSERC and the University of Texas at Austin (KV).

7. References

Allen, G. (1879). *The Colour-Sense: Its Origin and Development*, Trubner, ISBN 978-054-8102-93-0, London

Araújo, A.C.; Didonet, J.J.; Araújo, C.S.; Saletti, P.G.; Borges. T.R. & Pessoa, V.F. (2008). Colour vision in the black howler monkey (*Alouatta caraya*). *Visual Neuroscience*, Vol.25, No.3, pp. 243-248, ISSN 0952-5238

Barclay, R.M.R. (1995). Does energy or calcium availability constrain reproduction by bats? *Symposia of the Zoological Society of London*, Vol.67, pp. 245-258, ISSN 0084-5612

Barton, R.A. (2006). Olfactory evolution and behavioural ecology in primates. *American Journal of Primatology*, Vol.68, No.6, pp. 545-558, ISSN 0275-2565

Boubli, J.P. (1999). Feeding ecology of black-headed uacaris (*Cacajao melanocephalus melanocephalus*) in Pico da Neblina National Park, Brazil. *International Journal of Primatology*, Vol.20, No.5, pp. 719-749, ISSN 0164-0291

Bronstein, J.L. & Hoffman, K. (1987). Spatial and temporal variation in frugivory at a neotropical fig, *Ficus pertusa*. *Oikos*, Vol.49, pp. 261-268, ISSN 0030-1299

Caine, N.G.; Osorio, D. & Mundy, N.I. (2010). A foraging advantage for dichromatic marmosets (*Callithrix geoffroyi*) at low light intensity. *Biology Letters*, Vol.6, pp. 36-38, ISSN 1744-957X

Caine, N.G.; Surridge, A.K. & Mundy, N.I. (2003). Dichromatic and trichromatic *Callithrix geoffroyi* differ in relative foraging ability for red-green colour-camouflaged and non-camouflaged food. *International Journal of Primatology*, Vol.24, No.6, pp. 1163-1175, ISSN 0164-0291

Chapman, C.A.; Chapman, L.J.; Naughton-Treves, L.; Lawes, M.J. & McDowell, L.R. (2004). Predicting folivorous primate abundance: validation of a nutrition model. *American Journal of Primatology*, Vol.62, pp. 55-69, ISSN 0275-2565

Chapman, C.A. & Onderdonk, D.A. (1998). Forests without primates: primate/plant codependency. *American Journal of Primatology*, Vol.45, No.1, pp. 127-141, ISSN 0275-2565

Coley, P.D. & Aide, T.M. (1989). Red colouration of tropical leaves: A possible anti-fungal defence? *Journal of Tropical Ecology*, Vol.5, pp. 293-300, ISSN 0266-4674

Conklin, N.L. & Wrangham, R.W. (1994). The value of figs to a hind-gut fermenting frugivore: A nutritional analysis. *Biochemical Systematics and Ecology*, Vol.22, No.2, pp. 137-151, ISSN 0305-1978

di Fiore, A. (2004). Diet and feeding ecology of woolly monkeys in a western Amazonian rain forest. *International Journal of Primatology*, Vol.25, No.4, pp. 767-801, ISSN 0164-0291

Dierenfeld, E.S.; Mueller, P.J. & Hall, M.B. (2002). Duikers: native food composition, micronutrient assessment, and implications for improving captive diets. *Zoo Biology*, Vol.21, pp. 185-196, ISSN 0733-3188

Dominy, N.J. & Lucas, P.W. (2001). Ecological importance of trichromatic vision to primates. *Nature*, Vol.410, No.6826, pp. 363-366, ISSN 0028-0836

Dominy, N.J.; Lucas, P.W.; Ramsden, L.W.; Riba-Hernandez, P.; Stoner, K.E. & Turner, I.M. (2002). Why are young leaves red? *Oikos* Vol.98, No.1, pp. 163-176, ISSN 0030-1299

Dominy, N.J.; Svenning, J.C. & Li, W.H. (2003). Historical contingency in the evolution of primate colour vision. *Journal of Human Evolution*, Vol.44, No.1. pp. 25-45, ISSN 0047-2484

Felton, A.M.; Felton, A.; Wood, J.T.; Foley, W.J.; Raubenheimer, D.; Wallis, I.R. & Lindenmayer, D.B. (2009). Nutritional Ecology of *Ateles chamek* in lowland Bolivia: how macronutrient balancing influences food choices. *International Journal of Primatology*, Vol.30, No.5, pp. 675-696, ISSN 0164-0291

Fleagle, J.G. (1999). *Primate Adaptation and Evolution*, Academic Press, ISBN 978-012-2603-41-9, New York.

Frank, S.A. (1989). Ecological and evolutionary dynamics of fig communities. *Experientia*, Vol.45, pp. 674-680, ISSN 0014-4754

Ganzhorn, J.U. (1988). Food partitioning among Malagasy primates. *Oecologia*, Vol.75, pp. 436-450, ISSN 0029-8549

Ganzhorn, J.U. (1992). Leaf chemistry and the biomass of folivorous primates in tropical forests. *Oecologia*, Vol.91, No.4, pp. 540-547, ISSN 0029-8549

Ganzhorn, J.U.; Arrigo-Nelson, S.; Boinski, S.; Bollen, A.; Carrai, V.; Derby, A.; Donati, G.; Koenig, A.; Kowalewski, M. & Lahann, P. (2009). Possible fruit protein effects on primate communities in Madagascar and the neotropics. *PloS One*, Vol.4, No.12, pp. 1-8, ISSN 1932-6203

Garber, P.A. & Lambert, J.E. (1998). Primates as seed dispersers: ecological processes and directions for future research. *American Journal of Primatology*, Vol.45, pp. 3-8, ISSN 0275-2565

Gautier-Hion, A.; Duplantier, J.M.; Quris, R.; Feer, F.; Sourd, C.; Decoux, J.P.; Dubost, G.; Emmons, L.; Erard, C. & Hecketsweiler, P. (1985). Fruit characters as a basis of fruit choice and seed dispersal in a tropical forest vertebrate community. *Oecologia*, Vol.65, No.3, pp. 324-337, ISSN 0029-8549

Gautier-Hion, A. & Michaloud, G. (1989). Are figs always keystone resources for tropical frugivorous vertebrates? A test in Gabon. *Ecology*, Vol.70, No.6, pp. 1826-1833, ISSN 0012-9658

Goodman, S.M.; Ganzhorn, J.U. & Wilme, L. (1997). Observations at a *Ficus* tree in Malagasy humid forest. *Biotropica*, Vol.29, No.4, pp. 480-488, ISSN 0006-3606

Gould, K.S.; Kuhn, D.N.; Lee, D.W. & Oberbauer, S.F. (1995). Why leaves are sometimes red. *Nature*, Vol.378, pp. 241-242, ISSN 0028-0836

Hedrick, P.W. (2007). Balancing selection. *Current Biology*, Vol.17, No.7, pp. R230-R231, ISSN 0960-9822

Herbst, L.H. (1986). The role of nitrogen from fruit pulp in the nutrition of a frugivorous bat, *Carollia perspicillata*. *Biotropica*, Vol.18, pp. 39-44, ISSN 0006-3606

Hiramatsu, C.; Melin, A.D.; Aureli, F.; Schaffner, C.M.; Vorobyev, M.; Matsumoto, Y. & Kawamura, S. (2008). Importance of achromatic contrast in short-range fruit foraging of primates. *PLoS ONE*, Vol.3, No.10, pp. 3356, ISSN 1932-6203

Hladik, C.M.; Hladik, A.; Bousset, J.; Valdebouze, P.; Viroben, G. & Delort-Laval, J. (1971). Le regime alimentaire de primate de l'il de Barro Colourado (Panama). *Folia Primatologica*, Vol.16, pp. 85-122, ISSN 0015-5713

Hunt, D.M.; Carvalho, L.S.; Cowing, J.A. & Davies, W.L. (2009). Evolution and spectral tuning of visual pigments in birds and mammals. *Philosophical Transactions of the Royal Society of London B*, Vol.364, No.1531, pp. 2941-2955, ISSN 1364–503X

Interlandi, S.J. & Kilham, S.S. (2001). Limiting resources and the regulation of diversity in phytoplankton communities. *Ecology*, Vol.82, No.5, pp. 1270-1282, ISSN 0012-9658

Jacobs, G.H. (2007). New world monkeys and colour. *International Journal of Primatology*, Vol.28, pp. 729-759, ISSN 0164-0291

Jacobs, G.H. (2008). Primate colour vision: a comparative perspective. *Visual Neuroscience*, Vol.25, pp. 1-15, ISSN 0952-5238

Jacobs, G.H.; Neitz, M.; Deegan, J.F. & Neitz, J. (1996). Trichromatic colour vision in New World monkeys. *Nature*, Vol.382, No.6587, pp. 156-158, ISSN 0028-0836

Janson, C.H. (1983). Adaptation of fruit morphology to dispersal agents in a neotropical forest. *Science*, Vol.219, No.4581, pp. 187-189, ISSN 0036-8075

Janzen, D.H. (1979). How to be a fig. *Annual Review of Ecology and Systematics*, Vol.10, No.1, pp. 13-51, ISSN 0066-4162

Jordano, P. (1983). Fig-seed predation and dispersal by birds. *Biotropica*, Vol.15, No.1, pp. 38-41, ISSN 0006-3606

Kay, R.F. (1984). On the use of anatomical features to infer foraging behaviour in extinct primates. In: *Adaptations for Foraging in Nonhuman Primates*, Rodman, P.S. & Cant, J.G.H. (Eds.), 21-53, Columbia University Press, New York, ISBN 978-023-1052-27-6

Kendrick, E.L.; Shipley, L.A.; Hagerman, A.E. & Kelley, L.M. 2009. Fruit and fibre: the nutritional value of figs for a small tropical ruminant, the blue duiker (*Cephalophus monticola*). *African Journal of Ecology*, Vol.47, pp. 556-566, ISSN 0141-6707

Knogge, C.; Herrera, E.R.T. & Heymann, E.W. (2003). Effects of passage through tamarin guts on the germination potential of dispersed seeds. *International Journal of Primatology*, Vol.24, No.5, pp. 1121-1128, ISSN 0164-0291

Kunz, T.H. & Diaz, C.A. (1995). Folivory in fruit-eating bats, with new evidence from *Artibeus jamaicensis* (Chiroptera: Phyllostomidae). *Biotropica*, Vol.27, pp. 106-120, ISSN 0006-3606

Lambert, F.R. (1989). Fig-eating by birds in a Malaysian lowland rain forest. *Journal of Tropical Ecology*, Vol.5, pp. 401-412, ISSN 0266-4674

Lambert, J.E.; Chapman, C.A.; Wrangham, R.W. & Conklin-Brittain, N.L. (2004). The hardness of cercopithecine foods: Implications for the critical function of enamel thickness in exploiting fallback foods. *American Journal of Physical Anthropology* Vol.125, pp. 363-368, ISSN 0002-9483

Lee, D.W.; Brammeier, S. & Smith, A.P. (1987). The selective advantages of anthocyanins in developing leaves of mango and cacao. *Biotropica* Vol.19, No.1, pp. 40-49, ISSN 0006-3606

Link, A. & Stevenson, P.R. (2004). Fruit dispersal syndromes in animal disseminated plants at Tinigua National Park, Colombia. *Revista Chilena de Historia Natural* Vol.77, pp. 319-334, ISSN 0716-078X

Lucas, P.W.; Dominy, N.J.; Riba-Hernandez, P.; Stoner, K.E.; Yamashita, N.; Petersen-Pereira, W.; Salas-Pena, R.; Solis-Madrigal, S.; Osorio, D. & Darvell, B.W. (2003). Evolution and function of routine trichromatic vision in primates. *Evolution*, Vol.57, No.11, pp. 2636-2643, ISSN 1558-5646

Marshall, A.; Boyko, C.M.; Feilen, K.L.; Boyko, R.H. & Leighton, M. (2009). Defining fallback foods and assessing their importance in primate ecology and evolution. *American Journal of Physical Anthropology*, Vol.140, pp. 603-614, ISSN 0002-9483

Marshall, A.J. & Wrangham, R.W. (2007). Evolutionary consequences of fallback foods. *International Journal of Primatology*, Vol.28, pp. 1219-1235, ISSN 0164-0291

Melin, A.D. (2011). Polymorphic Colour Vision and Foraging in White-faced Capuchins: Insights from Field Research and Simulations of Monkey Vision. PhD Dissertation. Calgary: University of Calgary.

Melin, A.D.; Fedigan, L.M.; Hiramatsu, C.; Hiwatashi, T.; Parr, N. & Kawamura, S. (2009). Fig foraging by dichromatic and trichromatic *Cebus capucinus* in a tropical dry forest. *International Journal of Primatology*, Vol.30, No.6, pp. 753-775, ISSN 0164-0291

Melin, A.D.; Fedigan, L.M.; Hiramatsu, C. & Kawamura, S. (2008). Polymorphic colour vision in white-faced capuchins (*Cebus capucinus*): is there foraging niche divergence among phenotypes? *Behavioural Ecology and Sociobiology*, Vol.62, No.5, pp. 659-670, ISSN 1045-2249

Melin, A.D.; Fedigan, L.M.; Hiramatsu, C.; Sendall, C.L. & Kawamura, S. (2007). Effects of colour vision phenotype on insect capture by a free-ranging population of white-faced capuchins, *Cebus capucinus*. *Animal Behaviour*, Vol.73, No.1, pp. 205-214, ISSN 0003-3472

Melin, A.D.; Fedigan, L.M.; Young, H.C. & Kawamura, S. (2010). Can colour vision variation explain sex differences in invertebrate foraging by capuchin monkeys? *Current Zoology*, Vol.56, pp. 300-312, ISSN 1674-5507

Milton, K. (1979). Factors influencing leaf choice by howler monkeys: a test of some hypotheses of food selection by generalist herbivores. *American Naturalist*, Vol.114, No.3, pp. 362-378, ISSN 0003-0147

Mollon, J.D.; Bowmaker, J.K. & Jacobs, G.H. (1984). Variations of colour vision in a New World primate can be explained by polymorphism of retinal photopigments. *Proceedings of the Royal Society of London Series B, Biological Sciences*, Vol.222, pp. 373-399, ISSN 1364-5021

Molloy, L. & Hart, J.A. (2002). Duiker food selection: palatability trials using natural foods in the Ituri Forest, Democratic Republic of Congo. *Zoo Biology*, Vol.21, pp. 149-159, ISSN 0733-3188

Morgan, M.J.; Adam, A. &Mollon, J.D. (1992). Dichromats detect colour-camouflaged objects that are not detected by trichromats. *Proceedings of the Royal Society B, Biological Sciences*, Vol.248, No.1323, pp. 291-295, ISSN 0962-8452

O'Brien, T.G.; Kinnaird, M. & Dierenfeld, E.S. (1998). What's so special about figs? *Nature*, Vol.392, pp. 668, ISSN 0028-0836

Oftedal, O.T. (1991). The nutritional consequences of foraging in primates: the relationship of nutrient intakes to nutrient requirements. *Philosophical Transactions of the Royal Society of London Series B*, Vol.334, pp. 161-170, ISSN 0962-8436

Oliveira, L.C.; Hankerson, S.J.; Dietz, J.M. & Raboy, B.E. (2009). Key tree species for the golden-headed lion tamarin and implications for shade-cocoa management in southern Bahia, Brazil. *Animal Conservation*, Vol.13, pp. 60-70, ISSN 1367-9430

Osorio, D.; Smith, A.C.; Vorobyev, M. & Buchanan-Smith, H.M. (2004). Detection of fruit and the selection of primate visual pigments for colour vision. *American Naturalist*, Vol.164, No.6, pp. 696-708, ISSN 0003-0147

Osorio, D. & Vorobyev, M. (1996). Colour vision as an adaptation to frugivory in primates. *Proceedings of the Royal Society B*, Vol.263, pp. 593-599, ISSN 0962-8452

Palacios, E.; Rodriguez, A. & Defler, T.R. (1997). Diet of a group of *Callicebus torquatus lugens* (Humboldt, 1812) during the annual resource bottleneck in Amazonian Colombia. *International Journal of Primatology*, Vol.18, No.4, pp. 503-522, ISSN 0164-0291

Parr, N.; Melin, A.D. & Fedigan, L. In Press. Figs are not always fallback foods: The relationship between *Ficus* and *Cebus* in a tropical dry forest. *International Journal of Zoology*, ISSN 2231-3516

Polyak, S. (1957). *The Vertebrate Visual System*, University of Chicago Press, ISBN 978-022-6674-94-0, Chicago

Raboy, B.E.; Canale, G.R. & Dietz, J.M. (2008). Ecology of *Callithrix kuhlii* and a review of eastern Brazilian marmosets. *International Journal of Primatology*, Vol.29, No.2, pp. 449-467, ISSN 0164-0291

Regan, B.C.; Julliot, C.; Simmen, B.; Vienot, F; Charles-Dominique, P. & Mollon, J.D. (1998). Frugivory and colour vision in *Alouatta seniculus*, a trichromatic platyrrhine monkey. *Vision Research*, Vol.38, No.21, pp. 3321-3327, ISSN 0042-6989

Regan, B.C.; Julliot, C.; Simmen, B.; Vienot, F.; Charles-Dominique, P. & Mollon, J.D. (2001). Fruits, foliage and the evolution of colour vision. *Philosophical Transactions of the Royal Society of London B*, Vol. 356, pp. 229-283, ISSN 0962-8436

Riba-Hernández, P.; Stoner, K.E. & Lucas, P.W. (2005). Sugar concentration of fruits and their detection via colour in the Central American spider monkey (*Ateles geoffroyi*). *American Journal of Primatology*, Vol.67, No.4, pp. 411-423, ISSN 0275-2565

Rogers, M.E.; Maisels, F.; Williamson, E.A.; Fernandez, M. & Tutin, C.E.G. (1990). Gorilla diet in the Lope Reserve, Gabon: A nutritional analysis. *Oecologia*, Vol.84, pp. 326-339, ISSN 0029-8549

Serio-Silva, J.C.; Rico-Gray, V.; Hernández-Salazar, L.T. & Espinosa-Gómez, R. (2002). The role of *Ficus* (Moraceae) in the diet and nutrition of a troop of Mexican howler monkeys, *Alouatta palliata mexicana*, released on an island in southern Veracruz, Mexico. *Journal of Tropical Ecology*, Vol.18, No.6, pp. 913-928, ISSN 0266-4674

Shanahan, M.; So, S.; Compton, S.G. & Corlett, R. (2001). Fig-eating by vertebrate frugivores: a global review. *Biological Reviews of the Cambridge Philosophical Society*, Vol.76, No.4, pp. 529-572, ISSN 1464-7931

Smith, A.C.; Buchanan-Smith, H.M.; Surridge, A.K.; Osorio, D. & Mundy, N.I. (2003). The effect of colour vision on the detection and selection of fruits by tamarins (*Saguinus* spp.). *Journal of Experimental Biology*, Vol.206, pp. 3159-3165, ISSN 0022-0949

Snow, D.W. (1981). Tropical frugivorous birds and their food plants: a world survey. *Biotropica*, Vol.13, pp. 1-14, ISSN 0006-3606

Sterck, E.H.M.; Watts, D.P. & van Schaik, C.P. (1997). The evolution of female social relationships in nonhuman primates. *Behavioural Ecology and Sociobiology*, Vol.41, No.5, pp. 291-309, ISSN 1045-2249

Stone, B.C. (1979). Protective colouration of young leaves in certain Malaysian palms. *Biotropica*, Vol.11, pp. 126, ISSN 0006-3606

Stoner, K.E.; Riba-Hernández, P. & Lucas, P.W. (2005). Comparative use of colour vision for frugivory by sympatric species of platyrrhines. *American Journal of Primatology*, Vol.67, No.4, pp. 399-409, ISSN 0275-2565

Sumner, P. & Mollon, J.D. (2000a). Catarrhine photopigments are optimized for detecting targets against a foliage background. *Journal of Experimental Biology*, Vol.203, pp. 1963-1986, ISSN 0022-0949

Sumner, P. & Mollon, J.D. (2000b). Chromaticity as a signal of ripeness in fruits taken by primates. *The Journal of Experimental Biology*, Vol.203, pp. 1987-2000, ISSN 0022-0949

Surridge, A.K. & Mundy, N.I. (2002). Trans-specific evolution of opsin alleles and the maintenance of trichromatic colour vision in Callitrichine primates. *Molecular Ecology*, Vol.11, No.10, pp. 2157-2169, ISSN 0962-1083

Surridge, A.K.; Osorio, D. & Mundy, N.I. (2003). Evolution and selection of trichromatic vision in primates. *Trends in Ecology & Evolution*, Vol.51, pp. 198-205, ISSN 0169-5347

Tan, Y. & Li, W.H. (1999). Trichromatic vision in prosimians. *Nature*, Vol. 402, pp. 36, ISSN 0028-0836:

Tan, Y.; Yoder, A.D.; Yamashita, N. & Li, W.H. (2005). Evidence from opsin genes rejects nocturnality in ancestral primates. *Proceedings of the National Academy of Sciences*, Vol.102, No.41, pp. 14712-14716, ISSN 0027-8424

Taylor, W.P. (1934). Significance of extreme or intermittent conditions in distribution of species and management of natural resources, with a restatement of Liebig's law of minimum. *Ecology*, Vol.15, No.4, pp. 374-379, ISSN 0012-9658

Terborgh, J. (1983). *Five New World Primates: A Study in Comparative Ecology*, Princeton University Press, ISBN 978-069-1083-38-4, New York

Terborgh, J. (1986). Community aspects of frugivory in tropical forests. In: *Frugivores and Seed Dispersal*, Estrada, A. & Fleming, T.H. (Eds.), pp. 370-384, Kluwer Academy Publishers, ISBN 978-906-1935-43-8, New York

Thomas, D.W. (1984). Fruit intake and energy budgets of frugivorous bats. *Physiological Zoology*, Vol.57, pp. 457-467, ISSN 0031-935X

Tweheyo, M. & Babweteera, F. (2007). Production, seasonality and management of chimpanzee food trees in Budongo Forest, Uganda. *African Journal of Ecology*, Vol.45, No.4, pp. 535-544, ISSN 1365-2028

Urquiza-Haas, T.; Serio-Silva, J.C. & Hernandez-Salazar, L.T. (2008). Traditional nutritional analyses of figs overestimates intake of most nutrient fractions: A study of *Ficus perforata* consumed by howler monkeys (*Alouatta palliata mexicana*). *American Journal of Primatology*, Vol.70, pp. 432-438, ISSN 0275-2565

van Schaik, C.P. (1989). The ecology of social relationships amongst female primates. In: Comparative Socioecology: The Behavioural Ecology of Humans and Other Mammals, Standen, V. & Foley, R.A. (Eds.), Blackwell, Oxford

Veilleux, C.C. & Bolnick, D.A. (2009). Opsin gene polymorphism predicts trichromacy in a cathemeral lemur. *American Journal of Primatology*, Vol.71, pp. 86-90, ISSN 0275-2565

Vogel, E.; Neitz, M. & Dominy, N.J. (2007). Effect of colour vision phenotype in the foraging of white-faced capuchins, *Cebus capucinus*. *Behavioural Ecology*, Vol.18, pp. 292-297, ISSN 1045-2249

Wendeln, M.C.; Runkle, J.R. & Kalko, E.K.V. (2000). Nutritional values of 14 fig species and bat feeding preferences in Panama. *Biotropica*, Vol.32, No.3, pp. 489-501, ISSN 0006-3606

Wrangham, R.W. (1980). An ecological model of female-bonded primate groups. *Behaviour*, Vol.75, pp. 262-299, ISSN 0005-7959

Yamashita, N.; Stoner, K.E.; Riba-Hernández, P.; Dominy, N.J. & Lucas, P.W. (2005). Light levels used during feeding by primate species with different colour vision phenotypes. *Behavioural Ecology and Sociobiology*, Vol.58, No.6, pp. 618-629, ISSN 1045-2249

3

The Acoustic Behaviour as a Tool for Biodiversity and Phylogenetic Studies: Case of the *Rhammatocerus* Species Inhabiting Uruguay (Orthoptera, Acrididae, Gomphocerinae)

María-Eulalia Clemente[1], Estrellita Lorier[2], María-Dolores García[1] and Juan-José Presa[1]
[1]*Área de Zoología, Facultad de Biología, Universidad de Murcia,*
[2]*Sección de Entomología, Departamento de Biología Animal, Facultad de Ciencias, Universidad de la República,*
[1]*Spain*
[2]*Uruguay*

1. Introduction

Species diversity does have a pivotal role in the study and perception of biodiversity (Boero, 2010). One of the goals of Zoology is the study of the animal diversity, that is, the animal species, and Taxonomy is one of the basic disciplines to achieve it. Currently, biodiversity research requires a multidisciplinary approach (Boero, 2009), many different disciplines being involved, such as morphology, molecular biology, ecology, ethology..., that provide new characteristics to be considered. The study of biodiversity should proceed with the contribution of integrative taxonomy (Boero, 2010), taking into account, in addition to the traditional taxonomy, other disciplines of great utility, such as the study of the behaviour.

In this context, behaviour and sounds are relevant characteristics to discover new taxa (Valdecasas, 2011). Sound production in insects is widespread and has been recorded in different orders. It is involved in different behaviours, the most important of which are defence against predators, aggression and mating or sexual behaviour. Four types of senses are used by insects in their sexual behaviour: tactile, visual, chemosensory and acoustics. The acoustic sense is that generally have received more attention, this stimulus is often heard by humans and their production involves the movement of specialized structures that can be seen, usually directly. The types of sounds produced and producing mechanisms can be framed primarily in three categories: stridulation, vibration and percussion. The sounds of insects have been classified in the context of behaviour in several types: call or proclamation, courtship, aggregation, aggression, mating and sounds of interaction (Lewis, 1984; Ragge & Reynolds, 1998).

The sound on Orthoptera serves to promote social relations in the broadest sense of the term. This type of acoustic behaviour cannot be studied in isolation and cannot be understood except within the framework of the general behaviour of the species that is not just the sound production. From this point of view, we show that an exact knowledge of the physical elements of the sounds, as well as the morphology of these sound-producing organs is essential (Busnel, 1954). The sounds produced by the Orthoptera are important from a taxonomic point of view and has an important role as a mechanism of identification. They are of great value to establish the real status of local populations that may have few morphological differences (Blondheim, 1990; García et al., 1995).

The process of acoustic communication in reproduction is amply documented in Ensiferan insects (crickets and katydids): males of singing Ensifera (Orthoptera) emit specific calling songs used for species recognition, while courtship songs are generally less specific and could be mostly under the influence of sexual selection. In contrast, singing Caelifera (Orthoptera) perform more diverse behaviour prior to mating. In the subfamily Gomphocerinae (Acrididae) more specifically, many species emit calling songs, which are sufficiently specific to be used for species identification in the field, but they also perform very complex, often multimodal, courtship behaviour involving sequences of acoustic, vibrational and/or visual signals (Nattier et al., 2011; Ragge & Reynolds, 1998).

Neotropical Gomphocerinae form a group of grasshoppers whose taxonomy, systematics and biology are still poorly known (Otte, 1979; Otte & Jago, 1979). Within this group, *Rhammatocerus* Saussure, 1861 is widely distributed, from southern USA to central Argentine. Its species have a great economic importance; many of them are important crop and grazing pests (Carbonell et al., 2006; Cigliano & Lange, 1998; Salto et al., 2003), especially in Brazil and Colombia (Lecoq & Assis-Pujol, 1998). The genus *Rhammatocerus* is related to other genera such as *Parapellopedon* Jago, 1971 and *Cauratettix* Roberts, 1937.

Since its description, the most important studies on this genus are due to Jago (1971) and Carbonell (1995), but no taxonomical revisions of its entire species have been performed. At present, the genus is composed of 13 species (Assis-Pujol, 1997a 1997b, 1998; Carbonell, 1995), some of them still not clearly defined (Carbonell pers. comm.). The group is characterized by a high intraspecific variation and certain heterogeneity of its external morphology. So, other characters than morphological have been searched to clearly separate the species. As regards the species identification, in tribe Scyllinini, like in other Gomphocerinae, the genitalia allow differentiation between close species in only rare cases (Carbonell, 1995). The phallic complex is an important morphological character among Acrididae but, in Gomphocerinae, it has not a practical value (Carbonell, 1995). The female genitalia, especially the spermatheca, have revealed its utility in identifying some species (Assis-Pujol & Lecoq, 2000). Furthermore, molecular studies have not provided information helping species identification (Loreto et al., 2008).

As pointed before, the sounds produced by Orthoptera are of great taxonomic value. They play a well-known role as identification system during mating, and for this reason they are of great value for establishing the real status of local populations that display small morphological differences (Blondheim, 1990; García et al., 1995). The study of sounds produced by Gomphocerinae has repeatedly demonstrated its utility to solve species identification problems (Ragge & Reynolds, 1998); however the acoustic behaviour of

Neotropical Gomphocerinae has been studied to a very limited extent (COPR, 1982; García et al., 2003; Lorier, 1996; Lorier et al., 2010; Riede, 1987).

Rhammatocerus pictus (Brunner, 1900) and *Rhammatocerus brunneri* (Giglio-Tos, 1895) are robust species inhabiting wet high pastures in northern and central South America, between about latitudes 15 and 40 South. *Rhammatocerus pictus* is distributed in central and southern Brazil (Rio Grande do Sul), Paraguay, Chile (Malleco), northern and central Argentina, Bolivia and Uruguay (Assis-Pujol, 1998; Carbonell et al., 2006) and *Rhammatocerus brunneri* in central-southern Brazil, Paraguay, Bolivia and Uruguay (Assis-Pujol, 1997a). In Uruguay, both species occupy similar habitats (low humid areas covered with tall and dense grass and low hill slopes). They are distributed in the country (Assis-Pujol, 1998) and are more frequently found north of Río Negro. Both species are sympatric in Artigas, Rivera and Lavalleja Departments (Assis-Pujol, 1997a, 1998).

These two species are systematically close to each other, having been separated by Assis-Pujol (1998) just on the basis of morphological criteria: colour of hind femora and tibiae and the spermatheca shape. Their acoustic behaviour is up to date unknown.

Our objective is to establish the real status of these two taxa at the light of their sound production and acoustic behaviour, describing the songs produced in different behavioural situations, the behavioural units identified and the sound-morphological structures.

2. Materials and methods

This study on the sound production and stridulatory structures of *Rhammatocerus pictus* and *Rhammatocerus brunneri* was conducted with males and females proceeding from the "Colección de Entomología de la Facultad de Ciencias de la Universidad de la República", Uruguay, and specimens captured in Uruguay.

2.1 Microscopy techniques and characters considered in stridulatory file study

To study the structure of the stridulatory apparatus, 10 specimens of each sex and species (Table 1) were observed under a binocular microscope (Olympus SZH provided with 10X ocular lenses, 0.66-4X zoom objective, 2X lens and graduated eye piece) as well as with a Jeol 6100 scanning microscope, equipped with SEI (secondary electron images), working at an acceleration voltage of 15 kV and at 21 mm (working distance). Images were captured with the LINK ISIS program. Because the samples are hard and have not risk of being dehydrated, they only had to be cleaned as proposed by Clemente et al. (1989) and, then, coated with pure gold.

Measurements were taken by means of a sliding stage mounted on the stereoscopic binocular microscope, the displacement of which was measured by an attached dial calliper in combination with a graduated eye piece. The accuracy of the dial calliper was 0.05 mm.

For this study, the shape of the file and pegs have been taken into account, as well as the number of pegs in the file, femur length (HFL), file length (FL), peg density all along the file (PD) and in its middle area (PDM) and file length / femur length ratio (FLx100/HFL). Measurements are expressed in mm (Table 2).

	Specimens	Locality	Date	Collector
Rhammatocerus brunneri	4 males (nº 61, 62, 63, 65)	Cerro Chato Dorado. Rivera. Uruguay	12-II-2000	Clemente, García, Lorier
	1 male (nº 64) 2 females (nº 93, 94)	Sierra de la Aurora. Rivera. Uruguay	14-III-1961	C.S.C, A. Mesa, P. San Martín
	2 males (nº 67, 68) 1 female (nº 91)	Cuchilla de Cuñapirú. Rivera. Uruguay	21-I-1956	C.S.C.
	1 males (nº 69) 3 female (nº 86, 90, 95)	Ronda Alta. Río Grande do Sul. Brasil	24-II-1964	A. Mesa, M.A. Monné
	1 males (nº 70)	Bom Jesús.Río Grande do Sul. Brasil	26-II-1964	A. Mesa, M.A. Monné
	1 male (nº 71) 4 females (nº 87, 88, 89, 92)	40 km. N de Caaguazú. Paraguay	13-III-1965	C.S.C, A. Mesa, M.A. Monné
Rhammatocerus pictus	2 males (nº 56, 59)	Cerro Chato Dorado. Rivera. Uruguay	17-18-III-2001	E. Lorier
	2 males (nº 60, 66)	Lunarejo. Rivera. Uruguay	1-IV-1999	E. Lorier
	1 male (nº 57)	V.Lunarejo Rivera. Uruguay	13-II-2000	Clemente, García, Lorier
	1 male (nº 72) 1 female (nº 79)	Sª de la Aurora. Rivera. Uruguay	14-III-1961 11-III -1961	C.S.C, A. Mesa, R. San Martín
	1 male (nº 73) 1 female (nº 76) 1 female (nº 77)	Las Piedras. Canelones. Uruguay	5-II-1966 5-XI-1966 20-III-1964	A. Cármenes
	1 male (nº 74)	Pto. Pepeají. Paysandú. Uruguay	IV-1954	C.S.C.
	1 male (nº 58)	Buena Vista Agraciada. Soriano. Uruguay	8-II-2001	García, Lorier, Presa
	1 male (nº 75) 1 female (nº 83)	Lagoa Vérmelha. Rio Grande do Sul. Brasil	18-II-1964	C.S.C, A. Mesa, M.A. Monné
	1 female (nº78) 1 female (nº 80)	Ayo. Tres Cruces. Artigas. Uruguay	12-XI-1955 14-II-1955	Fac. Humanidades y Ciencias
	1 female (nº 81)	Tartagal. Salta. Argentina	29,31-I-1965	A. Mesa, R. Sandulski
	2 females (nº 82, 84)	San Lorenzo. Salta. Argentina	3-II-1965	A. Mesa, R. Sandulski
	1 female (nº 85)	Nonoai. Río Grande do Sul. Brasil	20-II-1964	A. Mesa, M.A. Monné

Table 1. Summary of the information concerning origin, collection date and number of specimens used to study the sound producing organs of *Rhammatocerus.*

		HFL (mm)	FL (mm)	Pegs	PD (mm^{-1})	PDM (mm^{-1})	FL x 100/HFL
R. brunneri	Males n=10	16.96 ± 0.76 (15.80-17.90)	5.09 ± 0.76 (4.5-5.70)	74.4 ± 12.4 (58-99)	14.60 ± 1.98 (11.60-18.68)	19.5 ± 3.2 (13-24)	29.96 ± 0.97 (28.61-31.84)
	Females n=10	23.86± 0.86 (22.70-25.25)	11.61 ± 0.73 (10.70-12.90)	84 ± 7.1 (72-95)	7.26 ± 0.70 (5.81-8.12)	12.6 ± 1.9 (10-16)	49.02 ± 3 (44.95-53.69)
R. pictus	Males n=10	17.92 ± 0.94 (16.50-19.40)	5.21 ± 0.62 (4.40-6.00)	73 ± 10.70 (58-90)	14.18 ± 2.61 (11.05-18.88)	18.6 ± 2.8 (14-22)	29.04 ± 2.78 (24.71-33.33)
	Females n=10	22.78 ± 1.32 (21.10-24.65)	10.78 ± 1.05 (8.95-12.05)	74.2 ± 3.29 (69-79)	6.94 ± 0.69 (6.09-8.16)	9.8 ± 1.8 (8-12)	47.57 ± 6.10 (36.30-54.50)

Table 2. Summarized data related to stridulatory apparatus of *Rhammatocerus brunneri* and *R. pictus*. HFL: hind femur length; FL: whole stridulatory file length; Pegs: total number of pegs in the whole file; PD: density of pegs in the whole file; PDM: density of pegs in the middle of the file. Values are expressed as mean ± SD and range in parenthesis.

Two-ways multivariate analysis of variance (MANOVA) was used to assess whether there were overall differences between study species regarding the stridulatory file, considering as dependent variables the abovementioned morphological traits. In this case separated analyses were run for each sex. For these and all other statistical analyses, software SPSS (v. 15.0) was used. All variables were log-transformed. The signification value, in all the cases, was $P \leq 0.05$.

2.2 Study of behaviour and sound production

This study was conducted with 9 males (Table 3), 6 males of *R. pictus* and 3 males of *R. brunneri*, and 312 different songs (126 from *R. pictus* and 186 from *R. brunneri*), registered in 5 different tapes (Table 3). The specimens were held in cages in the laboratory and fed mainly with grasses that were changed daily. Humidity was provided by daily watering and by cotton imbibed in water. Two types of cages were used: a glass cage, with net top, 20 x 11 x 14 cm, and a wooden cage with metal mesh top and glass front, 35 x 35 x 55 cm, both exposed to natural light or artificial light provided by a 40W bulb 12 hours per day (Table 3).

The sounds produced were recorded under different conditions, such as isolated specimens, a male together a female of the same species, a species together other different species, both species together.

Sound recordings were made in the laboratory (Table 3) using a Uher 4000 and a Uher 6000 analogical tape recorders (Uher Werke München, Barmseestrasse 11, 8000 München 71, Germany), at a tape speed of 9.5 cm/s, with a frequency response (in Hz) of 20-25000 and signal-to-noise ratio better than 66 dB A remote-control Uher M655 and a Uher M518 dynamic microphones, were located 10-20 cm from the specimens.

		Locality of capture	Date and collector	Recording date	Recording conditions	Tape	Type of sound and recording number
R. pictus	1 male	Cerro Batoví. Tacuaremb. Uruguay	1/IV/1999 E.Lorier	18/IV/1999	Wood cage, (bulb 25W) 20°C	EL6/1999	1 calling 4 courtship Rec.3/6/99
	1 male	Valle del Lunarejo. Rivera. Uruguay	13/II/2000 Clemente, García, Lorier	14/II/2000	Wood cage, (bulb 40W) 33°C	EL3/2000	1 calling Rec. 2/3/2000 Rec.3/3/2000
	1 male	Cerro Chato Dorado. Rivera. Uruguay	11/II/2000 Clemente, García, Lorier	22/II/2000	Wood cage, (2 bulbs 40W) 31-32°C	EL4/2000	4 calling 14 courtship Rec. 2/4/2000
	3 males	Rivera. Uruguay	18/III/2001 Lorier	20/III/2001 23/III/2001	Glass cage (bulb 40W) 27-28 °C	EL3/2001	96 disturbance 6 courtship Rec. 1/3/2002 Rec. 2/3/2001
R. brunneri	3 males	Cerro Chato Dorado. Rivera. Uruguay	12/II/2000 Clemente, García, Lorier	14/II/2000	Wood cage, (bulb 40W) 33°C	EL2/2000	36 disturbance 2 calling Rec. 8/2/2000
					Wood cage, (bulb 40W) 32°C	EL3/2000	118 disturbance 10 calling 7 courtship Rec. 1/3/2000 Rec. 3/3/2000
				22/II/2000	Wood cage, (2 bulbs 40W) 32 °C	EL4/2000	1 courtship 12 disturbance Rec. 1/4/2000

Table 3. Summary of the information concerning specimens used to study the acoustic behaviour of *Rhammatocerus* and studied recordings.

Observations of communicative and interactive behaviour and of the general activity of individuals were made in the laboratory and recorded with a JVC GR-AXM23 video-camera for subsequent analysis. Specimens were observed throughout the recordings, including mute periods, and the behaviour of the specimens in each situation was noted.

Sound recordings were analysed using a Mingograph 420 System attached to a digital oscilloscope (Tektronix 2211) and to a Krohn-Hite 3550 filter. To study the physical characteristics of the sound, the analogical signal was digitized with a Sound Blaster® AWE64 Gold at 8 bits and at a 44 kHz sampling frequency, and then studied using the Avisoft® SAS Lab Pro 3.8. PC software for MS-Windows. Oscillograms were obtained using the option One-dimensional functions, selecting the function Time-signal. The spectral characteristics were obtained using the same option than before, selecting, in the function Amplitude spectrum (linear), the Hamming evaluation window, bandwidth 0.526 Hz and resolution 0.336 Hz.

Males of both species produce different types of sound in different behavioural situations fitting with the categories from the literature (García et al., 2003; García et al., 2005; Ragge & Reynolds, 1998). The terminology used to describe songs follows that of Ragge & Reynolds (1998). The variables used for the study of the sounds are: echeme length, number of syllables, syllables length, rate of emission of syllables, peak of maximum amplitude, low quartile, middle quartile, upper quartile, minimum frequency, maximum frequency and band width. The following types of sounds were identified: (1) the calling song, produced spontaneously by a male; (2) the courtship song, produced by males when close to a female and (3) the disturbance song, produced by males when interacting with other individuals.

To explore the relationship among different types of sounds produced by the studied species, a principal component analysis (PCA) was performed considering all the variables used to study the sounds (see before). Since for disturbance songs not all response variables could be recorded, they were not included in the PCA. Two-ways MANOVA was used to formally assess the differences among type of sound (calling song and courtship song) and species (*R. pictus* and *R. brunneri*), considering the variables used for the study of sounds as dependent variables. When cases corresponding to different groups overlapped after running the PCA, a separate MANOVA was performed for these groups considering the same dependent variables and factors as for the overall MANOVA. Variables were log-transformed for both PCA and MANOVA and a significance level of 5% was selected.

Specimens and recordings are kept in the "Colección de Entomología de la Facultad de Ciencias de la Universidad de la República", Uruguay, and in the "Colección del Área de Zoología de la Universidad de Murcia", Spain. Recordings can be reviewed using the recording number (Table 3). Sample songs are also available at OSF (http://osf2.orthoptera.org) (Eades 2001).

3. Results

The sounds recorded, emitted by males in different behavioral situations, were all produced by femoro-tegminal mechanism, rubbing the pegs of the stridulatory file of hind femora against some specialized tegminal veins.

No differences between songs have been observed in relation to the recording conditions. No sound produced by females has been registered in any case, although they produce movements with their hind legs. These movements are of moderate amplitude, starting with the leg folded at about 45° respect the vertical and displacing about 45° from the starting position. This movement can be isolated, from resting position to the vertical, or be repeated several times, when close to other individuals, males or females.

3.1 Stridulatory file

In *Rhammatocerus pictus* the hind femora of both sexes have, along their inner surface, a stridulatory file that is almost linear and long in relation to femur length. The stridulatory pegs are well developed and regularly spread except at ends, where they are more irregularly and more separately distributed (Fig. 1A-D).

Male file, although shorter than that of the female, contains more pegs than that of the female. The peg density on the whole file is greater in males than in females (Table 2, Figs. 1A and C). The male pegs are conic shaped. They are inserted in the alveolus, which have a raised margin, and have a short peduncle (Fig. 1B). The female pegs, well developed, are also conic shaped with rounded apex, and have a short peduncle. They are also inserted in alveolus with raised margins (Fig. 1D).

In both sexes of *Rhammatocerus brunneri* the stridulatory file is linear and long in relation to femur length. Pegs are well developed and regularly spread except at ends, where they are more separated and irregularly disposed (Fig. 1 E-H).

Male file is half the length than that of female, but it has almost the same number of pegs and, so, the peg density is greater (Table 2, Fig. 1E and G). The male pegs are conic shaped and have a short peduncle. They are inserted by in the alveolus, which have a raised margin (Fig. 1F). The female pegs are also conic shaped, slightly irregularly spread and more separated than in males. Pegs are also inserted by a short peduncle in alveolus with raised margins (Fig. 1H).

The MANOVA revealed the lack of overall differences among species regarding the stridulatory file in the case of males ($F_{5, 14}$= 1.563; P= 0.234), although significant differences were detected for femur length ($F_{1, 18}$= 6.205; P=0.023). Females of different species showed significant overall differences ($F_{5, 14}$= 3.724; P= 0.024), as well as for the responses variables number of pegs in the file ($F_{1, 18}$= 15.947; P= 0.001) and peg density all along the file ($F_{1, 18}$= 11.920; P= 0.003).

3.2 Calling and courtship songs

The PCA performed extracted two components explaining most of the variance of the original data (79.5%). Component 1 was positively correlated with those variables dealing with frequency and negatively with duration of the echeme. Component 2 was positively correlated with both the number of syllables and the rate of emission of syllables and negatively with the duration of syllables (Fig. 2).

When scores for each case were plotted against component 1 and 2 three major groups were identified: 1) courtship and calling songs of *R. brunneri* (closely overlapped); 2) calling song of *R. pictus* and one of the types of courtship song of this species (pictus 1); 3) other type of courtship song of *R. pictus* (pictus 2) (Fig. 3). The MANOVA performed for both the courtship and calling songs of the studied species indicated the existence of overall significant differences among species ($F_{11, 23}$= 18.042; P<0.0001) and type of song ($F_{33, 68.47}$= 6.318; P<0.0001).

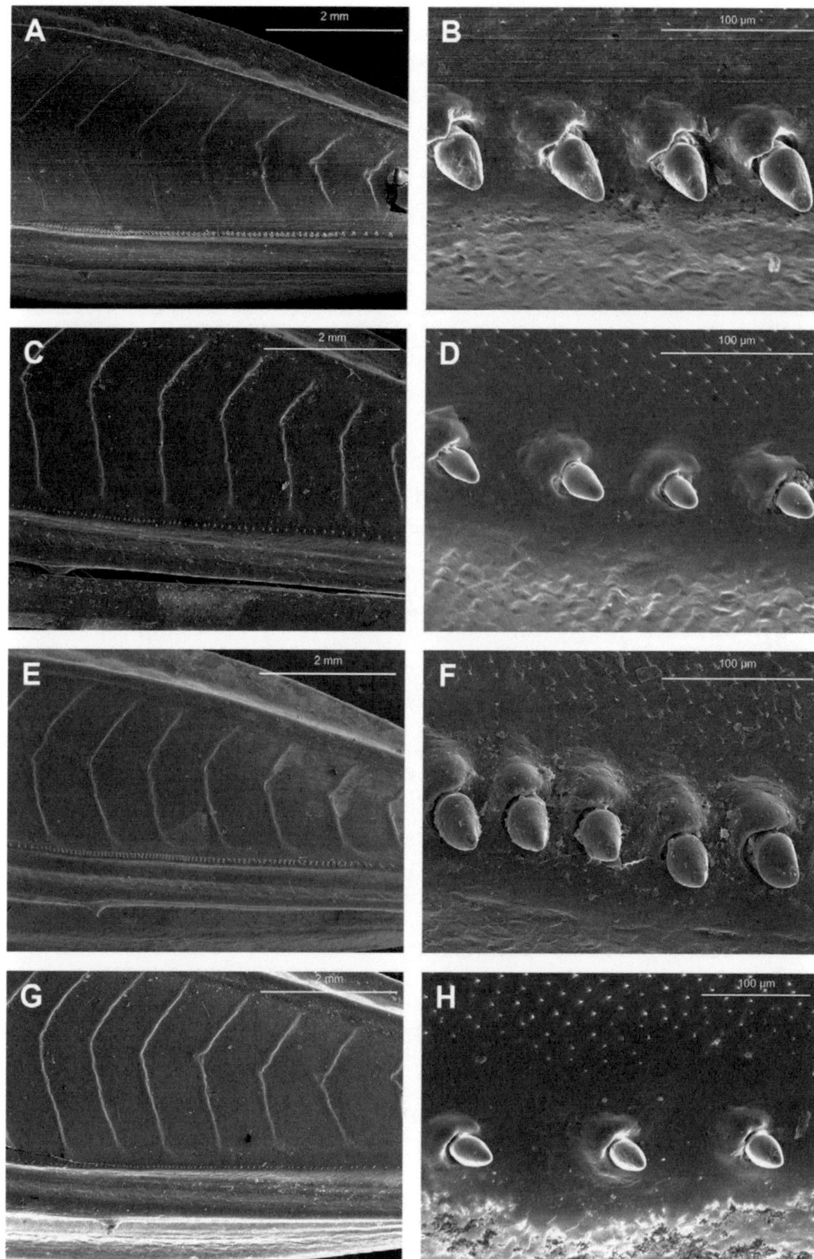

Fig. 1. Stridulatory files. A-D: *Rhammatocerus pictus*. A: male file general appearance, B: detail of male pegs of middle zone, C: female file general appearance, D: detail of female pegs of middle zone. E-H: *Rhammatocerus brunneri*. E: male file general appearance, F: detail of male pegs of middle zone, G: female file general appearance, H: detail of female pegs of middle zone.

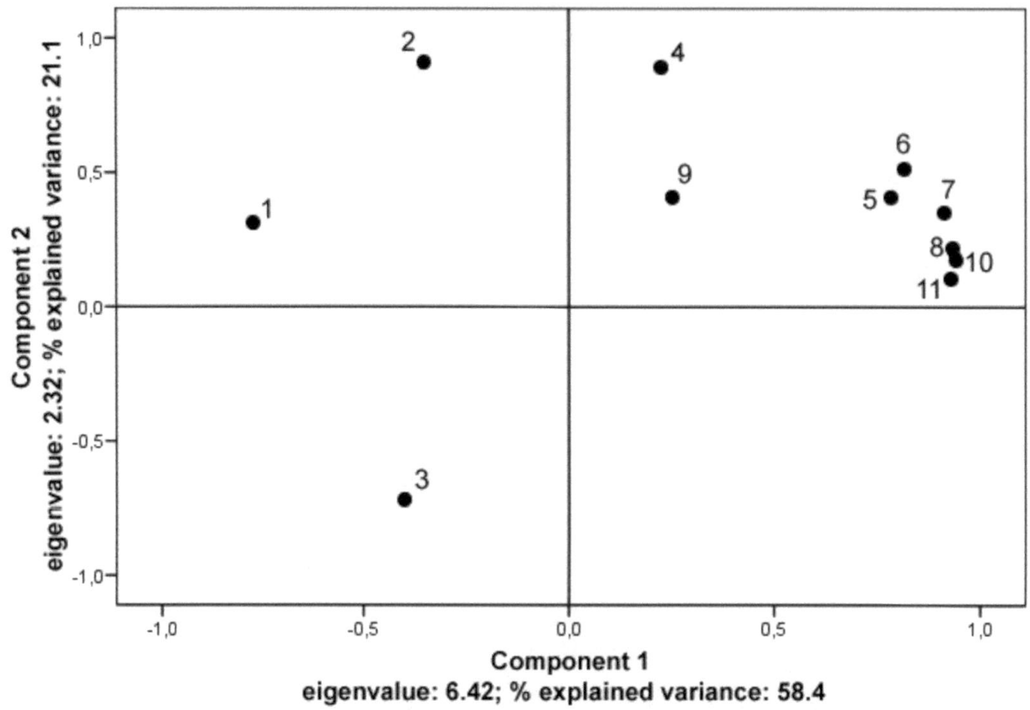

Fig. 2. Results of PCA performed to explore relationships among types of sound considering all the variables used: 1 echeme length; 2 number of syllables; 3 syllables length; 4 rate of emission of syllables; 5 peak of maximum amplitude; 6 low quartile; 7 middle quartile; 8 maximum frequency; 9 minimum frequency; 10 upper quartile; 11 bandwith

3.2.1 *Rhammatocerus pictus*

Calling song was a spontaneous song, consisting of echemes composed of a variable number of syllables (Table 4).

Its frequency spectrum occupied a broad band, with the peak of maximum amplitude at around 7 kHz (Table 4, Fig. 4C).

The echeme started with short and almost inaudible syllables. The sound increased in intensity after 1/3 of the echeme had been emitted. Syllables were clearly separated one from other by evident gaps (Figs. 4A and B).

It was produced by synchronously rubbing the two hind femora against the tegminae, both hind legs moving rapidly. At the beginning of the echeme the legs movement was hardly visible.

Courtship songs were composed of echemes of a variable number of syllables (Table 4). Two different types of song have been observed, the here called pictus 1 (Fig. 4D-F) emitted at a broad band, with the maximum amplitude at around 5 kHz, and the called pictus 2 (Fig. 4G-I) which occupied a much narrower band, with the maximum amplitude at around 2 kHz; both are significantly different (Fig. 3).

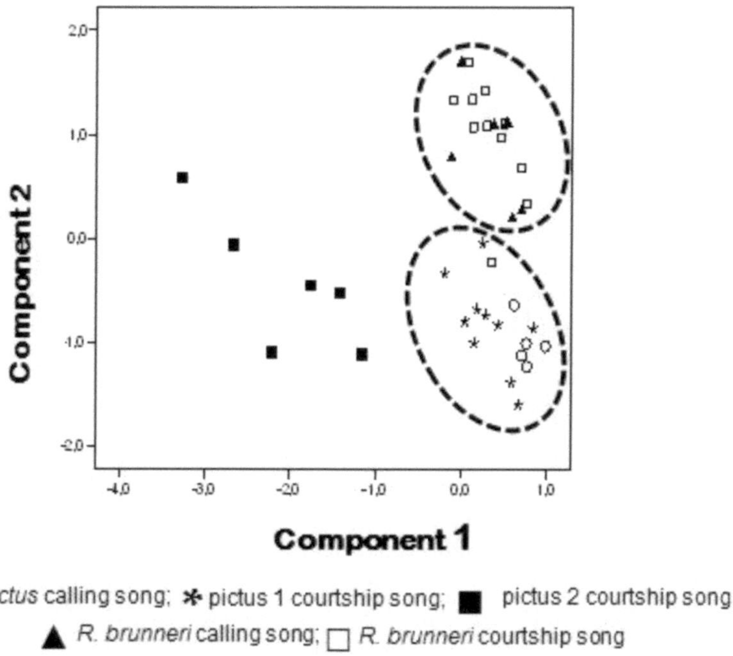

Fig. 3. Graphic showing the results of plotting scores for each cases against components 1 and 2 of PCA.

When the courtship started, the male moved walking towards the female; stopping when close and perpendicular to her. Then he moved his antennae up and down, synchronously at the beginning and, then, alternately, and started to sing. While singing, he directed his antennae towards the female, forming an around 150° angle. After the song, the male usually jumped suddenly on the female, trying to mate while touching her with his antennae. In most cases, the mating was not effective, the male being rudely rejected by the female by kicking the male, raising the hind legs or simply going away him.

While the courtship song, males performed two kinds of movement: 1) Mute movement: one to three slow and wide up and down movements of hind legs. They rose synchronously from the rest position (around 30° in relation to the main corporal axis) to around 75°, and then bent slightly asynchronously. 2). Stridulatory movement: a series of synchronous, quick and small up and down movements, of little amplitude (30-55°) respect to the corporal axis. In some cases, the movement 1 lacked, the courtship starting directly with the quick movement.

The songs started as imperceptible, almost inaudible, and sound increased in intensity until the end. At the final section, syllables were intense, similar in structure, with gaps between them (Fig. 4 D-E and G-H).

The MANOVA run separately for the group 2 identified by the PCA revealed that for this species, overall, calling song and pictus 1 courtship song were not significantly different ($F_{11, 3}$= 0.702; P= 0.712), existing differences only in the case of some the response variables dealing with frequency (P≤0.037): peak maximum amplitude ($F_{1, 13}$= 5.401; P= 0.037), low quartile ($F_{1, 13}$= 11.494; P= 0.005), upper quartile ($F_{1, 13}$= 6.026; P= 0.029), middle quartile ($F_{1, 13}$=7.279; P= 0.018) and minimum frequency ($F_{1, 13}$=11.527; P= 0.005).

3.2.2 *Rhammatocerus brunneri*

The calling song consisted of an echeme composed of syllables (Table 4, Fig. 5A and B), short and almost inaudible at the beginning and increasing in intensity further on. At the end, the echeme had a high intensity and the syllables are very close with almost no gap between them. The frequency spectrum of the sound occupies a broad band, the main peak being at around 8500 Hz (Table 4, Fig. 5C).

The leg movements to produce the sound are almost imperceptible at the beginning of the echeme. They became wider as the echeme went on and at the end legs seemed to move much quickly than before.

The courtship song was composed of echemes of a variable number of syllables (Table 4, Fig. 5D and E). As regards the spectral characteristics (Fig. 5F) they were similar to that of calling song there having not been found any statistically significant difference between this song and calling song.

The leg movements to produce the sound were also similar to that to produce calling song. When a male was performing a courtship, could be interrupted by other males, who started to perform the disturbance song.

During courtship the male follows the female, standing behind very close to her.

The MANOVA run separately for the group 1 identified by the PCA revealed that for *R. brunneri* there were not significant overall differences between courtship and calling songs ($F_{11, 5}$= 0.607; P=0.773), although the duration of the echeme (Table 4) was significantly ($F_{1, 15}$= 4.786; P= 0.045) different.

3.3 Disturbance song

3.3.1 *Rhammatocerus pictus*

Songs were composed of isolated syllables of variable duration (Table 4) irregularly emitted (Fig. 6A and B). Its frequency spectrum showed a quite broad band of emission; the maximum amplitude peak being at below 3 kHz (Fig. 6C). To produce song, the males moved rapidly and almost synchronously either hind legs or only one hind leg.

Males produced these sounds when the specimens were at near or in contact with each other. The signals were sometimes emitted in alternation. Some individuals were observed singing while walking. When a male was courting a female, other males near the couple started to sing the disturbance song that, in this case, played the role of rivalry song.

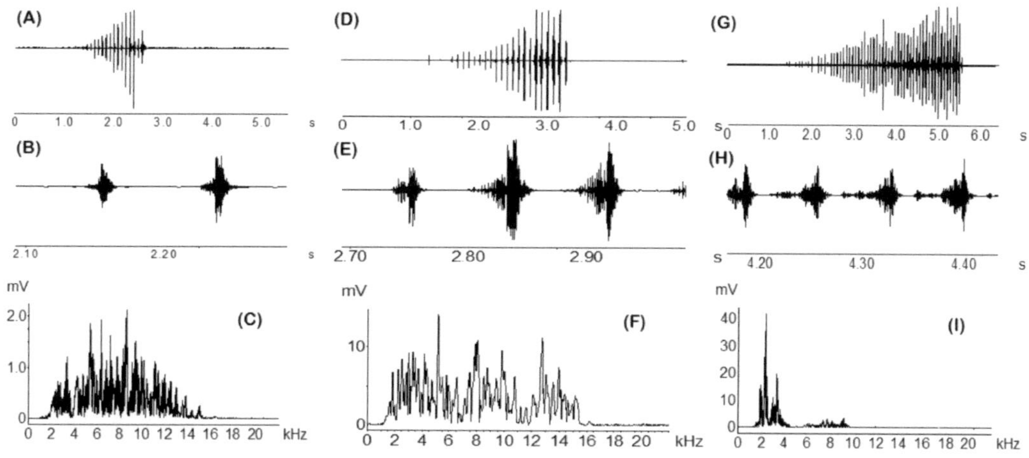

Fig. 4. *Rhammatocerus pictus*. Calling (A-C) and courtship (pictus 1 D-E and pictus 2 G-I) songs. Echeme (A, D and G); syllable detail (B, E and H), and frequency spectra (C, F and I).

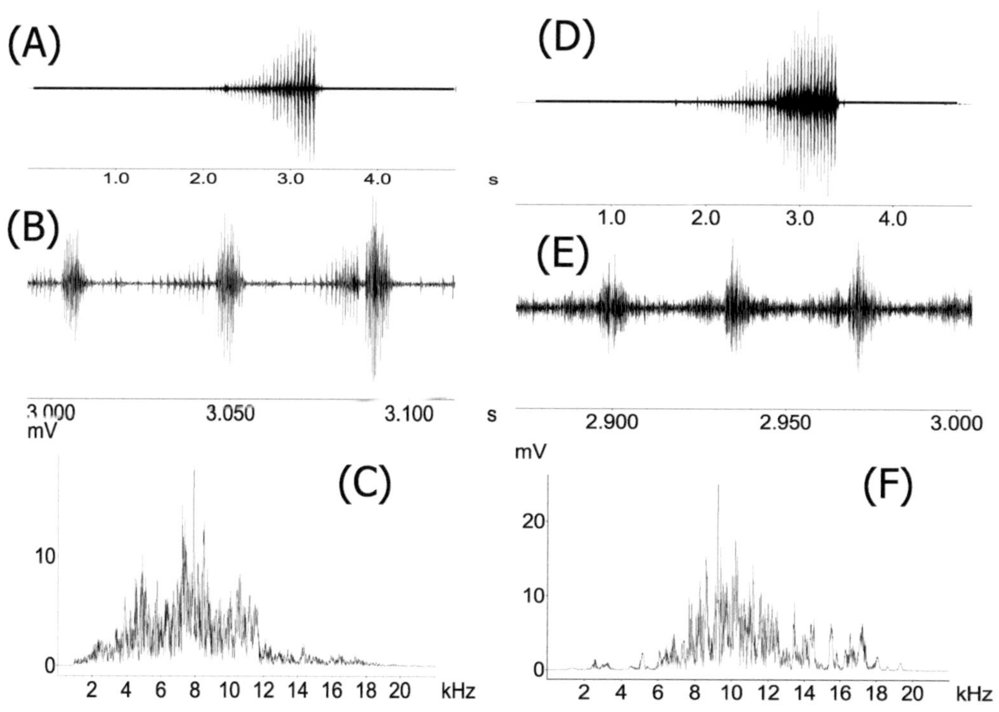

Fig. 5. *Rhammatocerus brunneri*. Calling song (A-C) and courtship song (D-F). Echeme (A and D); syllable detail (B and E) and frequency spectra (C and F).

3.3.2 *Rhammatocerus brunneri*

Songs were composed of isolated syllables (Table 4, Fig. 6D and E) emitted alternately by different individuals. The frequency spectrum occupied a broad band, with the peak of maximum amplitude at around 10 kHz (Fig. 6F). The leg movements to produce this song were similar to that performed by *R. pictus*, usually one or two quick asynchronous up and down movements.

This song was emitted in different circumstances, such as coinciding with or at the end of a calling song of other male or as rivalry song.

For disturbance songs, indicated the existence of overall significant differences among species (MANOVA: $F_{8, 15}$= 53.073; $P<0.001$).

No disturbance songs produced by females have been recorded. One or two mute up and down movements of their hind legs were observed after a male interaction or courtship song or when interacting with other female. These movements have been also observed performed by males when close to other individuals or interacting with them.

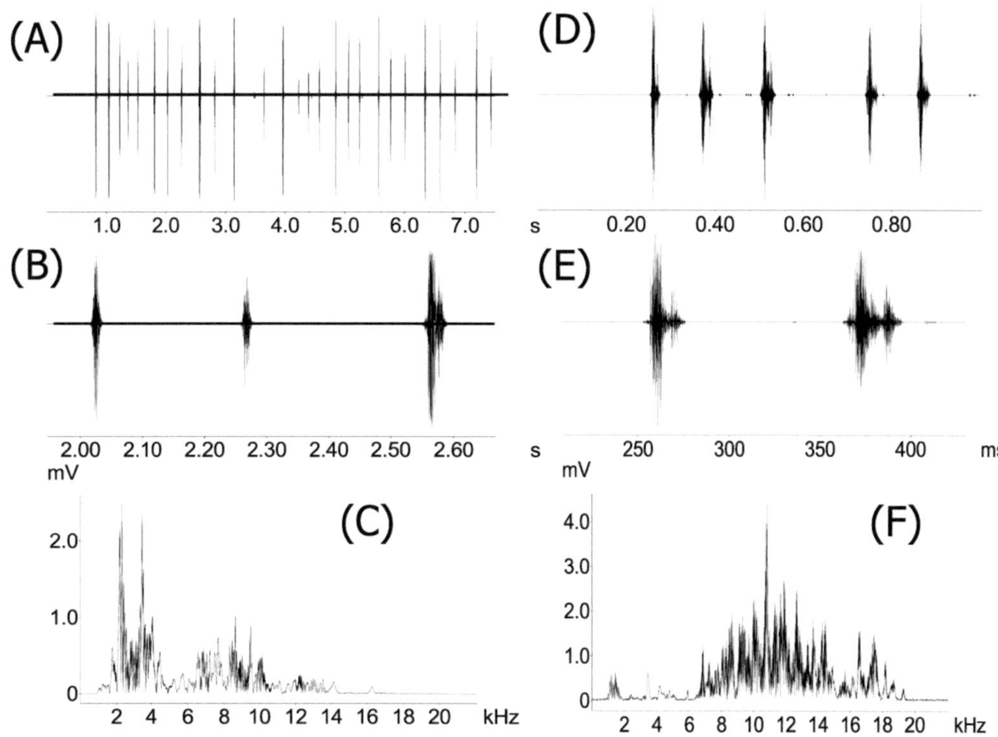

Fig. 6. Disturbance song. *Rhammatocerus pictus* (A-C) and *Rhammatocerus brunneri* (D-F). Sequence of syllables (A and D); syllable detail (B and E) and frequency spectra (C and F).

4. Discussion

The stridulatory file of *R. pictus* and *R. brunneri* are well developed in both males and females. The peg spread, regular all along the file except at the ends, fits with that of other Gomphocerinae (Clemente et al., 1989; Pitkin, 1976). It can be pointed that the metrical characteristics referred to the stridulatory file are not species specific for males of *R. pictus* and *R. brunneri* but can serve to differentiate the females of both species. Nevertheless, there seem to be some morphological differences between stridulatory pegs of males. In both cases pegs are conic shaped, but *R. pictus* have pegs clearly more elongated, with a more acute apex, and the alveolus have the raised margin more irregular than *R. brunneri*.

The acoustic repertoire of *R. pictus* and *R. brunneri* is similar to that of other neotropical Gomphocerinae species, such as *Parapellopedon instabilis* (Rehn, 1906), *Euplectrotettix ferrugineus* Bruner, 1900 and *Fenestra bohlsi* Giglio-Tos, 1895 (García et al., 2003; Lorier, 1996; Riede, 1987), being made up of:

1. Calling song, as defined by Bailey (1991), Dumortier (1963), Ewing (1989), García et al. (1995), Green (1995), Helversen & Helversen (1983), Lorier et al. (2010), Ragge (1987), Ragge & Reynolds (1998), Reynolds (1986), Riede (1987), among others, since the songs were produced by males being apart from other specimens.
2. Courtship song, as defined by Ragge & Reynolds (1998) among others, that is, the special song produced by a male when close to a female.
3. Disturbance song, characterized by the physical structure and context in which it is emitted, that is the sound produced when the specimens were near or in contact with each other (García et al., 2003, among others). In the literature, reference exists to a "contact cry" (Kontaktlaut). All these signals are brief and sometimes emitted in alternation (Dumortier, 1963; Faber, 1953; Jacobs, 1953) and it is said are of little or no importance in taxonomy or identification (Ragge & Reynolds, 1998). These sounds can be interpreted in an aggressive context as rivalry song when produced by several males as an answer to a courting male, as described for *Fenestra bolshi* (Lorier et al., 2010). From females, despite having a stridulatory file, no sound has been registered.

The function of these different types of signals linked to the reproductive behaviour in Orthoptera and other insects has been analysed and interpreted in the context of the sexual selection theory. Sound can evolve, among other possibilities, through sexual selection. Evolutionary pressures have to act towards a minor energetic cost and a minimum risk of predation in the mating (Bailey, 1991). The significance of the male calling songs of Orthoptera lies in the fact that they are believed to provide the main means of mater recognition and hence of reproductive isolation of sympatric species (Ragge, 1987; Ragge & Reynolds, 1998). Acoustic signals provide enough relevant information on species identification and sexual selection. So, sound is considered essential to solve taxonomical problems at the specific level. Closely related, sympatric species often use strikingly different acoustic criteria to discriminate the same set of conspecific and heterospecific signals (Gehardt & Huber, 2002). A marked difference between songs is a strong indicative of different species (Ragge & Reynolds, 1998). When two morphologically similar populations of Orthoptera have consistently different calling song, it is likely that they belong to different species. Conversely, when two populations show small morphological differences but have exactly the same calling song, they are probably forms of a single species (Ragge, 1987).

In most Gomphocerinae some acoustic signals identify the species and can avoid the crossing between sympatric close species. This is the case of *Rhammatocerus* species here studied, which show differences in physical characteristics of all types of song. Both species have overall different songs, as shown in Table 4 and Figures 4, 5 and 6. Differences are more or less evident depending on the type of song.

The calling song of *R. brunneri* is very similar to the courtship song (Fig. 5). Thus, this species uses a similar song, with little differences (Table 4), in two different behavioural situations. In *R. pictus* a similar situation exists between the calling song and one of the courtship songs recorded (pictus 1) (Fig. 4A-F). At the light of the recorded songs, this species may start courting with a song similar to the calling song (pictus 1) changing afterwards to a particular courtship song (pictus 2), accompanied by legs and antennae movements. Although the whole sequence has not been registered, it would be not surprising since the use of a courtship song similar to calling song has been observed in other Gomphocerinae species, such as *Chorthippus binotatus binotatus* (Charpentier, 1924) (García et al., 1995) and *Omocestus antigai* (Bolívar, 1987) (Clemente et al., 1999). In the last case, it could be observed that the courtship started with the calling song and, later, the courtship song was emitted.

This kind of behaviour fits with that observed, for example, in genus *Stenobothrus* Fischer, 1853, the ancestor of which should have had produced simple and likely identical calling and courtship songs. The complex courtship songs observed in many species resulted from the addition of new acoustic traits derived from the common calling song pattern and from the addition of visual signals produced by the movement of legs and antennae (Berger, 2008 as cited in Nattier et al., 2011). In *R. pictus,* during the courtship new traits related to the spectral characteristics of the sound and, on the other hand, visual signals by legs and antennae movements would have been added.

These similarity of calling and courtship songs in *R. brunneri* and, partly, in *R. pictus* could be explained by the signals flexibility observed in acridids by which the same signal can be used in different contexts, as here occurred. This would imply following the opinion of Otte (1977), a further stage in an evolutionary process not ended still, probably accompanying a sympatric speciation process. Probably, sounds should not be the unique barrier to avoid hybridizing but there could also take part other kind of signals, such as those visual or chemical (Otte, 1977), although behavioural traits are often evolutionary labile and hence communication systems often diverge more rapidly than do morphological and molecular characteristics (Gehardt & Huber, 2002). Nevertheless, the question of which of the two song types calling or courtship song is the phylogenetic older one remains open (Berger & Gottsberger, 2010) but a simplification of courtship repertoires in some species suggests that the evolution of courtship song, and that of mating behaviour as a whole, could also result from a dynamic process (Berger, 2008 as cited in Nattier et. al., 2011).

As regards the disturbance song, in some species phonotaxis is reduced when other males interfere, diminishing the efficiency of calling or courtship song. It could be pointed, for male disturbance songs – either spacing or answering other male courtship –, that the answering male, when alternating its song, uses its interference potential of acoustic channel

	Calling song		Courtship song			Disturbance song	
	R. pictus	*R. brunneri*	*R. pictus 1*	*R. pictus 2*	*R. brunneri*	*R. pictus*	*R. brunneri*
Echeme length (sec)	1.509 ± 0.164 (1.241-1.658)	2.130 ± 0.480 (1.618-3.22)	2.022 ± 0.588 (1.21-3.23)	3.02 ± 0.968 (2.151-4.807)	1.651 ± 0.380 (1.233-2.291)	-	-
Number of syllables	20.166 ± 1.602 (18-23)	49.416 ± 12.101 (34-67)	24.38± 7.65 (14-40)	38.85± 12.33 (28-61)	38± 12.23 (25-56)	-	-
Syllables length (sec)	0.049 s ±0.012 (0.028-0.064)	0.028 ± 0.002 (0.024-0.032)	0.041± 0.008 (0.031-0.070)	0.052 ± 0.005 (0.043-0.059)	0.027 ± 0.003 (0.023-0.032)	0.035 ±0.008 (0.020-0.066)	0.058 ± 0.023 (0.023-0.127)
Rate of emission (syll/sec)	13.471 ± 1.772 (12.06-16.61)	22.684 ± 3.542 (13.63-27.47)	12.07 ± 1.682 (9.14-15.39)	13 ± 0.328 (12.68-13.56)	22.9 ± 4.67 (13.63-29.29)	-	-
N	6	12	18	7	8	96	166
Peak of maximum amplitude	7198.08 ± 2962.745 (2790-11040)	8570.63 ± 2201.574 (1571-11864)	5014 ± 2239.707 (1297-10680)	2196.71 ± 422.754 (1884-3474)	7772.93 ± 2307.929 (3466-11170)	2752.14 ± 644.780 (2210-3830)	10065.38 ± 2145.442 (7550-14443)
Low quartile	5316.54 ± 796.125 (3760-6540)	6687.78 ± 1343.213 (3334-8645)	4340.39 ± 771.349 (2530-5351)	2236.79 ± 330.158 (1913-3213)	6129.47 ± 1524.788 (3639-9000)	2844.29 ± 332.883 (2150-3550)	8388.63 ± 1043.882 (6610-9619)
Middle quartile	8419.23 ± 803.845 (6990-9880)	9081.74 ± 1357.565 (6193-11148)	6879.75 ± 1361.498 (3924-8893)	3416.43 ± 1426.617 (2330-7380)	8720.00 ± 1293.614 (6341-10675)	4696.43 ± 1097.841 (3570-6930)	10899.75 ± 1054.661 (9560-12758)
Upper quartile	11354.62 ± 801.969 (8910-12270)	12040.63 ± 1507.873 (9146-14499)	10159.11 ± 1635.277 (6346-12917)	7058.50 ± 3157.601 (2853-14136)	11790.47 ± 1228.933 (8694-13339)	9090.00 ± 1350.504 (7080-11640)	13516.00 ± 1555.696 (10740-15500)
Minimum frequency	1864.62 ± 281.769 (940-2340)	1865.52 ± 1207.976 (963-7068)	1349.04 ± 327.541 (931-1920)	1301.29 ± 292.857 (1014-1838)	2078.73 ± 1178.164 (807-5307)	1542.14 ± 200.851 (1030-1930)	3237.38 ± 603.467 (2540-4430)
Maximum frequency	15971.92 ± 1312.951 (13130-18820)	17174.26 ± 2293.155 (12220-20916)	14593.64 ± 2035.698 (9735-17520)	7315.57 ± 3018.844 (3441-12973)	16105.13 ± 2529.865 (10874-18809)	11807.86 ± 1944.448 (8220-13970)	17744.50 ± 1357.726 (14470-18680)
Band width	14102.69 ± 1365.625 (11140-17070)	15306.85 ± 2635.109 (10150-19902)	13243.75 ± 1932.220 (8780-15993)	6013.64 ± 3030.971 (1606-11929)	14025.87 ± 3022.831 (8640-17463)	10259.29 ± 1980.810 (6710-12460)	14248.38 ± 1372.988 (11920-15860)
N	26	27	28	14	15	14	8

Table 4. Data summary concerning the physical characteristics of the songs emitted by the males of *Rhammatocerus pictus* and *R. brunneri* on the time and frequency domains (data of which are expressed in Hz)

(Cade, 1985; Greenfield, 1997). Thus, the emitters compete for the receipt's attention, in this case the silent female. Thereby it could be understood, for *R. brunneri*, the courtship interruption provoked by rivalry songs (McGregor & Peake, 2000). Differences between disturbance songs of the two species, *R. pictus* and *R. brunneri*, support the specific communicative value even of this type of song.

Anyway, since the acoustic communication involves a particular signal recognisable by the conspecifics to avoid the interspecific mating, its characteristics have to be specific. In the studied cases, it has been proved that the sound characteristics, both in the time and in the frequency domains, allow clearly discriminating cryptic species, and offer useful elements for phylogenetic analysis. Up to now, the only characters to separate between *R. pictus* and *R. brunneri* was the colour of hind femora and tibiae and the spermatheca shape (Assis-Pujol, 1998), but the real status of both taxa remained obscure. So, our results can have consequences on taxonomy since they bring new elements to the genus revision, justifying the separation of the two species, *R. pictus* and *R. brunneri*, as valid species (Carbonell pers. comm.).

5. Conclusion

Exploration of biodiversity is imperative (Boero, 2009) but is needed of features helping to the specific identification and differentiation especially among cryptic species. Currently, in addition to the morphological characters, easily observed and absolutely useful when well defined, other characteristics are available, such as the genetic or the behavioural ones. The sound production in insects provides excellent traits allowing discriminating species on the basis of the sound characteristics, in the time and in the frequency domains, as well as in the behavioural traits of the context in which sound is produced.

Rammathocerus brunneri and *R. pictus* are sympatric Neotropical Gomphocerinae species with very similar morphological traits, even those related to the stridulatory file, that is, the sound producing organ. Nevertheless, they produce songs that result species specific, offering new features to be considered for their actual specific status. In addition to that, the sounds and its accompanying behaviour provide new evidences to be taken into account in the phylogenetic history of the genus.

Thus, the acoustic behaviour, when present, is worth of being considered in taxonomic revisions due to its demonstrated utility. The job of naming animals is far from having been carried out (Boero, 2009) and sound can help to complete that job.

6. Acknowledgment

The authors are grateful to C.S. Carbonell for providing the study material and useful comments, and to Ministerio de Educación y Cultura of the Spanish Government for the "Proyecto de Investigación Conjunta", belonging to the "Programa de Cooperación Científica con Iberoamérica", to study "Bioacústica de Acrídidos: su aplicación en sistemática". We are also grateful to Drs. Ana Ruiz, Andrés Egea and David Verdiell, of the University of Murcia, for their invaluable help in the statistical treatment of data.

7. References

Assis-Pujol, C.V. de (1997a). Notas Sinonímicas e Redescrições de Duas Espécies de *Rhammatocerus* Saussure, 1861 (Orthoptera, Acrididae, Gomphocerinae, Scyllinini). *Boletim do Museu Nacional* (N.S.), Río de Janeiro, Vol. 376, pp. 1-12, ISSN 0080-312X.

Assis-Pujol, C.V. de (1997b). Duas Novas Espécies Brasileiras de *Rhammatocerus* Saussure, 1861. (Acrididae, Gomphocerinae, Scyllinini). *Boletim do Museu Nacional* (N.S.), Río de Janeiro, Vol. 380, pp. 1-10, ISSN 0080-312X.

Assis-Pujol, C.V. de (1998). Aspectos Morfológicos, Taxonómicos e Distribuição Geográfica de Cinco Espécies de *Rhammatocerus* Saussure, 1861 (Acrididae, Gomphocerinae, Scyllinini). *Boletim do Museu Nacional* (N.S, Río de Janeiro, Vol. 387, pp. 1-27, ISSN 0080-312X.

Assis-Pujol, C.V. de & Lecoq, M. (2000). Comparative study of spermathecae in eleven *Rhammatocerus* Saussure, 1861 grasshopper species (Orthoptera: Acrididae: Gomphocerinae: Scyllini). *Proceedings of the Entomological Society of Washington,* Vol. 120, No. 1, pp. 120-128, ISSN 0013-8797.

Bailey, W.J. (1991). *Acoustic behaviour of insects. An evolutionary perspective,* Chapman and Hall, ISBN 0-412-31980-2, London.

Berger, D. & Gottsberger, B. (2010). Analysis of the courtship of *Myrmeleotettix antennatus* (Fieber, 1853) - with general remarks on multimodal courtship behaviour in gomphocerine grasshoppers. *Articulata,* Vol. 25, No. 1, pp. 1-21, ISSN 0171-4090.

Blondheim, S.A. (1990). Patterns of reproductive isolation between two sibling grasshopper species *Dociostaurus curvicercus* and *D. jagoi jagoi* (Orthoptera: Acrididae: Gomphocerinae). *Transactions of the American Entomological Society,* Vol. 116, No. 1, pp. 1-64, ISSN 0002-8320.

Boero, F. (2009). Zoology in the era of biodiversity. *Italian Journal of Zoology,* Vol. 76, No. 3, pp. 239, ISSN 1125-0003

Boero, F. (2010). The study of species in the era of biodiversity: a tale of stupidity. *Diversity,* Vol. 2, pp. 115-126, ISSN 1424-2818.

Busnel, R.G. (Ed.) (1954). *Colloque sur l´Acoustique des Orthoptères,* Institut National de la Recherche Agronomique, Paris.

Cade, W. (1985). Insect mating and courtship behaviour. In: *Comprehensive insect physiology biochemistry and pharmacology,* Kerkut, G.A. & Gilbert, L.I. (Eds.), pp. 591-619, Pergamon Press, ISBN 0-08-030812-0, Oxford and New York.

Carbonell, C.S. (1995). Revision of the tribe Scyllinini, nov. (Acrididae: Gomphocerinae), with descriptions of new genera and species. *Transactions of the American Entomological Society,* vol. 121, No. 3, pp.87-152, ISSN 0002-8320.

Carbonell, C.S.; Cigliano, M.M. & Lange, C.E. (2006). *Especies de Acridomorfos (Orthoptera) de Argentina y Uruguay,* CD ROM, Publications on Orthopteran Diversity, The Orthopterist´s Society at the Museo de La Plata, ISBN 987-05-0546-5, Argentina.

Cigliano, M.M. & Lange, C.E. (1998). Orthoptera. In: *Biodiversidad de Artrópodos,* Morrone, J.J. & Coscarón, S. (Eds.), pp. 67-83,. A. Ediciones Sur, ISBN 950-9715-42-5, La Plata, Argentina.

Clemente, M.E.; García, M.D. & Presa, J.J. (1989). Estudio comparativo de la fila estriduladora de las especies de los géneros *Stenobothrus* Fischer, 1853, *Omocestus* Bolívar, 1878 y *Myrmeleotettix* Bolívar, 1914 presentes en la Península Ibérica (Orthoptera, Caelifera, Gomphocerinae). *Boletín de la Real Sociedad Española de Historia Natural* (Sección Biológica), Vol. 84, No. 3-4, pp. 343-361, ISSN 0366-3272.

Clemente, M.E.; García, M.D.; Arnaldos, M.I.; Romera, E. & Presa, J.J. (1999). Confirmación de las posiciones taxonómicas específicas de *Omocestus antigai* (Bolívar, 1897) y *Omocestus navasi* Bolívar, 1908 (Orthoptera, Acrididae). *Boletín de la Real Sociedad Española de Historia Natural* (Sección Biológica), Vol. 95, no. 3-4, pp. 27-50, ISSN 0366-3272.

C.O.P.R. (1982). *The locust and grasshopper agricultural manual,* Centre for Overseas Pest Research, ISBN 9780851351209, London.

Dumortier, B. (1963). Ethological and physiological study of sound emissions in Arthropoda. In: *Acoustic Behaviour of Animals,* Busnel, R.G. (Ed.), pp. 583-654, Elsevier, New York, Amsterdam.

Eades, D. (2001). Version 2 of the Orthoptera species file online. *Journal of Orthoptera Research,* Vol. 10, pp. 153-163. ISSN 1082-646.

Ewing, A.W. (1989). *Arthropod Bioacoustics. Neurobiology and Behaviour,* Comstock Publishing Associates, Cornell University Press, ISBN 08-014-2478X, Ithaca, New York.

Faber, A. (1953). *Laut-und Gebärdensprache bei Insekten. Orthoptera (Geradflügler). I,* Mitteilungen Staatliches Museum für Naturkunde in Stuttgart, Stuttgart.

García, M.D.; Clemente, M.E. & Presa, J.J. (1995). Manifestaciones acústicas de *Chorthippus binotatus binotatus* (Charpentier, 1825) (Orthoptera: Acrididae). Su estatus taxonómico y su distribución en la Península Ibérica. *Boletín de la Asociación española de Entomología,* Vol. 19, pp. 229-242, ISSN 0210-8984.

García, M.D.; Lorier, E.; Clemente, M.E. & Presa, J.J. (2003). Sound production in *Parapellopedon instabilis* (Rehn, 1906) (Orthoptera: Gomphocerinae). *Annales de la Société entomologique de France* (n.s), Vol. 39, No. 4, pp. 335-342, ISSN 0037-9271.

García, M.D.; Larrosa, E.; Clemente, M.E. & Presa, J.J. (2005). Contribution to the knowledge of genus *Dociostaurus* Fieber, 1853 in the Iberian Peninsula, with special reference to its sound production (Orthoptera: Acridoidea). *Anales de Biología,* Vol. 27, pp. 155-189, ISSN 0213-3997.

Gehardt, H.C. & Huber, F, (2002). *Acoustic communication in insects and anurans,* The University of Chicago Press, ISBN 0-226-28833-1, Chicago.

Green, S.V. (1995). Song characteristics of certain Namibian grasshoppers (Orthoptera: Acrididae: Gomphocerinae). *African Entomology,* Vol. 3, pp. 1-6, ISSN 1021-3589.

Greenfield, M.D. (1997). Acoustic communication in Orthoptera. In: *The bionomics of grasshoppers, katydids and their kin,* Gangwere, S.K., Muralirangan, M.C. & Muralirangan, M. (Eds.), pp. 197-230, Cab International, ISBN 0-85199-141-6, Wallingford, United Kingdom.

Helversen, D.v. & Helversen, O.v. (1983). II.3. Species recognition and acoustic localization in acridid grasshoppers: a behavioural approach. In: *Neuroethology and behavioral physiology,* Huber, F. & Markl, H. (Eds.), pp. 95-107, Springer - Verlag, ISBN 0387126449, Berlin, Heidelberg.

Jacobs, W. (1953). Vergleichende Verhaltensstudien an Feldheuschrecken (Orthoptera, Acrididae) und einigen anderen Insekten. *Zoologische Anzeiger,* Suppl. 17, pp.115-138, ISSN 0044-5231.

Jago, N.D. (1971). A review of the Gomphocerinae of the world, with a key to the genera (Orthoptera, Acrididae). *Proceedings of the Academy of Natural Sciences of Philadelphia,* Vol. 123, No. 8, pp. 205-343, ISSN 0097-3157.

Lecoq, M. & Assis-Pujol, C.V. (1998). Identity of *Rhammatocerus schistocercoides* (Rehn, 1906) from South and North of the Amazonian rain forest and new hypotheses on the outbreaks determinism and dynamics. *Transactions of the American Entomological Society,* Vol. 124, No. 1, pp. 13-23, ISSN 0002-8320.

Lewis, T. (Ed.) (1984). *Insect Communication,* 12 th. Symposium of the Royal Entomological Society of London, Academic Press Inc, ISBN 0-12-447175-7, London.

Loreto, V.; Cabrero, J.; López-León, M.D.; Camacho, J.P.M. & de Souza, M.J. (2008). Comparative analysis of r-DNA location in five neotropical Gomphocerinae grasshopper species. *Genetica,* Vol. 132, pp. 95-101, ISSN 0016-6707.

Lorier, E. (1996). *Estudio de la comunicación acústica en especies de Euplectrottettix (Acrididae, Gomphocerinae),* Repertorios de tesis de maestría y doctorado 1986-1994, Ministerio de Educación y Cultura, PEDECIBA, UDELAR, PNUD, Uruguay, Oficina del libro AEM, Montevideo, Uruguay.

Lorier, E.; Clemente, M.E.; García, M.D. & Presa, J.J. (2010). El comportamiento acústico de *Fenestra bohlsii* Giglio-Tos, 1895 (Orthoptera, Acrididae, Gomphocerinae). *Neotropical Entomology,* Vol. 39, No. 6, pp. 839-853, ISSN 1519-566X.

McGregor, P.K.& Peake, T.M. (2000). Comunication networks: social environments for receiving and signaling behaviour. Acta Ethologica, Vol. 2, pp. 71-81, ISSN 0873-9749.

Nattier, R.; Robillard, T.; Amedegnato, C.; Couloux, A.; Cruaud, C. & Desutter-Grandcolas, L. (2011). Evolution of acoustic communication in the Gomphocerinae (Orthoptera: Caelifera: Acrididae). *Zoologica Scripta,* Vol. 40, No. 5, pp. 479-497, ISSN 0300-3256.

Otte, D. (1977). Comunication in Orthoptera. In: *How animals communicate,* Sebeok, T. (Ed.), pp. 334-361, Indiana University Press, ISBN 0-253-32855-1, Bloomington, Indiana.

Otte, D. (1979). Revision of the grasshopper tribe Orphulellini (Gomphocerinae, Acrididae). *Proceedings of the Academy of Natural Sciences of Philadelphia,* Vol. 131, pp. 52-58, ISSN 0097-3157.

Otte, D. & Jago, N.D. (1979). Revision of the grasshopper genera *Silvitettix* and *Compsacris* (Gomphocerinae, Acrididae). *Proceedings of the Academy of Natural Sciences of Philadelphia,* Vol. 131, pp. 257-288, ISSN 0097-3157.

Pitkin, L.M. (1976). A comparative study of the stridulatory files of the British Gomphocerinae (Orthoptera: Acrididae). *Journal of Natural History,* Vol. 10, pp.17-28, ISSN 0022-2933.

Ragge, D.R. (1987). The songs of the western European grasshoppers of the genus *Stenobothrus* in relation to their taxonomy (Orthoptera: Acrididae). *Bulletin of the British Museum (Natural History)* Entomology series, Vol. 55, pp. 393-424, ISSN 0524-6431.

Ragge, D.R. & Reynolds, W.J. (1998). *The song of the Grasshoppers and Crickets of Western Europe,* Harley Books, ISBN 0-946589-49-6, Colchester.

Reynolds, W.J. (1986). A description of the song of *Omocestus broelemannnni* (Orthoptera: Acrididae) with notes on its taxonomic position. *Journal of Natural History*, Vol. 20, pp. 111-116, ISSN 0022-2933.

Riede, K. (1987). A comparative study of mating behaviour in some Neotropical grasshoppers (Acridoidea). *Ethology*, Vol. 76, pp. 265-296, ISSN 0179-1613.

Salto, C.; Primo, J. & Luiselli, S. (2003). Preferencias alimenticias de *Rhammatocerus pictus* (Bruner) y *Aleuas lineatus* Stål (Orthoptera: Acrididae) en condiciones semicontroladas. INTA Rafaela. Anuario 2003. Available from: http://www.inta.gov.ar/rafaela/info/documentos/anuario2003/a2003_p148.pdf (accessed November 9, 2010).

Valdecasas, A.G. (2011). Una disciplina científica en la encrucijada: la Taxonomía. *Memorias de la Real Sociedad Española de Historia Natural, Segunda época*, Vol. 9, pp. 9-17. ISSN 1132-0869.

4

Detecting Non-Local Japanese Pine Sawyers in Yunnan, Southwestern China via Modern Molecular Techniques

Shao-ji Hu, Da-ying Fu and Hui Ye
Laboratory of Biological Invasion and Transboundary Ecosecurity,
Yunnan University, Kunming,
P. R. China

1. Introduction

The term "biological invasion" comes with two aspects, on one hand it refers to the introduction of an exotic species, and/or non-local populations of any species to a given geographical area, while on the other hand it refers to the ecological and/or the economical consequences of such activity (Perrings et al., 2002). With accelerating socioeconomic development and globalization, issues of biological invasion are of increasing concern, as some invasive alien species are capable of causing catastrophe to local environments and the economies (Xu et al., 2006; Meyerson & Mooney, 2007).

Intentionally or accidentally, human beings constantly introduce organisms to new habitats. For those cases of intentional introduction, some were for agronomic purposes and have been proved to be beneficiary, such as potatoes, maize, peanuts, and sunflowers (Pope et al., 2001; S.M. Wang, 2004). However, others like *Eichharnia crassipes* (Martius) had caused a diversity of problems (Harley et al., 1996; Julien, 2001; T.J. Hu et al., 2009). For accidental introduction, the consequences are often more negative, as these organisms often go undetected until they have expanded their population considerably. The invasion and expansion of *Eupatorium adenophorum* Sprengel in western China is a typical example (R. Wang & Y.Z. Wang, 2006; R. Wang et al., 2011).

Yunnan Province in southwestern China is situated in the low-latitudinal plateau region (LLPR) connecting the Indochinese peninsula and the western portion of mainland China, where the terrain and climate are complex (Z.Y. Chen, 2001; S.Y. Wang & W. Zhang, 2005). Such distinct geographical position and natural environment entitle Yunnan a wide scope of ecosystems except desert and ocean in the northern hemisphere (Guo & Long, 1998). Apart from being the representative of terrestrial biodiversity, such complexity also makes Yunnan very vulnerable to invasive alien species.

The western portion of Yunnan is generally known as the longitudinal range-gorge region (LRGR) where great mountains and deep valleys run parallel through Yunnan and connect the Indochinese peninsula and the inner area of western China, which has been considered as the terrestrial pathway of biological invasion in Yunnan (He et al., 2005). The eastern portion of Yunnan is the western margin of the Yunnan-Guizhou altiplano, where most of the terrain are plains with low hills with an altitude lower than the central portion of Yunnan as well as most of the LRGR (S.Y. Wang & W. Zhang, 2005). The subdued slope of this area also allows non-local organisms to enter Yunnan from the adjacent areas of China. Meanwhile, Yunnan also plays an important role in preventing invasive alien species from entering China, as the mountains lie in northern portion are natural barriers to obstruct the populations of non-local organisms from expanding. Therefore, Yunnan bears a vital strategic function of regional ecosecurity defined by the combination of terrestrial pathway and frontier prevention of biological invasion.

To enhance relevant research in detecting exotic species and/or non-local populations in Yunnan, research on invasive alien species has been conducted to reveal origin, expansion, and the mechanisms of invasion. For cases of invasive insects, researches on two exotic tephritid pests, *Bactrocera dorsalis* (Hendel) and *B. correcta* (Bezzi) have achieved the goals of population recognition and route reestablishment via modern molecular techniques (Shi et al., 2010; Liu, unpublished data). These studies demonstrated that the invasion of the two fruit flies represented the mode of natural and long-term invasion, inhabitation, and expansion. Moreover, the merits of these important research also provided ideas for the research on biological invasion in the new era. We used molecular techniques to detect non-local populations of a forest pest in southwestern Yunnan which strongly suggested another possible mode of biological invasion.

2. Background information

2.1 The Japanese pine sawyer

The Japanese pine sawyer, *Monochamus alternatus* Hope (Coleoptera: Cerambycidae), is a stem-boring beetle widely distributed in eastern Asia and the northern portion of the Indochinese peninsula (Davis et al., 2008). This polyphagous beetle feeds on conifers throughout the life history, including many unrelated species of *Pinus, Picea, Abies, Cedrus, Larix,* and *Cupressus* (Ning et al., 2004; Davis et al., 2008). *M. alternatus* is a univoltine species, which produces only one generation per year (Togashi, 1989; L.P. Wang, 2004; Zhao et al., 2004). After copulation, the female adults gnaw ovipositing wounds in the bark of host trees, and lay eggs in the space between the phloem and the xylem (Anbutsu & Togashi, 2000). The newly hatched larvae ingest wood tissue from both the phloem and the xylem, and start to excavate "U" shaped tunnels from the third instar. The final instar larvae build oval pupal chambers at the end of the tunnels to pupate. The newly eclosed adults feed on shoots, needles, and bark to obtain nutrients for maturation (Shibata, 1987). Copulation usually takes place five to ten days after emergence, each adult is able to mate more than once (L.P. Wang, 2004; H. Yang et al., 2006). As a typical secondary stem-boring species, the female adults tend to select stressed hosts for oviposition (S.J. Hu et al., 2009), which is induced by the volatile chemicals (i.e., α-pinene, β-pinene, 3-carnine, and ethanol) emitted from the hosts (Ikeda et al., 1986; Yamasaki et al., 1989), but often influenced and deterred by bark thickness, branch diameter, ovipositing scars from other female adults, and larval frass (Nakamura et al., 1995a, 1995b; Anbutsu & Togashi, 2000; Li & Z.N. Zhang, 2006; S.J. Hu et al., 2009; Z.X. Yang et al., 2010).

2.2 Association with the pine wood nematodes

The pine wood nematode, *Bursaphelenchus xylophilus* (Steiner et Buhrer) (Nematoda: Aphelenchoididae), is a quarantine phytopathogenic organism of conifers indigenous to northern America but casually spread to Eurasia and southern Japan in last century (CABI & EPPO, 1997). The nematode, like other species of genus *Bursaphelenchus*, lives in the vascular tissue of their coniferous hosts, which decreases the transportation of water and resin, and subsequently weakens the hosts and causes a syndrome named the pine wilt disease (Mamiya, 1983). The pathogenicity of *B. xylophilus* varies significantly. In North America, reports on its damages are rarely seen (Wingfield et al., 1986). However, in the vast majority of Japan and China, where *B. xylophilus* is considered as an invasive species, the infestation is often fatal, and has caused catastrophic timber loss (Mamiya, 1988; X.B. Hu et al., 1997). In nature, the relocation of *Bursaphelenchus* nematodes depends on coleopterous vectors, but the association between nematodes and vectors differs with taxa (Linit, 1988). For *B. xylophilus*, the long-horned beetles of genus *Monochamus* are the major vector (Linit et al., 1983; Evans et al., 1996; Kulinich & Orlinskii, 1998), and research has confirmed that *M. alternatus* is the key vector of *B. xylophilus* in eastern Asia (Mamiya & Enda, 1972; Kobayashi et al., 1984). In infested hosts, the immature pine wood nematodes are able to locate the pupae of *M. alternatus* and board into the tracheae, and the newly eclosed adults carry the nematodes when seeking food and new hosts. During the course of feeding and ovipositing, the nematodes detach from the beetles and enter the new hosts through the feeding and ovipositing wounds to initiate a new round of infestation (Edwards & Linit, 1992; Naves et al., 2007).

2.3 Economic importance of *M. alternatus*

There are basically four aspects of economic importance of *M. alternatus*. As a stem-boring coleopteran of the conifers, the direct feeding and ovipositing of the adults damages the tender shoots, needles, and bark of the host, which may weaken the host when a sizable population occurs in a stand. The tunnels inside the host built by the larvae often cause more damage which eventually kills the host and substantially reduces the economic value of the timber (L.P. Wang, 2004). *M. alternatus* spends the entire immature stage in the host, which makes it very difficult to be detected from timber and/or wood packaging materials and can be easily relocated to new habitats via transportation in trading (Haack, 2006). As timber and/or wood packaging materials with *B. xylophilus* can also be relocated (Braasch et al., 2001; Gu et al., 2006), the concealed *M. alternatus* would subsequently introduce the nematodes into new habitats via its dispersal. Hence, *M. alternatus* has been recognized as a dangerous forest pest by forestry authorities worldwide.

2.4 The pine wilt disease in China

In mainland China, the pine wilt disease was reported for the first time in a small patch of *Pinus thunbergii* near the Sun Yat-sen Mausoleum in Nanjing, Jiangsu Province in 1982 (Y.Y. Wang et al., 1991). Afterwards, the infested area expanded quickly throughout eastern China. From the bulletins released by the SFA (State Forestry Administration) of China, a clear tendency of westward expansion can be observed (SFA, 2004 ~ 2007). During the past two years, some counties in Henan, Shaanxi, and Sichuan provinces in northern, northwestern, and southwestern China also reported infested areas of the pine wilt disease (SFA, 2011) (Fig. 1).

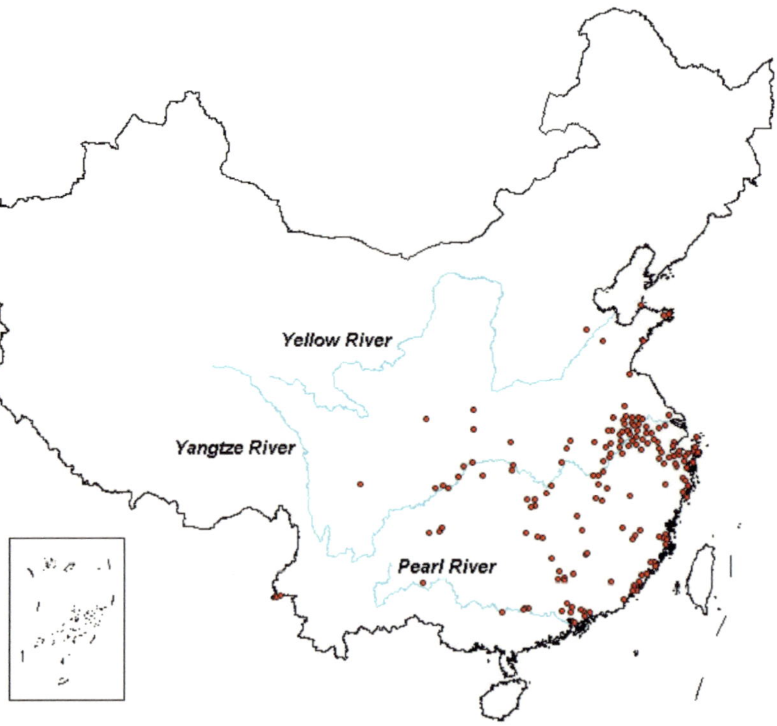

Fig. 1. The distribution pattern of the pine wilt disease (red dots) in mainland China based on the historical data from 2004 ~ 2011, data source: SFA (2004 ~ 2011).

With the expansion of pine wilt disease, an interesting distribution pattern was revealed. In eastern and southern China, the distribution concentrated near harbors, like the deltas of the Yangtze River and the Pearl River, while in the central portion of China, the infested areas are mostly along the Yangtze River. Such a distribution pattern suggested that the dispersal of *B. xylophilus* was closely related to transportation.

The occurrence of the pine wilt disease in Yunnan Province, southwestern China was distinct. Not only because it was one of the latest infested areas after the massive expansion, but also due to its unique geographical position. Unlike eastern China where many harbors are distributed, the first, single, and isolated pine stand (*Pinus kesiya* var. *langbianensis*) in Yunnan infested by *B. xylophilus* was located in the westernmost corner, a remote township named Wanding near the Sino-Burmese border (SFA, 2007).

3. The hypothesis

Given that there had been no report on the occurrence of *B. xylophilus* in Burma (CABI & EPPO, 1997), it was not logical to speculate the pine wood nematode in Wanding came from any Burmese source. Moreover, Wanding, the only site of infestation, lies in the west portion of the LRGR, which is regarded as natural barriers obstructing organisms from expanding by natural means (He et al., 2005). Therefore, it was also not logical to speculate that the nematodes came from adjacent areas in Yunnan. However, after a careful patrol of the initial infested pine stand, investigators discovered that the infested trees was centered an under-constructing telecommunication facility, which consumed a considerable amount of electronics manufactured in eastern China (Z.Q. Li, pers. comm.). Hence, a reasonable hypothesis was conceived that the nematodes in Wanding were casually introduced with non-local *M. alternatus* populations hidden in the wood-packaging materials. Providing that *M. alternatus* from different localities possess different genetic profiles, the non-local individuals can then be distinguished by analyzing the genetic differences on population level, and the source can be traced by comparing the genetic profiles with the samples taken from the suspected region.

4. Validating the hypothesis – Theory and experiment

4.1 Molecular achievements on *Monochamus*

Although *Monochamus* species are important forest pests across the Eurasia, research employing molecular techniques were quite limited compared to other arthropod pests. Cesari et al. (2005) started the first molecular taxonomy research on seven *Monochamus* species by using the combined mtDNA data of *cox1* and *12s* genes. Afterward, such research were quickly developed in Japan, where *M. alternatus* had become a threat to forestry. Kawai et al. (2006) published a paper on the genetic structure of 27 populations of *M. alternatus* from Japan and China (with 25 populations from Japan and two populations from China). Shoda-Kagaya (2007) published paper on microsatellite markers. Then, Koutroumpa et al. (2009) conducted a criticizing research which revealed *Numts* resembling the *cox1* and *cox2* genes of *M. galloprovincialis* and *M. sutor*, alerting scientists that precautions are necessary when applying the gene markers from the mitogenome. There was no such research on *M. alternatus* in China published by the end of 2009.

4.2 Designing the experiment

Appropriate sampling strategy requires samples to be taken from populations scattered in the entire research range, and for each population, sufficient individuals should be collected to represent it (S.Y. Chen & Y.P. Zhang, 2006). Based on this, seven populations representing *M. alternatus* from central, southern, southwestern, and northern Yunnan were selected, in which three populations were chosen from the southwestern Yunnan where the pine wood nematodes were reported. In an attempt to determine the source of the pine wood nematode, two reference populations from outside Yunnan were set, with one from Hubei Province representing central China and the other from Zhejiang Province representing eastern China (Fu et al., 2010; Fig. 2). Adult *M. alternatus* were collected by flight traps baited with barkborer lure (Chinese Academy of Forestry, Zhejiang, China).

Choosing the correct genetic marker can minimize problems in data analysis (Simon et al., 1994) and maximize the future potential of cross references with data published by other researchers. Population phylogeny requires gene markers with moderate evolutionary rate, free of recombination, which means that mtDNA, a maternal heritage genome without recombination, is the ideal choice. Considering that there has been no report on *Numts* in *M. alternatus*, which can also be detected and excluded by strict data examination when encountered. The prefunding research by Kawai et al. (2006) resolved a clear population phylogeny using a gene fragment of *cox2*; the research discussed here chose the same *cox2* marker.

The same PCR primers developed by Roehrdanz (1993) and Kawai et al. (2006) were applied, and the PCR reaction was performed by the protocols described by Kawai et al. (2006). Bioinformatic and statistic software like DAMBE 5.0.7 (Xia & Xie, 2001), MEGA 4.0 (Tamura et al., 2007), SAMOVA 1.0 (Dupanloup et al., 2002), AMOVA 3.1 (Excoffier et al., 2005), and SSPS 13.0 (SPSS Inc., Illinois, US) were applied to analyze haplotype assemblage, Kimura two-parameter (K2P) distance (Kimura, 1980), NJ phylogenetic reconstruction (Saitou & Nei, 1987), population grouping, and multidimensional scaling (MDS) (Lessa, 1990). The mapping of haplotype distribution was performed in AcrView 3.3 (ESRI, USA).

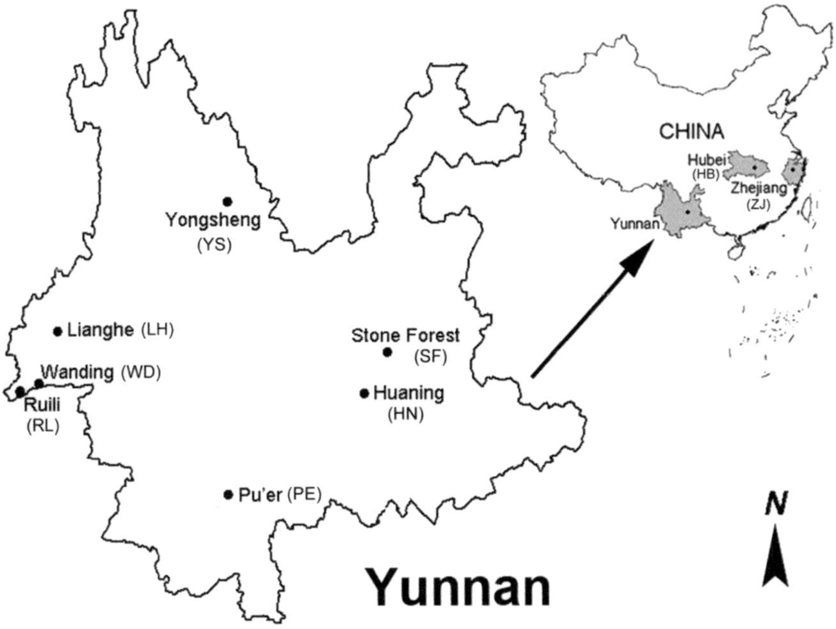

Fig. 2. The sampling sites of *M. alternatus* in Yunnan, Hubei, and Zhejiang. Dots on the map of China represent the capital cities of the three provinces. After Fu et al. (2010).

5. Validating the hypothesis – Molecular evidences

5.1 Haplotype assemblage

Eighteen haplotypes were defined by 22 polymorphic sites in the 565 bp *cox2* sequence, with haplotypes 1, 2, and 5 showing much higher frequencies. The distribution pattern of these 18 haplotypes was locality related: haplotypes 1 to 4 were widely distributed, dominating most of the central, northern, and southern Yunnan; haplotypes 6 to 18 were found within single localities. Haplotype 5 was shared by populations from southwestern Yunnan and Zhejiang; however, haplotypes 1 to 4 were neither shared by population from Hubei nor population from Zhejiang; and haplotypes 12 to 15 were only found in Hubei population. Further analysis of the haplotypes derived from the three populations in southwestern Yunnan discovered a high ratio of haplotype 5: with 69.2%, 70.0%, and 71.4% in Ruili, Wanding, and Lianghe populations, respectively. The matrix of shared haplotypes showed a boundary between the three populations from southwestern Yunnan and the remaining populations, with the shared haplotypes being confined within each population group and the unique haplotypes being presented only in two of the populations from southwestern Yunnan (Table 1). When mapping haplotype frequencies into pie charts, a more obvious population boundary as well as the existence of non-local individuals in three populations from southwestern Yunnan can be observed (Fig. 3).

Population	PE	RL	WD	HN	SF	LH	YS	ZJ	HB
Pu'er									
Ruili	0								
Wanding	0	1							
Huaning	1	0	0						
Stone Forest	2	0	0	2					
Lianghe	2	1	1	1	2				
Yongsheng	1	0	0	1	1	1			
Zhejiang	0	1	1	0	0	1	0		
Hubei	0	0	0	0	0	0	0	0	
Unique haplotypes	0	3	3	0	0	0	0	3	4

Table 1. Matrix of shared haplotypes of the nine *M. alternatus* populations (below diagonal) and the numbers of unique haplotypes (last row). Population codes correspond to those in Fig. 2.

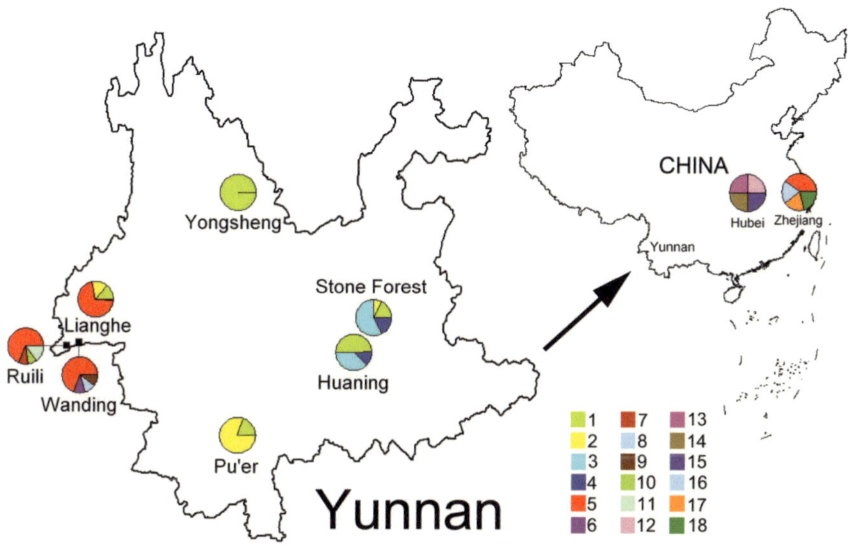

Fig. 3. Mapping of haplotype frequencies of all nine populations of *M. alternatus*.

5.2 Genetic distance

The Kimura two-parameter (K2P) distance varied from 0.0014 to 0.0132, with the distance between Pu'er (southern Yunnan) and Yongsheng (northern Yunnan) being the minimum and that between Stone Forest (central Yunnan) and Hubei being the maximum (Table 2). The K2P distances among populations from Pu'er, Huaning, Stone Forest, and Yongsheng and among populations from Ruili, Wanding, and Lianghe showed close genetic connection; but the K2P distances between any given population from Pu'er, Huanian, Stone Forest, or Yongsheng and that from Ruili, Wanding, or Lianghe suggested otherwise (Table 2). Hence, the seven populations sampled from Yunnan could be divided into two groups, hereafter designated as the SPG, the southwestern population group containing populations from Ruili, Wanding, and Lianghe, and the RPG, the remaining population group containing populations from Pu'er, Huaning, Stone Forest, and Yongsheng. The grouping result was then supported by the multidimensional scaling (MDS) analysis (Fig. 4). Notably, both the K2P distances and the MDS analysis showed a close relationship between the population from Zhejiang and all populations from the SPG, however, the relationship between the population from Hubei and any population from both the SPG and the RPG was much greater (Table 2; Fig. 4).

Site	PE	RL	WD	HN	SF	LH	YS	ZJ	HB
PE		0.0037	0.0035	0.0012	0.0015	0.0028	0.0013	0.0037	0.0040
RL	0.0084		0.0007	0.0036	0.0040	0.0012	0.0034	0.0008	0.0024
WD	0.0084	0.0018		0.0035	0.0038	0.0012	0.0032	0.0009	0.0024
HN	0.0020	0.0090	0.0090		0.0013	0.0029	0.0012	0.0037	0.0040
SF	0.0022	0.0101	0.0101	0.0022		0.0032	0.0020	0.0040	0.0042
LH	0.0064	0.0027	0.0031	0.0071	0.0081		0.0026	0.0013	0.0025
YS	0.0014	0.0070	0.0070	0.0020	0.0031	0.0053		0.0034	0.0038
ZJ	0.0093	0.0021	0.0026	0.0098	0.0109	0.0035	0.0079		0.0027
HB	0.0122	0.0070	0.0077	0.0123	0.0132	0.0084	0.0107	0.0086	

Table 2. Matrix of the Kimura two-parameter (K2P) distance (below diagonal) and standard errors (above diagonal) of between populations of *M. alternatus*. Population codes correspond to those in Fig. 2. Data source: Fu et al. (2010).

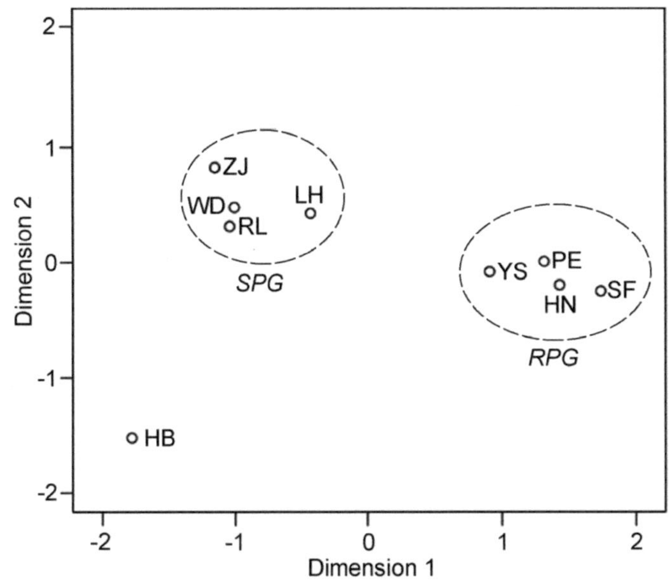

Fig. 4. The multidimensional scaling (MDS) plots of *M. alternatus* populations at different locations based on Kimura two-parameter (K2P) distances. SPG, the southwestern population group; RPG, the remaining population group. Population codes correspond to those in Fig. 2. After Fu et al. (2010).

5.3 Phylogenetic cladistics

The neighbor-joining (NJ) phylogenetic tree based on the K2P distance of the 18 haplotypes resolved three major clades designated from A to C (Fig. 5). Clade A contained six populations from Yunnan, with four from northern, central, and southern portion, and only one sample from Wanding, southwestern portion was included. Clade B contained the three populations from southwestern Yunnan and the population from Zhejiang, eastern China. However, no sample from other portions of Yunnan grouped in clade B. Clade C only contained the population from Hubei, central China. The topological structure of the NJ phylogenetic tree suggests that the seven populations consisted of two groups, the RPG and the SPG, regardless of the connection between them by a few samples from Wanding and Lianghe. It also suggested that the genetic profile of the population from Hubei is different from the remaining ones involved in the research.

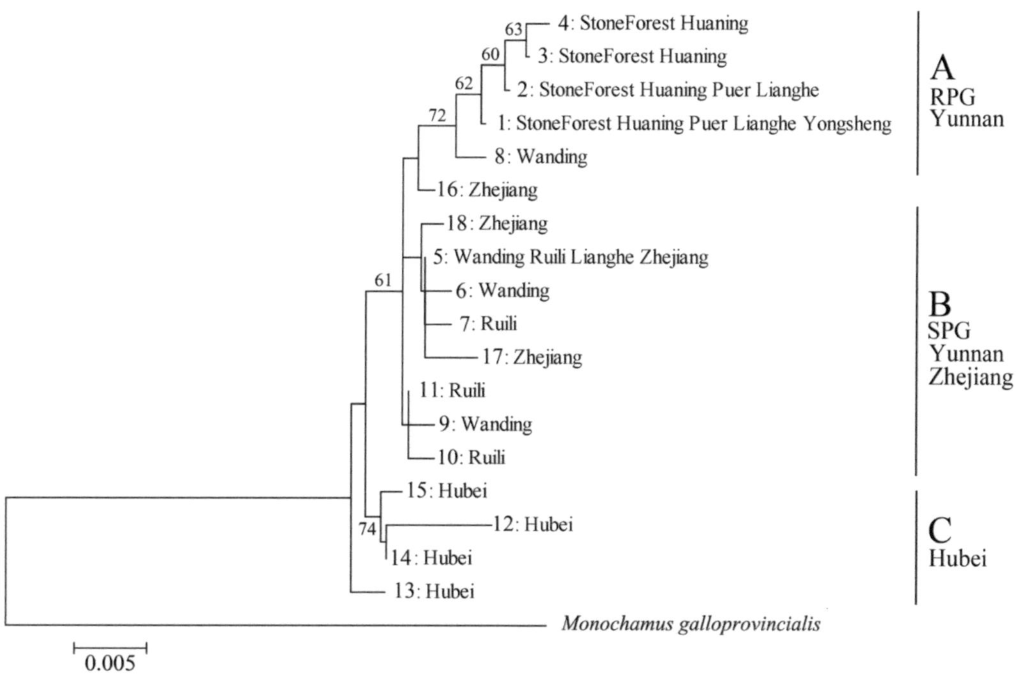

Fig. 5. The neighbor-joining (NJ) phylogenetic tree of the 18 haplotypes of *M. alternatus*, numbers above the branches represent the bootstrap values (> 50). After Fu et al. (2010).

5.4 Population genetic structure

SAMOVA and AMOVA analyses were applied to verify the genetic structure of the seven populations from Yunnan. When the number of groups (K) grows and the F_{CT} value reaches the plateau, the SAMOVA analysis yields the optimal result. In this research, the F_{CT} value reached the maximum ($F_{CT} = 0.73$) when $K = 2$, with populations from Pu'er, Huaning, Stone Forest, and Yongsheng being one group and populations from Ruili, Wanding, and Lianghe being the other (Fig. 6), supporting the grouping result mentioned previously. The grouping was then tested by the AMOVA analysis, which indicated that 72.54% of the variation occurred between the two groups, 21.86% occurred within populations, and only 5.60% occurred among populations within groups (Table 3). The relatively high variation within populations was also mirrored by the haplotype assemblage that most of the populations possessed multiple haplotypes, especially the three populations from southwestern Yunnan rich in unique haplotypes (Table 1; Fig. 3).

Source of variation	*d. f.*	Variance components	Percentage of variation	*F*	*P*
Between groups	1	1.725	72.54	$F_{CT} = 0.73$	0.025
Among populations within groups	5	0.133	5.60	$F_{SC} = 0.21$	0.001
Within populations	54	0.520	21.86	$F_{ST} = 0.78$	0.000
Total	60	2.378	100	--	--

Table 3. AMOVA analysis results. Data source: Fu et al. (2010).

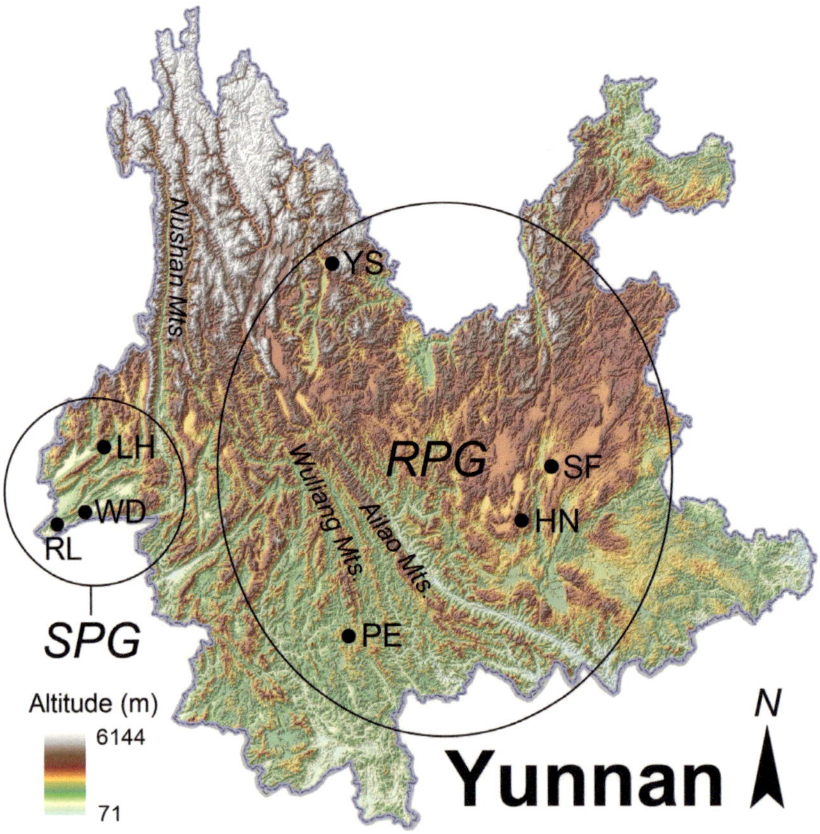

Fig. 6. The terrain of Yunnan and the grouping result of seven populations of *M. alternatus* determined by SAMOVA and AMOVA analyses with major geographical barriers marked (Ailao Mountains, Wuliang Mountains, and Nushan Mountains). SPG, the southwestern population group; RPG, the remaining population group. Population codes correspond to those in Fig. 2.

6. Determination of non-local *M. alternatus*

The experiment provided molecular evidence strongly supporting the following two hypotheses: 1) the genetic structure divides the seven *M. alternatus* populations from Yunnan into two groups, the southwestern population group (SPG) and the remaining population group (RPG), and 2) it is most likely that there were non-local individuals of *M. alternatus* in the three populations sampled from southwestern Yunnan, as these populations shared one of the dominant haplotypes of the reference population from Zhejiang. However, in an attempt to finally determine the identity of non-local *M. alternatus*, the relationship between such genetic divergence and the natural geographical characteristics of Yunnan should be considered thoroughly.

Genetic divergence has been frequently discussed in terms of natural geographical barriers (Yagi et al., 2001; Shoda-Kagaya, 2007; Shi et al., 2010). Yunnan is a typical LLPR with complex terrain, where numerous population phylogenetic studies on different insects have revealed the role that geographical barriers play in obstructing the gene flow among populations from various localities (Shi & Ye, 2004; Liu et al., 2007; P. Chen & Ye, 2008; Shi et al., 2010). Basically, the mountain ranges in the LRGR portion of Yunnan are considered effective geographical barriers, which restrict gene flow in the latitudinal direction.

Since the natural environment of Yunnan is able to cause genetic divergence, the separation of two population groups could be either the result of natural geographical barriers or the consequence of introduced non-local sources. The seven sampling sites in Yunnan were separated by great mountain ranges. Ruili, Wanding, and Lianghe were separated from other populations by Nushan Mountains, Pu'er was separated from Stone Forest, Huaning, and Yongsheng by Wuliang and Ailao Mountains (Fig. 6). But it is interesting to note that the genetic distances among populations from the RPG were limited to 0.0014 to 0.0031 (Table 2), the while genetic variation among them comprised only 5.6% of the total divergence (Table 3), and they shared all of the haplotypes defined by the samples taken from them (Table 1; Fig. 3). Hence, the geographical barriers such as Wuliang and Ailao Mountains had not caused sufficient genetic divergence among these populations. Judging from the phylogenetic tree (Fig. 5), these four populations were latest diverged, which may represent a short inhabitation history of *M. alternatus* in Yunnan under natural condition. To the contrast, the genetic distances between populations from the SPG and the RPG were much higher than that within each group, varied from 0.0064 to 0.0101 (Table 2), the genetic variation reached 72.54% (Table 3), and most of the unique haplotypes were defined (Table 1). Given the discussion above, the geographical barriers can not explain such divergence.

Assuming that the SPG and the RPG were two genetically independent sources, there must be gene flows among these populations in the course of evolution, which will be detected by shared haplotypes. It is noticeable that population from Lianghe shared two haplotypes with populations from the RPG; however, no haplotype from the SPG was shared by any other populations. Given that the individuals from the RPG were able to "migrate" into the SPG, the same phenomenon should have happened the other way around. Another assumption was that the genetic divergence between the SGP and the RPG was caused by introduction of non-local *M. alternatus*. The haplotype distribution demonstrated that about 90% individuals of the southwestern populations shared haplotype 5 with population from Zhejiang (Table 1; Fig. 3), and the phylogenetic tree indicated a close genetic connection between haplotype 5 and those unique haplotypes (Fig. 5), which had been regarded as a symbol of establishment of non-local populations (Slatkin & Hudson, 1991; J. Hu et al., 2008).

As mentioned previously, the pine wilt disease in Wanding was reported four years after the arrival of electronics with wood packaging materials from eastern China where the pine wood disease was most severe. Considering the characteristics of the immature stages of *M. alternatus* and its association with *B. xylophilus*, it was possible that the un-treated, infested wood packaging materials disposed casually near the construction site became the source of infestation. This is not only would help to explain the significant genetic divergence between the SPG and the RPG in Yunnan, but also can explain the genetic connection between the SPG and the population from Zhejiang instead of Hubei.

7. Conclusion

The research proved the existence of non-local *M. alternatus* in southwestern Yunnan by utilizing the modern molecular techniques. It is the first, yet not only an isolated study on the population phylogeny of the *Monochamus* longhorn beetles in China, but also another exemplar of applying the molecular techniques to the research of biological invasion. Apart from demonstrating the feasibility of detecting non-local populations, it also provided an insight that, unlike the dispersal of other invasive alien species, such as fruit flies, the spread of pine wood nematodes in mainland China was mostly due to human-aided relocation of infested, un-treated wood packaging materials. Therefore, preventing the infested wood materials from circulating or being transported may become the most important checkpoint of preventing the pine wilt disease from expanding.

8. Future research and objectives

In the research discussed here, the modern molecular techniques provided strong evidence which proved the dispersal of the pine wood nematodes in southwestern Yunnan was caused by the casual introduction of non-local Japanese pine sawyer, *Monochamus alternatus*, and even helped to identify the possible origin of those non-local *M. alternatus* by combining both of the molecular data and the historical facts of telecommunication construction in the infested area. The most advanced technology nowadays allows the scientists to identify the pine wood nematodes directly from either the wood sample or the beetles (X.R. Wang et al., 2010; X.R. Wang et al., 2010; Y.Q. Hu et al., 2011). By utilizing such innovative methods, the foresters and plant quarantine staff are able to turn the identification of invasive organisms from sheer speculation to solid evidence. By promoting such concept along with the modern molecular techniques, not only will future law enforcement benefit, but also the regional and international trading embargos surrounding quarantine and/or invasive organisms shall be solved in a proper manner.

Apart from identifying non-local populations, population phylogeny can also be used to study the historical dispersal (including both natural dispersal and human-aided relocation) of insects, as suggested by research mentioned earlier (Kawai et al., 2006; Shoda-Kagaya, 2007; Shi et al., 2010). Since China has a 28-year dispersal history of the pine wilt disease, the dispersal pattern of the pine wood nematode as well as its key vector, *M. alternatus*, should be an interesting field to research. However, to date, there is only one paper on the dispersal pattern of the nematodes in China (Sun et al., 2008). Currently, the senior author and his colleagues are conducting a panoramic research on the population phylogeny of *M. alternatus* sampled from multiple localities in mainland China. The result of this research is to partially answer the long-asked question regarding the dispersal pattern of *M. alternatus* during the spread of the pine wilt disease in China. And as it is now possible to test the genetic profiles of *B. xylophilus* and *M. alternatus* together, future studies on the dispersal pattern and original sources of the pine wood nematodes in China can be revealed more thoroughly.

9. Acknowledgements

The related researches were conducted under a joint cooperation of the following persons and faculties: Z. Zhang (Chinese Academy of Forestry), Y.Z. Pan (Southwest Forestry University), H.P. Liu (Forest Disease and Pest Control and Quarantine Bureau of Yunnan Province), P.Y. Zhou (FDPCQB of Dehong Prefecture), R.J. Rui and W.C. Cai (Forest Disease and Pest Control and Quarantine Station of Yongsheng), Z.Q. Li and B.C. Liu (FDPCQS of Wanding), Z. Nong (FDPCQS of Ruili), H. Zhou (Forestry Bureau of Pu'er), D.D. Chen, X.Y. Ma, S.N. Ge, and C. Wu (Yunnan University), and X. Zhang (Kunming Institute of Zoology). The related research programs were conducted with the financial supports of the Key Program of State Forestry Administration (2006BAD08A19105), the International S&T Cooperation Program of China (2006DFA31790), the Key Scientific Project Fund of the Science and Technology Bureau of Yunnan Province (2005NG03), and the "211" Key Project of Yunnan University (21134025).

10. References

Anbutsu, H. & Togashi, K. (2000). Deterred ovipostion response of *Monochamus alternatus* (Coleoptera: Cerambycidae) to oviposition scars occupied by eggs. *Agricultural and Forest Entomology*, 2 (3): 217-223.

Braasch, H., Tomiczek, C., Metge, K., Hoyer, U., Burgermeister, W., Wulfert, I., & Schönfeld, U. (2001). Records of *Bursaphelenchus* spp. (Nematoda, Parasitaphelenchidae) in coniferous timber imported from the Asian part of Russia. *Forest Pathology*, 31 (3): 129-140.

CABI & EPPO. (1997). *Quarantine Pests for Europe (2nd edition)*. Wallingford: CAB International.

Cesari, M., Marescalchi, O., Francardi, V., & Mantovani, B. (2005). Taxonomy and phylogeny of European *Monochamus* species: first molecular and karyological data. *Journal of Zoological Systematics and Evolutionary Research*, 43 (1): 1-7.

Chen, P. & Ye, H. (2008). Relationship among five populations of *Bactrocera dorsalis* based on mitochondrial DNA sequences in western Yunnan, China. *Journal of Applied Entomology*, 132 (7): 530-537.

Chen, S.Y. & Zhang, Y.P. (2006). The genetic approaches and applications in the research of the origin of domesticated animals. *Chinese Science Bulletin*, 51 (21): 2469-2475. [in Chinese]

Chen, Z.Y. (2001). *The Pandect of Yunnan Climate*. Beijing: China Meteorological Press. [in Chinese]

Davis, E.E., Albercht, E.M., & Venette, R.C. (2008). *Monochamus alternatus*. In: Venette, R.C. *Exotic Pine Pests: Survey Reference*, USDA Forest Service.

Dupanloup, I., Schneider, S., & Excoffier, L. (2002). A simulated annealing approach to define the genetic structure of populations. *Molecular Ecology*, 11 (12): 2571-2581.

Edwards, O.R. & Linit, M.J. (1992). Transmission of *Bursaphelenchus xylophilus* through oviposition wounds of *Monochamus carolinensis* (Coleoptera: Cerambycidae). *Journal of Nematology*, 24 (1): 133-139.

Evans, H.F., McNamara, D.G., Braasch, H., Chadoeuf, J., & Magnusson, C. (1996). Pest risk analysis for the territories of the European Union (as the PRA area) on *Bursaphelenchus xylophilus* and its vectors in the genus *Monochamus*. *EPPO Bulletin*, 26: 199-249.

Excoffier, L., Lava, G., & Schneider, S. (2005). Arlequin ver. 3.0: an integrated software package for population genetics data analysis. *Evolutionary Bioinformatics Online*, 1: 47-50.

Fu, D.Y., Hu, S.J., Ye, H., Haack, R.A., & Zhou, P.Y. (2010). Pine wilt disease in Yunnan, China: evidence of non-local pine sawyer *Monochamus alternatus* (Coleoptera: Cerambycidae) population revealed by mitochondrial DNA. *Insect Science*, 17 (5): 439-447.

Gu, J., Braasch, H., Burgermeister, W., & Zhang, J. (2006). Records of *Bursaphelenchus* spp. intercepted in imported packaging wood at Ningbo, China. *Forest Pathology*, 36 (5): 323-333.

Guo, H.J. and Long, C.L. (1998). *The Biodiversity of Yunnan*. Kunming: Yunnan Science & Technology Press: 1-10. [in Chinese]

Haack, R.A. (2006). Exotic bark- and wood-boring Coleoptera in the United States: recent establishments and interceptions. *Canadian Journal of Forest Research*, 36 (2): 269-288.

Harley, K.L.S., Julien, M.H., & Wright, A.D. (1996). Water hyacinth: a tropical worldwide problem and methods for its control. In: Cussans, B.H., Devine, G.W., Duke, M.D., Fernandez-Quintanilla, S.O., Helweg, C., Labrada, A., Landes, R.E., Kudsk, M., & Streibig, P. (eds.) *Proceedings of the 2nd International Weed Control Congress*. Copenhagen: Slagelse: 630-644.

He, D.M., Wu, S.H., Peng, H., Yang, Z.F., Ou, X.K., and Cui, B.S. (2005). A study of ecosystem changes in Longitudinal Range-Gorge Region and transboundary eco-security in Southwest China. *Advances in Earth Science*, 20 (3): 932-943. [in Chinese with English abstract]

Hu, J., Zhang, J.L., Nardi, F., & Zhang, R.J. (2008). Population genetic structure of the melon fly, *Bactrocera cucurbitae* (Diptera: Tephritidae), from China and Southeast Asia. *Genetica*, 134 (3): 319-324.

Hu, S.J., Fu, D.Y., Li, Z.Q. & Ye, H. (2009). Factors effecting the distribution of the oviposition scars of *Monochamus alternatus* in *Pinus kesiya* var. *langbianensis*. *Forest Pest and Disease*, 6 (4): 1-2, 11. [in Chinese with English abstract]

Hu, T.J., Wang, Y.C., & Lian, Q.P. (2009). Advances on hazards and controls of common water hyacinth, *Eichhornia crassipes*. *Fisheries Science and Technology*, 25 (6): 27-30. [in Chinese with English abstract]

Hu, X.B., Qu, T., & Zheng, H. (1997). On the control strategies of pine wilt disease in China. *Forest Pest and Disease*, 16 (3): 30-32. [in Chinese]

Hu, Y.Q., Kong, X.C., Wang, X.R., Zhong, T.K., Zhu, X.W., Mota, M.M., Ren, L.L., Liu, S., & Cai, M. (2011) Direct PCR-based method for detecting *Bursaphelenchus xylophilus*, the pine wood nematode in wood tissue of *Pinus massoniana*. *Forest Pathology*, 41 (2): 165-168.

Ikeda, T., Yamane, A., Enda, N., Oda, K., Makihara, H., Ito, K., & Ôkochi, I. (1986). Attractiveness of volatile components of felled pine trees for *Monochamus alternatus* (Coleoptera: Cerambycidae). *Journal of the Japanese Forestry Society*, 68 (1): 15-19.

Julien, M.H. (2001). Biological control of water hyacinth with arthropods: a review to 2000. In: Julien, M.H., Hill, M.P., Center, T.D., and Ding, J.P. (eds). *ACIAR Proceedings*, 102: 8-20.

Kawai, M., Shoda-Kagaya, E., Maehara, T., Zhou, Z.L., Lian, C.L., Iwata, R., Yamane, A., & Hogetsu, T. (2006). Genetic structure of pine sawyer *Monochamus alternatus* (Coleoptera: Cerambycidae) populations in Northeast Asia: consequences of the spread of pine wilt disease. *Environmental Entomology*, 35 (2): 569-579.

Kimura, M. (1980). A simple method for estimating evolutionary rates of base substitutions through comparative studies of nucleotide sequences. *Journal of Molecular Evolution*, 16 (2): 111-120.

Kobayashi, F., Yamane, A., & Ikeda, T. (1984). The Japanese pine sawyer beetle as the vector of pine wilt disease. *Annual Review of Entomology*, 29: 115-135.

Koutroumpa, F.A., Lieutier, F., & Roux-Morabito, G. (2009). Incorporation of mitochondrial fragments in the nuclear genome (*Numts*) of the longhorned beetle *Monochamus galloprovincialis* (Coleoptera, Cerambycidae). *Journal of Zoological Systematics and Evolution Research*, 47 (2): 141-148.

Kulinich, O.A. & Orlinskii, P.D. (1998). Distribution of conifer beetles (Scolytidae, Curculionidae, Cerambycidae) and wood nematodes (*Bursaphelenchus* spp.) in European and Asian Russia. *EPPO Bulletin*, 28: 39-52.

Lessa, E.P. (1990). Multidimensional analysis of geographical genetic structure. *Systematic Zoology*, 39 (3): 242-252.

Li, S.Q. & Zhang, Z.N. (2006). Influence of larval frass extracts on the oviposition behaviour of *Monochamus alternatus* (Col., Cerambycidae). *Journal of Applied Entomology*, 130 (3): 177-182.

Linit, M.J. (1988). Nematode-vector relationships in the pine wilt disease system. *Journal of Nematology*, 20 (2): 227-235.

Linit, M.J., Kondo, E., & Smith, M.T. (1983). Insects associated with the pinewood nematode, *Bursaphelenchus xylophilus* (Nematoda: Aphelenchoididae), in Missouri. *Environmental Entomology*, 12 (2): 467-470.

Liu, J.H., Shi, W., and Ye, H. (2007). Population genetics analysis of the origin of the oriental fruit fly, *Bactrocera dorsalis* Hendel (Diptera: Tephritidae), in northern Yunnan Province, China. *Entomological Science*, 10 (1): 11-19.

Mamiya, Y. (1983). Pathology of pine wilt disease caused by *Bursaphelenchus xylophilus*. *Annual Review of Phytopathology*, 21: 201-220.

Mamiya, Y. (1988). History of pine wilt disease in Japan. *Journal of Nematology*, 20 (2): 219-226.

Mamiya, Y. & Enda, N. (1972). Transmission of *Bursaphelenchus lignicolus* (Nematoda: Aphelenchoididae) by *Monochamus alternatus* (Coleoptera: Cerambycidae). *Nematologica*, 18 (2): 159-162.

Meyerson, L.A. & Mooney, H.A. (2007). Invasive alien species in an era of globalization. *Frontiers in Ecology and the Environment*, 5 (4): 199-208.

Nakamura, H., Tsutsui, N., & Okamoto, H. (1995a). Oviposition habit of the Japanese pine sawyer, *Monochamus alternatus* Hope (Coleoptera, Cerambycidae) I. Factors affecting the vertical distribution of oviposition scars in a pine tree. *Japanese Journal of Entomology*, 63 (3): 633-640.

Nakamura, H., Tsutsui, N., & Okamoto, H. (1995b). Oviposition habit of the Japanese pine sawyer, *Monochamus alternatus* Hope (Coleoptera, Cerambycidae) II. Effect of bark thickness on making oviposition scars. *Japanese Journal of Entomology*, 63 (4): 739-745.

Naves, P.M., Camacho, S., de Sousa, E.M., & Quartau, J.A. (2007). Transmission of the pine wood nematode *Bursaphelenchus xylophilus* through feeding activity of *Monochamus galloprovincialis* (Col., Cerambycidae). *Journal of Applied Entomology*, 131 (1): 21-25.

Ning, T., Fang, L.Y., Tang, J., & Sun, J.H. (2004). Advances in research on *Bursaphelenchus xylophilus* and its key vector *Monochamus* spp. *Entomological Knowledge*, 41 (2): 97-104. [in Chinese with English abstract]

Perrings, C., Williamson, M., Barbier, E.B., Delfino, D., Dalmazzone, S., Shogren, J., Simmons, P., & Watkinson, A. (2002). Biological invasion risks and the public good: an economic perspective. *Conservation Ecology*, 6 (1): 1-7.

Pope, K.O., Pohl, M.E.D., Jones, J.C., Lentz, D.L., von Nagy, C., Vega, F.J., & Quitmyer, I.R. (2001). Origin and environmental setting of ancient agriculture in the lowlands of Mesoamerica. *Science*, 292 (5520): 1370-1373.

Roehrdanz, R.L. (1993). An improved primer for PCR amplification of mitochondrial DNA in a variety of insect species. *Insect Molecular Biology*, 2 (2): 89-91.

Saitou, N. & Nei, M. (1987). The neighbor-joining method: a new method for reconstructing phylogenetic trees. *Molecular Biology and Evolution*, 4 (4): 406-425.

SFA (State Forestry Administration). (2004). *The Bulletin of State Forestry Administration of P. R. China, (2004-2).* http://www.forestry.gov.cn/distribution/2004/04/15/zwgk-2004-04-15-5965.html. [in Chinese]

SFA (State Forestry Administration). (2005). *The Bulletin of State Forestry Administration of P. R. China, (2005-2).* http://www.forestry.gov.cn/distribution/2005/01/31/zwgk-2005-01-31-5962.html. [in Chinese]

SFA (State Forestry Administration). (2006). *The Bulletin of State Forestry Administration of P. R. China, (2006-1).* http://www.forestry.gov.cn/distribution/2006/02/22/zwgk-2006-02-22-5960.html. [in Chinese]

SFA (State Forestry Administration). (2007). *The Bulletin of State Forestry Administration of P. R. China, (2007-4).* http://www.forestry.gov.cn/distribution/2007/07/11/zwgk-2007-07-11-5949.html. [in Chinese]

SFA (State Forestry Administration). (2009). *The Bulletin of State Forestry Administration of P. R. China, (2009-3).* http://www.forestry.gov.cn/portal/main/govfile/13/govfile_1635.html [in Chinese]

SFA (State Forestry Administration). (2011). *The Bulletin of State Forestry Administration of P. R. China. (2011-2).* http://www.forestry.gov.cn/portal/main/govfile/13/govfile_1792.htm. [in Chinese]

Shi, W., Kerdelhué, C., & Ye, H. (2010). Population genetic structure of the oriental fruit fly, *Bactrocera dorsalis* (Hendel) (Diptera: Tephritidae) from Yunnan province (China) and nearby sites across the border. *Genetica*, 138: 377-385.

Shi, W. & Ye, H. (2004). Genetic differentiation in five geographic populations of oriental fruit fly, *Bactrocera dorsalis* (Hendel) (Diptera: Tephritidae) in Yunnan province. *Acta Entomologica Sinica*, 47 (3): 384-388. [in Chinese with English abstract]

Shibata, E. (1987). Oviposition schedules, survivorship curves, and mortality factors within trees of two cerambycid beetles (Coleoptera: Cerambycidae), the Japanese pine sawyer, *Monochamus alternatus* Hope, and sugi bark borer, *Semanotus japonicus* Lacordaire. *Researches on Population Ecology*, 29 (2): 347-367.

Shoda-Kagaya, E. (2007). Genetic differentiation of the pine wilt disease vector *Monochamus alternatus* (Coloeptera: Cerambycidae) over a mountain range --- revealed from microsatellite DNA markers. *Bulletin of Entomological Research*, 97 (2): 167-174.

Simon, C., Francesco, F., Beckenbach, A., Crespi, B., Liu, H., & Flook, P. (1994). Evolution, weighting, and phylogenetic utility of mitochondrial gene sequences and a compilation of conserved polymerase chain reaction primers. *Annals of the Entomological Society of America*, 87 (6): 651-701.

Slatkin, M. & Hudson, R.R. (1991). Pairwise comparisons of mitochondrial DNA sequences in stable and exponentially growing populations. Genetics, 129 (2): 555-562.

Sun, J., Yang, S.Y., Cui, C.L., Zhang, C.X., Lin, M.S., & Zhang, K.Y. (2008). Possible transmission routes of *Bursaphelenchus xylophilus* in China based on molecular data. *Journal of Nanjing Agricultural University*, 31 (2): 55-60. [in Chinese with English abstract]

Tamura, K., Dudley, J., Nei, M., & Kumar, S. (2007). MEGA 4: Molecular Evolutionary Genetics Analysis (MEGA) Software Version 4.0. *Molecular Biology and Evolution*, 24 (8): 1596-1599.

Togashi, K. (1989). Development of *Monochamus alternatus* Hope (Coleoptera: Cerambycidae) in relation to oviposition time. *Japanese Journal of Applied Entomology and Zoology*, 33 (1): 1-8.

Wang, L.P. (2004). Study on the biological characteristic of *Monochamus alternatus* Hope. *Journal of Fujian Forestry Science and Technology*, 31 (3): 23-26. [in Chinese with English abstract]

Wang, R., Wang, J.F., Qiu, Z.J., Meng, B., Wan, F.H., & Wang, Y.Z. (2011). Multiple mechanisms underlie rapid expansion of an invasive alien plant. *New Phytologist*, 191 (3): 828-839.

Wang, R. & Wang, Y.Z. (2006). Invasion dynamics and potential spread of the invasive alien plant species *Ageratina adenophora* (Asteraceae) in China. Diversities and Distributions 12: 397–408.

Wang, S.M. (2004). Introduction of the American-originated crops and its influence on the Chinese agricultural production structure. *Agricultural History of China*, 24 (2): 16-27. [in Chinese with English abstract]

Wang, S.Y. & Zhang, W. (2005). *Yunnan Geography*. Kunming: The Nationalities Publishing House of Yunnan. [in Chinese]

Wang, X.R., Kong, X.C., Jia, W.H., Zhu, X.W., Ren, L.L., & Mota, M.M. (2010). A rapid staining-assisted wood sampling method for PCR-based detection of pine wood nematode *Bursaphelenchus xylophilus* in *Pinus massoniana* wood tissue. *Forest Pathology*, 40 (6): 510-520.

Wang, X.R., Zhu, X.W., Kong, X.C., & Mota, M.M. (2010). A rapid detection of the pinewood nematode, *Bursaphelenchus xylophilus* in stored *Monochamus alternatus* by rDNA amplification. *Journal of Applied Entomology,* 135 (1/2): 156-159.

Wang, Y.Y., Song, Y.S., Zang, X.Q., & Liu, Y. (1991). The ten-year review and future control measures of the pine wilt disease in China. *Forest Pest and Disease,* 10 (3): 39-42. [in Chinese]

Wingfield, M.J., Bedker, P.J., & Blanchette, R.A. (1986). Pathogenicity of *Bursaphelenchus xylophilus* on pines in Minnesota, Wisconsin. *Journal of Nematology,* 18 (1): 44-49.

Xia, X.H. & Xie, Z. (2001). DAMBE: Data analysis in molecular biology and evolution. *Journal of Heredity,* 92 (4): 371-373.

Xu, H.G., Ding, H., Li, M.Y., Qiang, S., Guo, J.Y., Han, Z.M., Huang, Z.G., Sun, H.Y., He, S.P., Wu, H.R., & Wan, F.H. (2006). The distribution and economic losses of alien species invasion to China. *Biological Invasions,* 8: 1495-1500.

Yagi, K., Katoh, T., Chichvarkhin, A., Shinkawa, T., & Omoto, T. (2001). Molecular phylogeny of butterflies *Parnassius glacialis* and *P. stubbendorfii* at various localities in East Asia. *Genes & Genetic Systems,* 76 (4): 229-234.

Yamasaki, T., Sakai, M., & Miyawaki, S. (1989). Oviposition stimulants for the beetle, *Monochamus alternatus* Hope, in inner bark of pine. *Journal of Chemical Ecology,* 15 (2): 507-516.

Yang, H., Wang, J.J., Zhao, Z.M., Yang, D.M., & Zhang, H. (2006). Effects of multiple mating on quantitative depletion of spermatozoa, fecundity, and hatchability in *Monochamus alternatus. Zoological Research,* 27 (3): 286-290. [in Chinese with English abstract]

Yang, Z.X., Wang, J.M., Chen, X.M., Duan, Z.Y., & Ye, S.D. (2010). The vertical distribution characteristics of *Monochamus alternatus* on *Pinus yunnanensis* trunks. *Forest Research,* 23 (4): 607-611. [in Chinese with English abstract]

Zhao, Y.X., Dong, Y., & Xu, Z.H. (2004). Bionomics and geographical distribution of *Monochamus alternatus* Hope (Coleoptera: Cerambycidae) in Yunnan Province. *Forest Pest and Disease,* 23 (5): 13-16. [in Chinese with English abstract]

5

Development of an Individual-Based Simulation Model for the Spread of Citrus Greening Disease by the Vector Insect *Diaphorina citri*

Youichi Kobori[1], Fugo Takasu[2] and Yasuo Ohto[3]
[1]*Japan International Research Center for Agricultural Sciences,*
[2]*Nara Women's University,*
[3]*National Agricultural Research Center,*
Japan

1. Introduction

To establish an Integrated Pest Management (IPM) system for crop pest insects and disease, risk assessment of relevant hindrances is necessary. Typically, the data for such risk assessment analysis is collected through a field survey or field experiments. However, the clarification of important factors often requires assessment of the interaction between such factors. Modelling provides a basis for the theoretical description of such interaction, based on field data, and for relevant simulations. We developed a simulation model which estimates the spread of citrus greening disease (Huanglongbing, HLB) by the vector insect *Diaphorina citri* Kuwayama. Then, we examined the parameters affecting the spread dynamics of the disease, using the model to inform risk assessment. We targeted the area of the Mekong Delta, Vietnam, one of the regions severely affected by the spread of HLB. At the beginning of this chapter, we describe the disease and the vector insect. Then we outline the developmental methodology for the HLB disease-spread estimation model. Finally, we provide suggestions aimed at preventing the spread of HLB, based on the disease-spread estimation results in several situations.

1.1 Citrus greening disease

Citrus greening disease is a limiting factor in citrus production worldwide (Bové, 2006). The symptoms of HLB are similar to those of nutritional stress (Halbert & Manjunath, 2004). A survey conducted over an eight-year period in Réunion Island, for example, indicated that 65% of the trees were badly damaged and rendered unproductive within seven years of planting (Aubert et al., 1996). In Thailand, citrus trees generally decline within 5-8 years after planting due to HLB (Roistacher, 1996). In his compilation of global infection statistics, Toorawa (1998) estimated that 50 million trees were infected in South and Southeast Asia, and 3 million were infected in Africa. In India and Saudi Arabia, there has been a marked decline in the citrus industry as a result of HLB.

The pathogens are phloem-inhabiting bacteria in the generous *Candidatus* Liberibacter (Halbert et al., 2004). Although these bacteria have hitherto not been sufficiently cultured for the application of Koch's postulates, some experimental results have strongly suggested that they are the pathogen in HLB (Su et al., 1986; Buitendag & von Broembsen, 1993). There are two principal means of transmission for a healthy tree: graft transmission, whose frequency has been estimated in many previous studies (e.g., Lin & Lin, 1990; van Vuuren, 1993), but with widely varying values; and transmission through vector insects. In Asian countries, HLB is borne by the Asian citrus psyllid, *D. citri*.

Detection of the HLB pathogen has been achieved by several methods. DNA identification through the PCR method was used to detect the bacteria both in citrus plants and vector insects (Bové et al., 1993; Tian et al., 1996). Wang et al. (2006) conducted a study which developed a reproducible conventional PCR method with several primer sets, and two quantitative real-time PCR methods for detection and monitoring of the pathogen. The HLB pathogen also can be detected with an electron microscope, ELISA (Garnier & Bové, 1993).

1.2 Life cycles of the vector insect and the disease

The *D. citri* is a Hemipteran insect, measuring 3 to 4 mm in length, with piercing-sucking mouthparts that allow this pest to feed on the phloem of citrus spp. and other related rutaceous plants. The eggs of *D. citri* are laid on the new leaf growth of expanding terminals, in the folds of unfurled leaves, and behind developing leaf buds (Chavan & Summanwar, 1993). There are five nymphal instars (Aubert, 1987). Adults may live several months, and females may lay as many as 800 eggs in a lifetime, under artificial rearing conditions (Mead, 1977); however, the longevity and fecundity in actual field conditions are not well known. Temperature-dependent development of *D. citri* has been estimated (Liu & Tsai, 2000; Yang et al., 2006; Nakata, 2006), and the results reveal a relatively consistent trend, even though the host plants used in the experiments varied. In general, nymphs grew faster at higher temperatures, except for 32.5 °C. Nakata (2006) estimated the developmental zero and effective accumulative temperature of the egg as 13.7°C and 46.9 degree-days, respectively. The developmental zero and effective accumulative temperature of the nymphs were 11.6°C and 192.3 degree-days, respectively (Nakata, 2006).

Fourth and fifth instar nymphs and adults of *D. citri* can acquire the pathogen, and emerged adults that have fed on infected plants as nymphs can transmit the pathogen to healthy plants (Capoor et al., 1974; Xu et al., 1988). Once the bacterium is acquired, the psyllid will retain and transmit the bacterium throughout its life. Multiplication of the pathogen in individual *D. citri* was investigated by Inoue et al. (2009). The efficiency of transmission to healthy plants by the virulent adult *D. citri* was estimated in several studies. The virulency of vectors grown on infected trees, and the transmission rate of virulent vectors to healthy host trees, were different in each report. Inoue et al. (2009) estimated a successful transmission of 67% (for test plants) by inoculative adult *D. citri* that were given acquisition feeding in the nymphal developmental period, suggesting that the pathogenic bacteria was present in the salivary glands of these psyllids. In another study, the transmission rate of virulent *D. citri* to healthy plants was estimated 1% (Huang et al., 1984). In our previous study, the virulence of vectors grown on infected trees, and the transmission rate of virulent vectors to healthy host trees, were almost 90% and almost 25%, respectively (Ohto & Kobori, unpublished data).

2. The model

Until quite recently, almost all models of pathogens and hosts were developed within the framework of mean-field models, which assumed that the respective individuals were uniformly distributed and that the respective interactions occurred uniformly. In recent years, however, the increased calculation speed of computers has enabled the development of an Individual-Based Model (IBM), wherein each individual, of the pathogen and the vector, behave individually, acting according to a predefined set of rules. The model is able to examine disease-spread dynamics in a simulation field, by calculating the cumulative results of the individual behaviors. Because it can treat the vector and host individually, the IBM offers a new and powerful tool in the study of insect borne plant disease. We therefore employed this technique to develop an HLB disease-spread model based on the *D. citri* vector, with reference to the pine wilt disease-spread model developed by Takasu (2009). The C language used in writing the source code for the model was based on the C language technology of simulated individuals, published electronically by Takasu (2008).

2.1 Framework of the model and estimation of the parameters

In this simulation, we targeted the region of the Mekong Delta, Vietnam, in order to establish the basic parameters. Some parameters were provided by previous reports and some by our own observations in the target area. The model is summarized in Fig. 1.

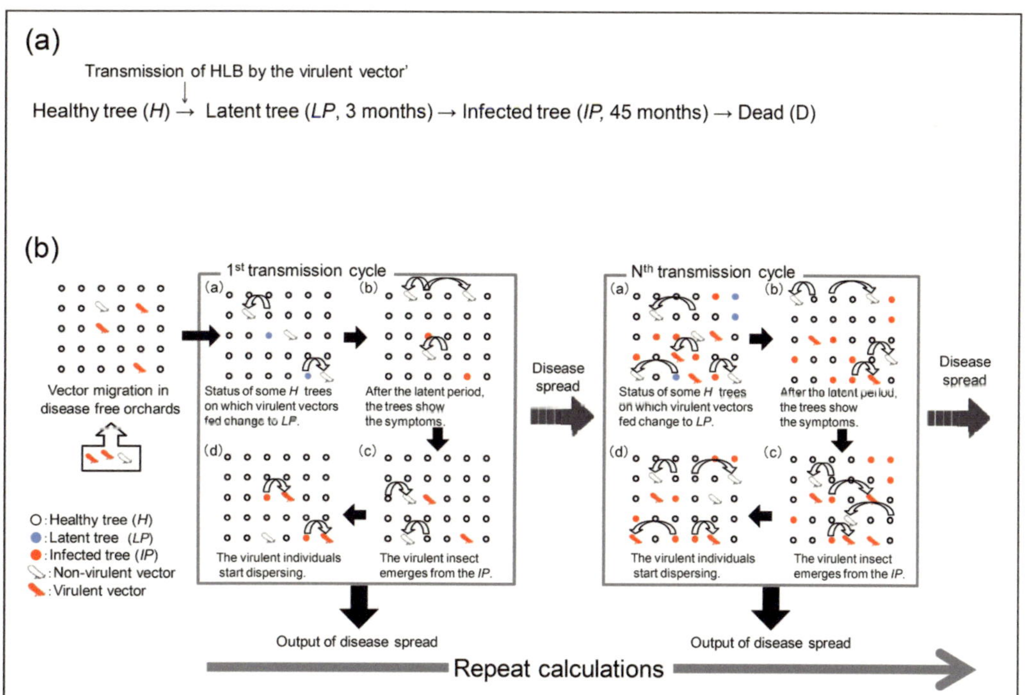

From Kobori et al. (2010).

Fig. 1. (a) Change in tree status over time. (b) Summary of citrus greening disease-spread simulation by our model.

We assumed that the citrus trees were arranged in a two-dimensional regular lattice, with the position of a given tree as '(x, y)', where x and y are integers, and $-L \leq x \leq L$ and $-M \leq y \leq M$ determines the spatial dimensions of the lattice. The distance between trees was assumed to be 2.5 m, in accordance with the common planting regime in the region. We assumed a discrete time progression defined by one month (*t*).

Tree status was defined as one of the following: (1) healthy period (*H*), when the tree is healthy and has the potential to be infected with HLB by the feeding of virulent *D. citri*; (2) latent period (*LP*), when the tree has been infected with HLB by *D. citri* but still does not have infective ability; (3) infectious period (*IP*), when the *D. citri* growing on the tree have the potential to transmit HLB with a probability of *V*; and (4) dead period (*D*), when the tree is dead or there is no tree at the given point in the lattice. The value *V* was estimated at 0.9 from our aforementioned previous study (Ohto & Kobori, unpublished data). Transition through these tree statuses is irreversible: $H \rightarrow LP \rightarrow IP \rightarrow D$. There have been few previous reports that estimated the latent and infectious periods of HLB. The growth stage, cultivar and environmental differences of the host plant may affect these, but we do not know the details. Hence, in this model, we estimated the default values, based on our own field observation and a survey questionnaire conducted in Vietnam, as the following: the *LP* tree essentially changes status to *IP* in 3*t*, and the *IP* to *D* in 45*t* (Fig. 1 (a)).

The vector insects are assumed to move around according to a given dispersal kernel. We discuss the determination of this kernel in Section 2.2. If the insect is virulent, it transmits HLB with a probability of *Tr* to tree *H*. In this report, we set *Tr* at 0.3, in accordance with our aforementioned previous study (Ohto & Kobori, unpublished data). A vector insect individual produces an average of *Rp* next-generation individuals according to Gaussian distribution. In this report we assumed an *Rp*-value of 10, in accordance with our life-table study in Japan (Kobori, un-published data) and similar observations in Vietnam. We also established a threshold number, of 100 individuals of the next generation on a given tree, from our observations in Vietnam.

To exclude edge and corner effects associated with a rectangular two-dimensional space, the model space was assumed to be a torus. In addition, we distributed enough *D* trees around the simulation to prevent adults from arriving at the edge.

2.2 Determination of the dispersal kernel

To determine the dispersal kernel of the *D. citri*, we carried out choice and no-choice tests in a glass house, and artificial release experiments in the experimental field in Japan.

In the glass house, we estimated factors affecting the dispersal activity of *D. citri*, first by means of a no-choice test. Movement from a 'growth' tree (that is, one on which the *D. citri* has grown from egg to adult) to another tree was observed, in time, for each situation (Kobori et al., 2011a). We examined high – low insect density situations, and new-bud – no-new-bud situations, for the growth tree. The dispersal movement of the adult *D. citri* was greater for trees with new buds than for trees without new buds (Fig. 2). The effects of *D. citri* density were not detected in this experiment. Additionally, we investigated host-plant selection by *D. citri* through choice tests (Fig. 3). Choice testing between trees with and without buds revealed that released adults preferred trees with buds during the testing period. This preference disappeared when both trees were treated with a sticky spray to prevent adult movement between trees. These results suggest that adult *D. citri* do not select trees with new buds from a distance but recognize them by means of random movement between trees.

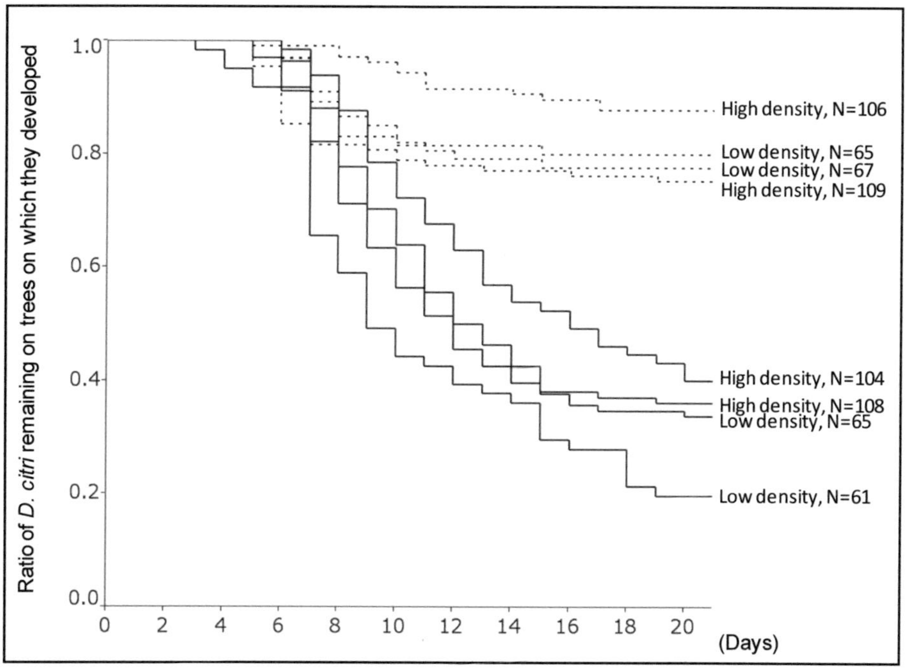

Solid lines: trees with buds; broken lines: trees without buds; N: initial number of adult *D. citri* on the tree. From Kobori et al. (2011a).

Fig. 2. Relationship between the number of days and the number of migrating adult *D. citri*.

Through artificial release experiments, we estimated mean dispersal distance, frequency of movement and dispersal direction (Kobori et al., 2011b). We artificially released, in separate experiments, 10,000 and 1,000 adult *D. citri*, marked in pink, at the center of the experimental field, in which 33 X 33 potted host plants were set out in a grid at 2.5 m intervals (for the marking method, see Nakata, 2008).

In the 10,000-adult experiment, the mean dispersal distance from the release point was 5-6 m, and in the 1,000-adult experiment it was 6-12 m. In each case, the proportion of *D. citri* found on different days over the experimental periods declined with increasing distance from the release point, and did not change substantially over time (Fig. 4). These results suggested that once the adult *D. citri* arrived on the host plants they hardly moved again. Moreover, the center of distribution would be expected to move with the wind direction, if each *D. citri* continued its random movement, because the same wind direction was observed throughout most of each experimental period; however, the center of distribution in fact stayed close to the release point throughout the experimental periods (Fig. 5). Individuals that moved more than 7.5 m from the release point were found on the lee side. These results suggested that once *D. citri*, here carried by the wind, had arrived on the host plants, they too hardly moved again.

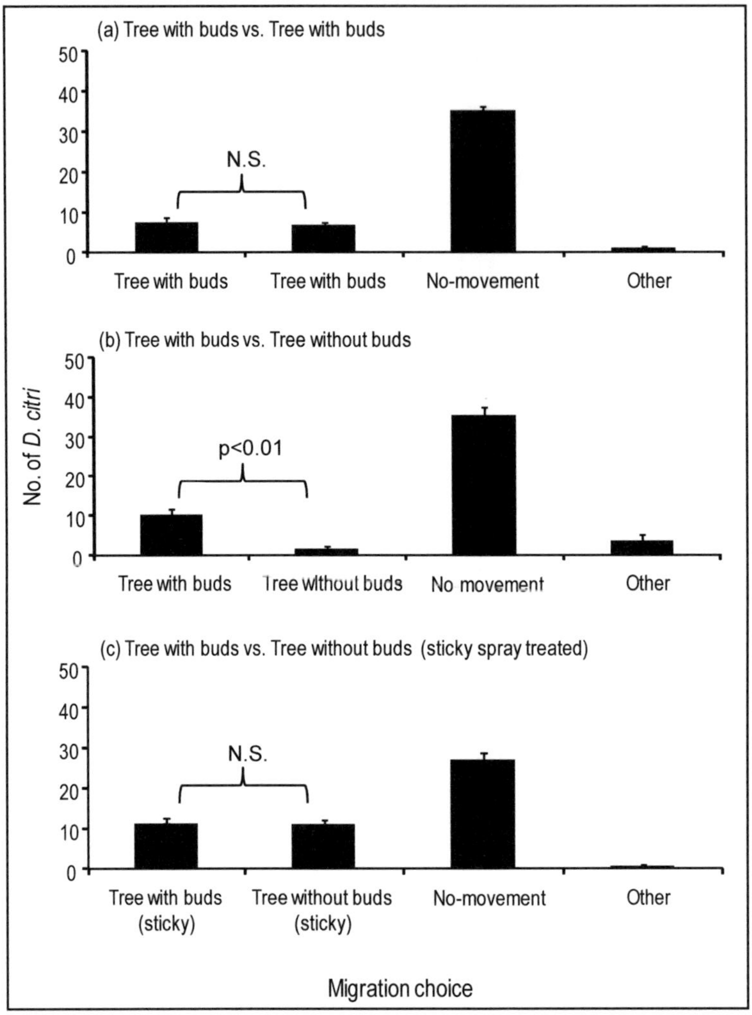

Tree with buds: 8 to 10 buds present on the tree; tree without buds: no buds on the tree; other: individuals found on the side wall of the glasshouse or missing inside the glasshouse. From Kobori et al. (2011a).

Fig. 3. Results of a choice test with adult *D. citri* (mean ± S.E.).

In light of the above results, we determined the dispersal kernel. We did not include the effects of vector insect density in the frequency of movement in the kernel. On the other hand, it is highly possible that the existence of new buds affects movement behavior. However, in the Mekong Delta, Vietnam, our targeted area, there are new buds on the host trees almost the entire year. Thus, we did not include bud effects in our model. The direction of movement was isotropic, and the distance of a given movement, r, follows a probability distribution w(r). For the dispersal kernel, we approximated mobility by means of the formula:

$$w(r) = \gamma^{e^{-\gamma r}} \tag{1}$$

γ was estimated at 0.8 from previous experiments (Fig. 4). As a result of a given movement, the insect lands on the nearest-neighbour host tree. If the insect lands on a *D* tree, the individual moves again without feeding or reproducing, in accordance with the kernel.

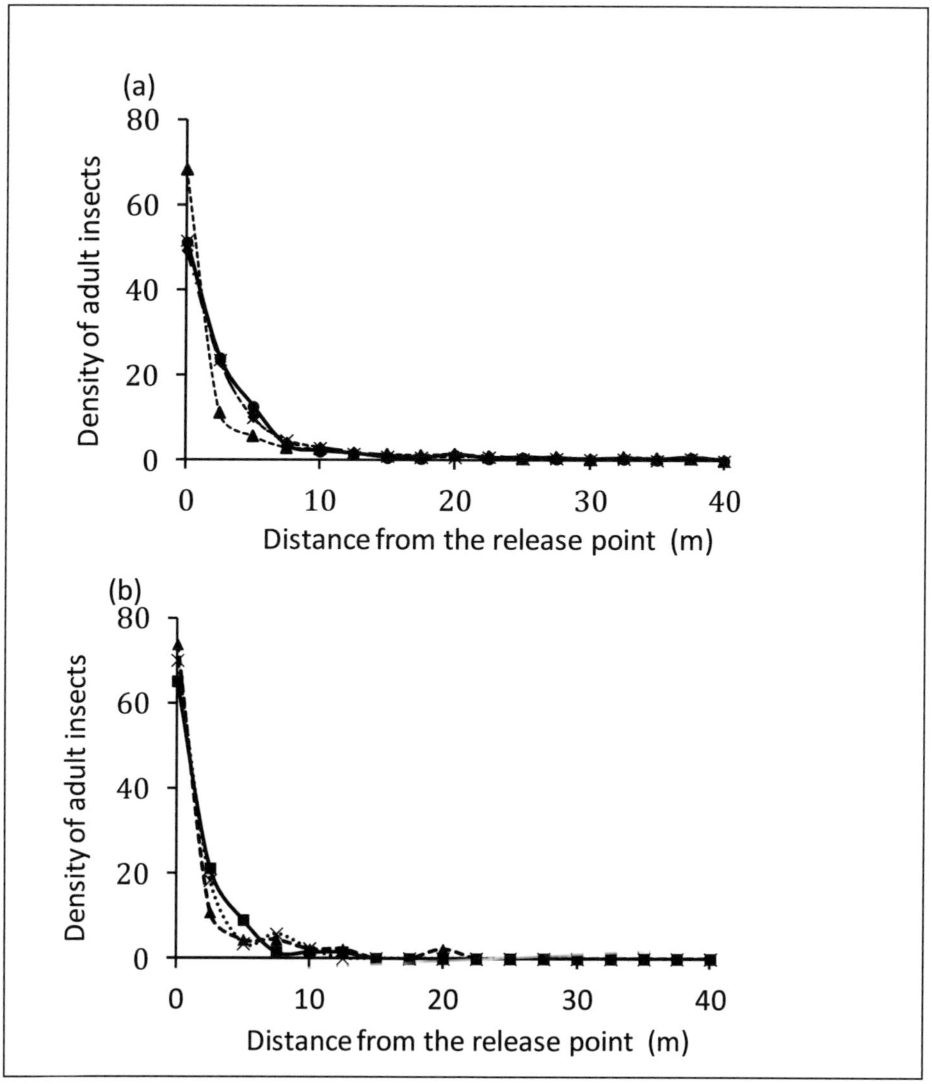

These values were calculated in a one-dimensional cross-section from south to north in the field.
Density of adult insects = (number of individuals / total number of the individuals in field) × 100.
▲: 3days, ■: 7 days, ×: 14 days, ●: (a) 20 days, (b) 21 days after release. From Kobori et al. (2011b).

Fig. 4. The density of *D. citri* (*D*) found at different time periods, plotted against distance from the release point; (a) 10,000 adults released, (b) 1,000 adults released.

When the individual lands on an *H*, *LP* or *IP* tree, it does not move again. In the process of moving, individuals have a certain probability (*Dm*) of dying. After reproduction, individuals have a certain probability (*Dl*) of dying; individuals remaining alive (probability of 1-*Dl*) are integrated into the next generation. There is insufficient previous research to provide estimates for these parameters, and we could not estimate their value by certain observation or experiment. Thus, we roughly estimated these parameters at *Dm*: 0.7 and *Dl*: 0.9 by speculative inference from life-table analysis and our own limited observation in this report (e.g. Mead, 1977; Tsai et al., 2002).

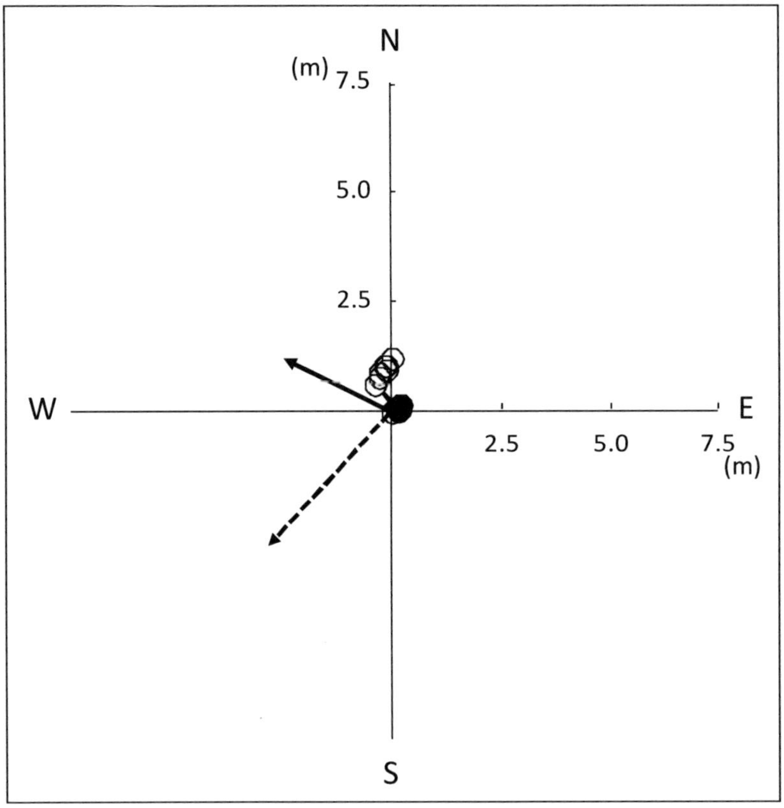

Empty and solid circles indicate 10,000 adults and 1,000 adults released, respectively. The two arrows (solid: 10,000 adults; dashed: 1,000 adults) indicate the composite wind vectors over the experimental periods, as calculated from the daily dominant wind directions. The average velocities were 2.42 m/s in the case of 10,000 adults released, and 3.98 m/s in the case of 1,000 adults released. From Kobori et al. (2011b).

Fig. 5. Center of distribution, over the experimental period, of insects released at the zero coordinate.

3. HLB disease-spread simulation using the model

3.1 Spread of HLB in a field under default conditions

We simulated HLB spread under default conditions. Field size was set at *L*, *M* = 100. Healthy trees (*H*) were distributed at the center of the field (*L*, *M* = 20). The remaining trees were *D*. Then, 100 virulent vector insects were distributed at the center of the *H* field. We set the model calculation period for 84 months.

Within 12*t*, more than 10 newly infected trees had appeared in the given field (Fig. 6 (a)). The spread speed increased with time (Fig. 6 (b)). By the end of the calculation period, more than half of the *H* trees had changed status to *LP*, *IP* or *D*.

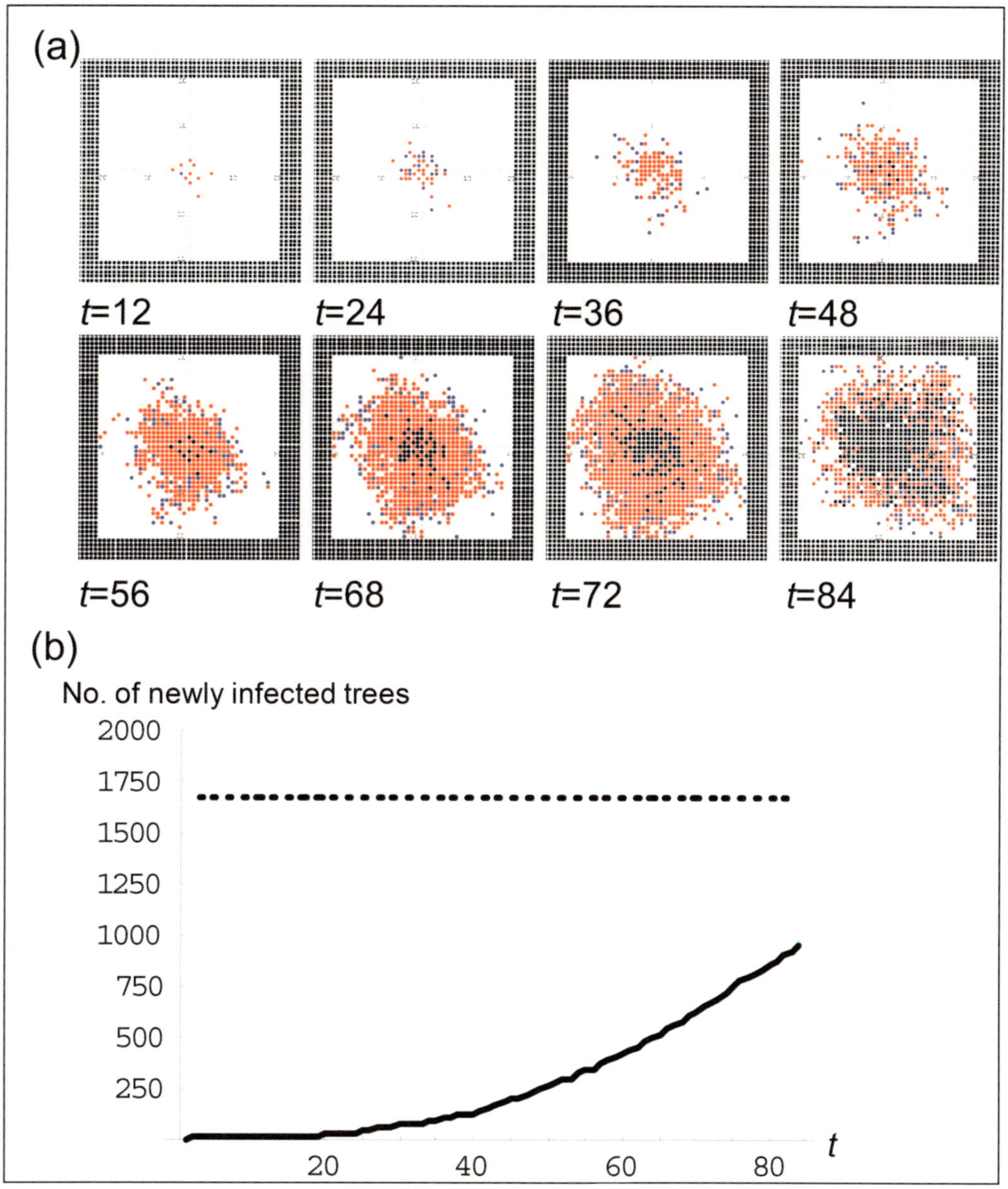

Blue dot indicates Latent Period (*LP*); red: Infectious Period (*IP*), black: Dead (*D*). Healthy (*H*) trees are not shown. Solid line: number of trees newly infected; dotted line: number of initial *H* trees in the simulated field (1,681 trees).

Fig. 6. (a) Snapshots of tree-status distribution for *t*=12, 24, 36, 48, 56, 68, 72, 84. (b) Number of trees newly infected in the field over time.

3.2 Effectiveness of delaying the latent period (LP), and of removing infectious (IP) trees, in suppressing the spread rate of HLB in an orchard

We estimated the effectiveness of delaying the latent period (*LP*) and removing infectious (*IP*) trees in suppressing the spread rate of HLB in an orchard, compared with the default situation. The field was defined according to the default conditions above. Then, 100 virulent vector insects were distributed at the center of the *H* field. We set the model calculation period for 84 months.

First, we changed the number of time steps required to transition from *LP* to *IP* from the default (3*t*) to 6, 12 and 24*t*, respectively, in order to estimate the effects of delaying the latent period. The model estimation suggested that the spread rate of HLB in the orchard slowed as the time from *LP* to *IP* increased (Fig. 7). The theory that mean generation time has more affect on the intrinsic rate of natural increase than the reproductive rate, would account for this result. Our estimation suggested that cultivars in which the transition from *LP* to *IP* is delayed might show resistance to HLB and thus suppress the rate of disease spread.

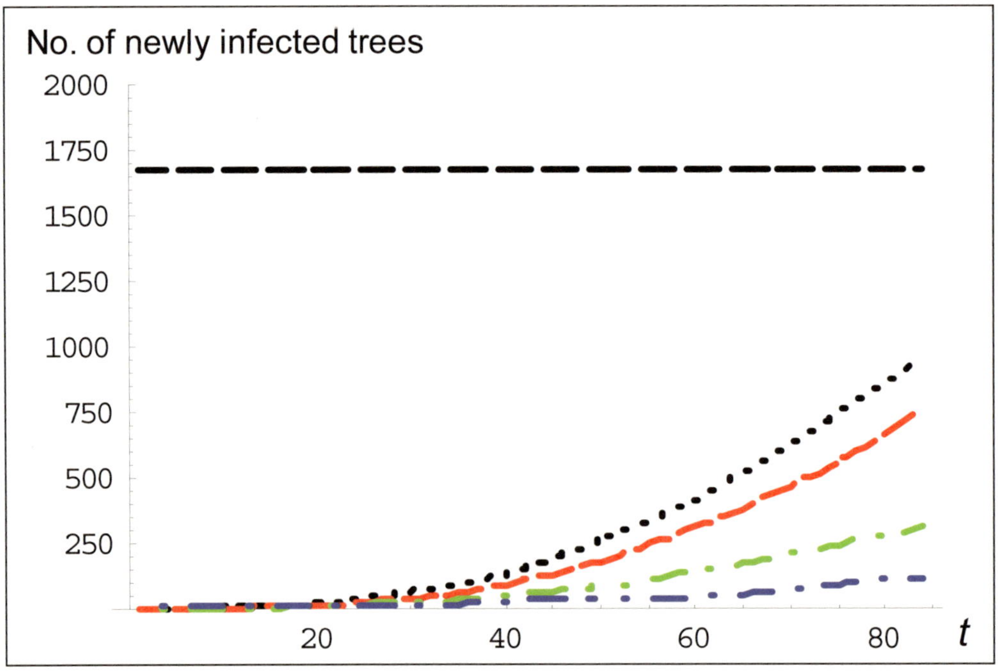

Black dotted line, 3*t* to change from *LP* to *IP* (default parameter); red dashed line: 6*t*; green dash-dotted line: 12*t*; blue dash-dotted line: 24*t*. Black dashed line is number of initial *H* trees in the simulated field (1,681 trees). No. of newly infected trees = *LP*+*IP*+*D* (excluding initial *D* trees).

Fig. 7. Effects of delaying the transition from latent period (*LP*) to infectious period (*IP*).

Second, we estimated the effects of removing infectious (*IP*) trees. We assumed that the *IP* trees were removed at 24, 12 and 6*t*, respectively, after their status had changed from *LP* (default: trees were removed at 45*t*, when their status changed to *D*), and we compared the spread rates of HLB. Under the no-removal scenario (default), the number of newly infected trees was more than 1,000 (Fig. 8). When we removed the *IP* trees 24*t* after their status change, the number of newly infected trees was almost same as under the default condition. However, when we removed the *IP* trees 9*t* after their status change, the number of newly infected trees decreased relative to the default condition. Moreover, when we removed the *IP* trees 6*t* after their status change, the number of newly infected trees barely increased. We may thus predict that the removal of infectious trees will be efficacious in preventing the spread of HLB in a field.

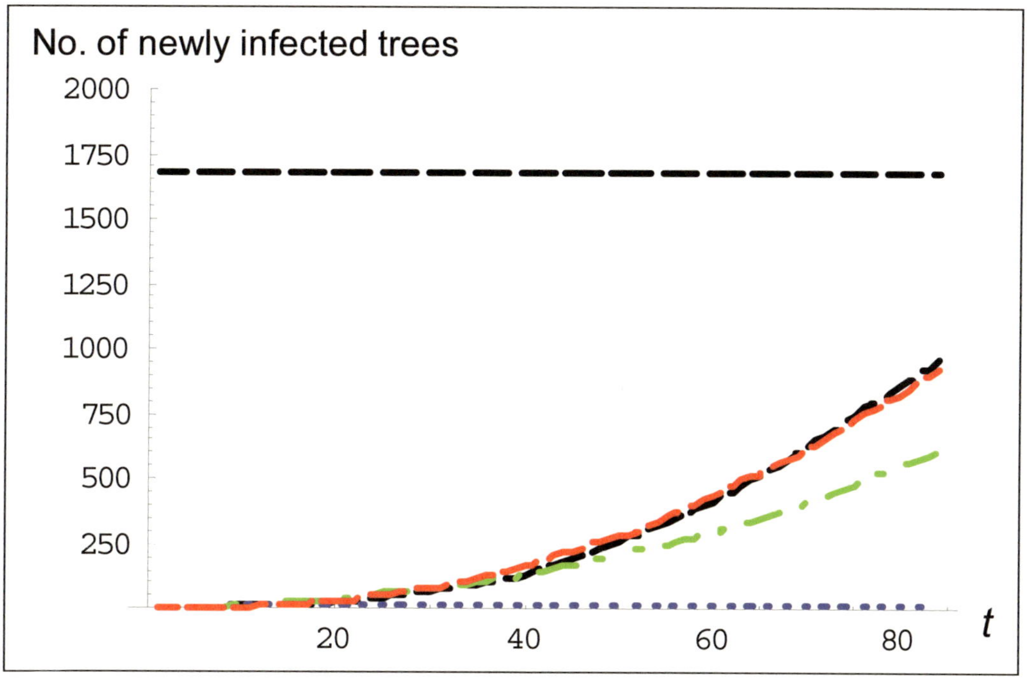

Black dashed line: no removal; red dashed line: removal 24*t* after infection; green dash-dotted line: 12*t*; blue dash-dotted line: 6*t*. Black dashed line is number of initial *H* trees in the simulated field (1,681 trees). No. of newly infected trees = *LP*+*IP*+*D* (excluding initial *D* trees).

Fig. 8. Effects of infectious tree (*IP*) removal in the field.

3.3 Attempts to simulate more realistic scenarios

The vector insect population follows seasonal trends (e.g., Nakahira et al., 2011). Thus, we established high/low reproductive periods and conducted the simulation. The number of newly infected trees increased with time without immigration of virulent vector insects. The model forecast that more than 80% of trees were infected with HLB by 84*t*. Hence, we suggest that the use of disease-free seedlings offers a fundamental technique for preventing HLB spread in an orchard. Additionally, we estimated the suppression of the HLB spread rate through systemic pesticide treatment in a newly planted orchard. Several reports have indicated that systemic pesticide treatment causes high mortality rates in *D. citri* (e.g., Ichinose et al., 2010). We assumed that pesticide applied in the new orchard would be effective for two years. Hence, we assumed that, until 24*t* after planting, 100% of the insects that had migrated from an infected orchard into the newly planted orchard would die after feeding (if an individual was virulent it might transmit HLB, but would not reproduce on the tree). After this period, individuals migrating from the infected orchard would reproduce in the newly planted orchard. The number of newly infected trees was lower in the plantings treated with systemic pesticide than in those with the negative controls. We concluded that treatment with systemic pesticide was effective in suppressing the spread rate of HLB in a newly planted orchard, and the effect was definitive when the virulent *D. citri* migrated from a nearby orchard (Fig. 9). Part of this study have been reported in Kobori et al. (2011c).

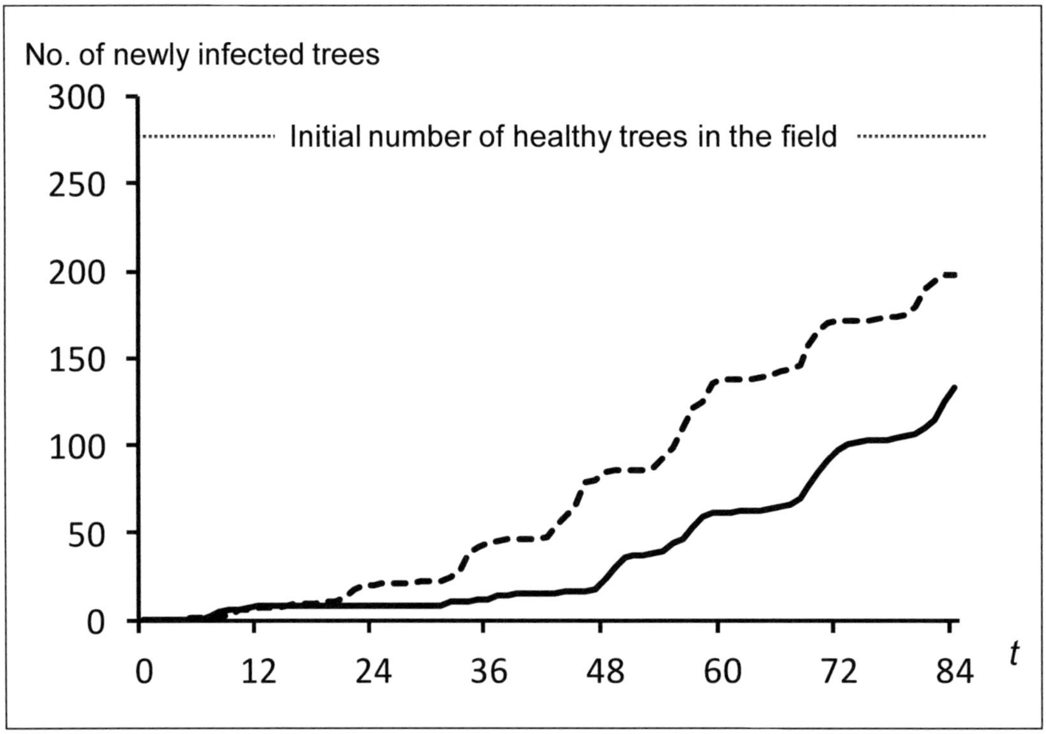

Dashed line, negative control (did not treat systemic pesticide); soiled line, treated with systemic pesticide (effective for 24*t*). Number of initial *H* trees in the simulated field was 256. No. of newly infected trees = *LP*+*IP*+*D* (excluding initial *D* trees). Modified from Kobori et al. (2011).

Fig. 9. Effects of systemic pesticide to suppress HLB in the field.

Although these results generated valuable suggestions, there was a problem in our simulated conditions. In a number of replications, extinction of the vector insect occurred. In actual observation, even insects not found in a given season could be found in later observations. One cause of this may be the closed nature of our simulation system. In reality, vector insects could immigrate into the field. Currently, we are expanding the model to incorporate this possibility. In addition, the dispersal kernel was developed for calculations of single-field behavior. We must revise the kernel to simulate larger scenarios, such as those involving multiple, separated fields.

4. Conclusion

This chapter described the development of an individual-based modelling technique to simulate disease-spread dynamics. Our model was able to provide parameters for each individual citrus tree and the respective vector insect, and thereby to examine disease-spread dynamics in the simulation field by calculating the cumulative results of the individual behaviors. This model can be applied to many diseases vectored by insects, although we developed the model for HLB.

Our simulated results suggested that both delaying the transition from latent period (*LP*) to infectious period (*IP*), and removing infectious (*IP*) trees, suppressed the spread rate of HLB in an orchard. Additionally, the results of our a preliminary trail suggested that cooperative control of *D. citri* on the part of orchard owners in a given area, to reduce the number of virulent individuals, may be effective in suppressing the spread rate of HLB.

5. Acknowledgment

Part of this study was conducted as part of the collaborative research project between the Japan International Research Center for Agricultural Sciences (JIRCAS) and the Southern Horticultural Research Institute of Vietnam (SOFRI). In addition, part of this work was supported by KAKENHI (23780052). We thank Mr. Masato Shimajiri, Mr. Masakazu Hirata, and the staff of the Technical Support Section of TARF/JIRCAS for their help in managing and maintaining the experimental field for estimation of the parameters. We are grateful to Ms. Aya Yano, Mr. Kenji Mishima, Mr. Hiroyuki Murota, Ms. Kiyomi Toume and Satoshi Kawate for their help in producing the released insects and controlling our plant conditions for the parameter estimation.

6. References

Aubert, B. (1987). *Trioza erytreae* del Guercio and *Diaphorina citri* Kuwayama (Homoptera: Psylloidea), the two vectors of citrus greening disease: Biological aspects and possible control strategies. *Fruits, Vol.* 42, pp.149-162, ISSN 0248-1294

Aubert, B., Grisoni, M., Villemin, M. & Rossolin. G. (1996). A case study of huanglongbing (greening)control in Réunion, *Proceedings of 13th Conference of the International Organization of Citrus Virologists (IOCV)*, pp. 276-278, China, November 2005

Buitendag, C. H. & von Broembsen, L. A.(1993). Living with Citrus Greening in South Africa, *Proceedings of 12th Conference of the International Organization of Citrus Virologists (IOCV)*, pp. 269-273, India, November 1992

Bové, J. M., Garnier, M., Ahlawat, Y. S., Chakraborty, N. K. & Varma, A. (1993). Detection of the Asian Strains of the Greening BLO by DNA-DNA Hybridization in Indian Orchard Trees and Malaysian *Diaphorina citri* Psyllids, *Proceedings of 12th Conference of the International Organization of Citrus Virologists (IOCV)*, pp. 258-263, India, November 1992

Capoor, S. P., Rao, D. G. & Viswanath, S. M. (1974). Greening disease of citrus in the Deccan Trap Country and its relationship with the vector, *Diaphorina citri* Kuwayama, *Proceedings of 6th Conference of the International Organization of Citrus Virologists (IOCV)*, pp. 43-49, Swaziland, August 1972

Chavan, V. M. & Summanwar, A. S. (1993). Population dynamics and aspects of the biology of citruspsylla, *Diaphorina citri* Kuw., in Maharashtra, *Proceedings of 12th Conference of the International Organization of Citrus Virologist (IOCV)*, pp. 286-290, India, November 1992

Garnier, M. & Bové, J. M. (1993). Citrus Greening Disease and the Greening Bacterium, *Proceedings of 12th Conference of the International Organization of Citrus Virologists (IOCV)*, pp. 212-219, India, November 1992

Halbert, S. E. & Manjunath, K. L. (2004). Asian citrus psyllids (Sternorrhyncha: Psyllidae) and greening disease of citrus: A literature review and assessment of risk in Florida. *Florida Entomologist,* Vol. 87, pp. 330-353, ISSN 0015-4040

Huang, C. H., Tsai, M. Y. & Wang, C.L. (1984). Transmission of citrus likubin by a psyllid, *Diaphorina citri. Journal of Agricultural Research of China*, Vol.33, pp..15-72.

Ichinose, K., Bang, D. V., Tuan, D. H. & Dien, L. Q. (2010). Effective use of neonicotinoids for protection of citrus seedlings from invasion by *Diaphorina citri* (Hemiptera: Psyllidae). *Journal of Economic Entomology,* Vol.103, pp. 127-35, ISSN 0022-0493

Inoue, H., Ohnishi, J., Ito, T., Tomimura, K., Miyata, S., Iwanami, T. & Ashihara, W. (2009). Enhanced proliferation and efficient transmission of *Candidatus* Liberibacter asiaticus by adult *Diaphorina citri* after acquisition feeding in the nymphal stage. *Annals of Applied Biology*, Vol. 155, pp. 29-36, ISSN 0003-4746

Kobori, Y., Nakata, T. & Ohto Y. (2011a). Estimation of dispersal pastern of adult Asian psyllid, *Diaphorina citri* Kuwayama (Hemiptera: Psyllidae). *Japanese journal of Applied Entomology and Zoology*, Vol. 55, pp. 177-181, ISSN: 0021-4914 (In Japanese with English summary, tables and figures)

Kobori, Y., Nakata, T., Ohto, Y. & Takasu, F. (2011b). Dispersal of adult Asian citrus psyllids, *Diaphorina citri* Kuwayama (Homoptera: Psyllidae), the vector of citrus greening disease, in artificial release experiments. *Applied Entomology and Zoology*, Vol. 46, pp. 27-30, ISSN 0003-6862

Kobori, Y., Ohto, Y., Nakata, T & Ichinose, K. (2011c). Spread risk estimation of citrus greening disease vectored by *Diaphorina citri* (Hemiptera: Psyllidae) in a citrus orchard using an individual-based model. *JIRCAS Working Report No.72*, pp. 39-43, ISSN 1341-710X

Kobori, Y., Ohto, Y., Nakata, T. & Takasu, F. (2010). Development of individual based-model for simulating the spread of Citrus greening disease by the vector insects. In: Research Highlights 2010., Available from http://www.jircas.affrc.go.jp/kankoubutsu/highlight/highlights2010/2010_30.html

Lin, K. H. & Lin, K. H. (1990). The citrus huanglongbing (greening) disease in China, *Rehabilitation of Citrus Industry in the Asia Pacific Region. Proceedings of Asia Pacific International Conference on Citriculture*, pp. 1-26, Thailand, February 1990

Liu, Y. H. & Tsai, J. H. (2000). Effects of temperature on biology and life table parameters of the Asian citrus psyllids, *Diaphorina citri* Kuwayama (Homoptera: Psyllidae). *Annals Applied Biology*, Vol. 137, pp. 201-216, ISSN 0003-4746

Mead, F. W. (1977). The Asiatic citrus psyllid, *Diaphorina citri* Kuwayama (Homoptera: Psyllidae). Entomology Circular 180. Florida Department of Agriculture and Consumer Services, Division of Plant Industry, pp. 4

Nakahira K., Kobori Y., Ohto Y. , Chau N. M. & Dien L. Q. (2011). Population dynamics of the Asian citrus psyllid (*Diaphorina citri* Kuwayama, Hemiptera) in a king mandarin (*Citrus nobilis* Loureiro) orchard in Mekong Delta. *JIRCAS Working Report No.72*, pp. 39-43, ISSN 1341-710X

Nakata, T. (2006). Temperature-dependent development of the citrus, *Diaphorina citri* (Homoptera: Psylloidea), and the predicted limit of its spread based on overwintering in the nymphal stage in temperature regions of Japan. *Applied Entomology and Zoology*, Vol. 41, pp. 383–387, ISSN 0003-6862

Nakata, T. (2008). Effectiveness of micronized fluorescent powder for marking citrus psyllid, *Diaphorina citri*. *Applied Entomology and Zoology*, Vol. 43, pp. 33-36, ISSN 0003-6862

Roistacher, C. N. (1996). The Economics of Living with Citrus Diseases: Huanglongbing (Greening) in Thailand, *Proceedings of 13th Conference of the International Organization of Citrus Virologists (IOCV)*, pp. 279-285, China, November 2005

Su, H.-J., Cheon, J.-U. & Tsai, M.-J. (1986). Citrus greening (Likubin) and some viruses and their control trials, pp. 143-147 In *Plant Virus Diseases of Horticultural Crops in the Tropics and Subtropics. FFTC Book Series No. 33.* (Asian and Pacific Council. Food & Fertilizer Technology Center ed.), Food and Fertilizer Technology Center for the Asian and Pacific Region, Taiwan

Takasu, F. (2009). Individual-based modelling of the spread of pine wilt disease: vector beetle dispersal and the Allee effect. *Population Ecology, Vol.* 51, pp. 399-409 ISSN 1438-3896

Takasu, F. (2008). An introduction to individual-based modelling - a framework to implement IBM. In: Research activities, 10. 4. 2008, Available from http://gi.ics.nara-wu.ac.jp/~takasu/research/IBM/ibm.html

Tian, Y., Ke, S. & Ke, C. (1996). Polymerase chain reaction for detection and quantitation of *Liberobacter asiaticum*, the bacterium associated with Huanglongbing (Greening) of citrus in China, *Proceedings of 13th Conference of the International Organization of Citrus Virologists (IOCV)*, pp. 252-257, China, November 2005

Toorawa, P. (1998). La maladie du huanglongbing (greening) des agrumes a L'Île Maurice. Detectionde "*Candidatus* Liberobacter asiaticum" et "Candidatus Liberobacter africanum" dans lesagrumes et les insects vecteurs. *Doctoral Thesis, L' University de Bordeaux*, pp. 186

Tsai, J. H., Wang, J. J. & Liu, Y. H. (2002). Seasonal abundance of the Asian citrus psyllid, *Diaphorina citri* (Homoptera: Psyllidae) in southern Florida. *Florida Entomologist*, Vol. 85, pp. 446-451, ISSN 0015-4040

Van Vuuren, S. P. (1993). Variable transmission of African greening to sweet orange, *Proceedings of 12th Conference of the International Organization of Citrus Virologists (IOCV)*, pp. 264-268, India, November 1992

Wang, Z., Yin, Y., Hu, H., Yuan, Q., Peng, G. & Xia, Y. (2006) Development and application of molecular-based diagnosis for '*Candidatus* Liberibacter asiaticus', the causal pathogen of citrus huanglongbing. *Plant Pathology*, Vol. 55, pp. 630–638, ISSN 1365-3059

Xu, C. F., Xia, Y. H., Li, K.-B. & Ke, C. (1988). Further study of the transmission of citrus huanglongbing by a psyllid, *Diaphorina citri* Kuwayama, *Proceedings of 10th Conference of the International Organization of Citrus Virologists (IOCV)*, pp. 243-248, Spain, November 1986

Yang, Y., Huang, M., Beattie, G. A.C., Xia, Y., Ouyang, G. & Xiong, J. (2006). Distribution, biology, ecology and control of the psyllid *Diaphorina citri* Kuwayama, a major pest of citrus: A status report for China. *International Journal of Pest Management*, Vol. 52, pp. 343-352, ISSN 0967-0874

6

Current Status of Entomopathogenic Fungi as Mycoinecticides and Their Inexpensive Development in Liquid Cultures

Abid Hussain[1,2*], Ming-Yi Tian[1], Sohail Ahmed[3] and Muhammad Shahid[4]

[1]Department of Entomology, College of Natural Resources and Environment, South China Agricultural University, Guangzhou,

[2]Department of Arid Land Agriculture, Faculty of Agriculture and Food Sciences, King Faisal University, Hofuf, Al-Hassa,

[3]Department of Agricultural Entomology, University of Agriculture, Faisalabad,

[4]Department of Chemistry and Biochemistry, University of Agriculture, Faisalabad,

[1]China

[2]Saudi Arabia

[3,4]Pakistan

1. Introduction

Synthetic chemical pesticides remained the mainstay of pest eradication for more than 50 years. However, insecticide resistance, pest resurgence, safety risks for humans and domestic animals, contamination of ground water, decrease in biodiversity, and other environmental concerns have encouraged researchers for the development of environmentally benign strategies for pest control including the use of biological control agents. Naturally occurring biological control agents are important regulatory factors in insect populations. Many species are employed as biological control agents of insect pests in glass-house and row crops, orchards, ornamentals, range, turf and lawn, stored products, and forestry and for the abatement of pest and vector insects of veterinary and medical importance (Burges, 1981; Lacey & Kaya, 2000; Tanada & Kaya, 1993).

The application of microorganisms for control of insect pests was proposed by notable early pioneers in invertebrate pathology such as Agostino Bassi, Louis Pasteur, and Elie Metchnikoff (Steinhaus, 1956, 1975). These biological control agents such as viruses, bacteria, protozoa, nematodes and most fungi exert considerable control of target populations.

Among micro-organisms, entomopathogenic fungi constitute the largest single group of insect pathogens. Generally, two groups of fungi are found to cause diseases in insects. Entomopathogenic fungi belong to the orders Entomophthorales and Hypocreales (formerly called Hyphomycetes). Several other entomopathogenic fungi from other taxonomic groups are also known. Until now, over 700 species of fungi are known to infest insects (Wraight et

*Corresponding Author

al, 2007). Such insect killing fungi present major advantages. Firstly, they are important natural enemies of arthropods (Chandler et al, 2000), capable of infecting them directly through the integument. Secondly, cultivation of those fungi and production of infective conidia are easy and fairly cheap (Roberts & Hajek, 1992). Finally, entomogenous fungi can be found under different ecological conditions (Ferron, 1978).

Unique among entomopathogens, fungi do not have to be ingested and can invade their hosts directly through the exoskeleton or cuticle. Therefore, entomopathogenic fungi can infect non-feeding stages such as eggs and pupae. The insect cuticle is the first barrier against biological insecticides. Insect cuticle mainly formed from three layers such as, epicuticle, procuticle and epidermis. Each layer has different chemical structure and properties (Juárez & Fernández, 2007). The epicuticle is very thin (0.1–3 μm) and multi-layered. The outermost surface layer of the epicuticle is the lipid layer, it is mostly resistant to enzyme degradation and exhibits characteristic such as water barrier properties (Hadley, 1981); unless physically disrupted, it can help to prevent passage of cuticle degrading fungal enzymes. The site of invasion among insects is often between the mouthparts, inter-segmental folds or through spiracles. At these sites, locally high humidity promotes conidial germination and the cuticle is non-sclerotised and more easily penetrated (Clarkson & Charnley, 1996; Hajek & St. Leger 1994).

Conidia upon landing on a potential host, initiates a series of steps that could lead to a compatible (infection) or a non-compatible (resistance) reaction. In a compatible reaction, fungal recognition and attachment proceed to germination on the host cuticle. Once the epicuticle is breached, progress by the penetration peg through the cuticle may be more or less direct via penetrant hyphae, penetrant structures may also extend laterally (Hajek & St. Leger, 1994). Fargues (1984) proposed adhesion to occur at three successive stages: (1) adsorption of the fungal propagules to the cuticular surface; (2) adhesion or consolidation of the interface between pre-germinant propagules and the epicuticle; (3) fungal germination and development at the insect cuticular surface, until appresorium is developed to start the penetration stage. Zacharuk (1970b) proposed an active adhesion process for *M. anisopliae* after detecting epicuticle dissolution and mucoid material penetrating the pore canals. Infection will proceed after a successful penetration has been achieved.

In terrestrial environment, fungal conidial germination proceeds with the formation of germ tube (Boucias & Pendland, 1991) or appressorium (Madelin et al, 1967; Zacharuk, 1970a), which forms a thin penetration peg that breaches the insect cuticle via mechanical (turgor pressure) or enzymatic means (proteases) (Zacharuk, 1970b). Exocellular mucilage, proposed to enhance binding to the host cuticle, is also secreted by several entomogenous fungi during the formation of infective structures (Boucias and Pendland, 1991). In *M. anisopliae*, appressorium formation, hydrophobins, and the expression of cuticle-degrading proteases are triggered by low nutrient levels (St. Leger et al, 1992), demonstrating that the fungus senses environmental conditions or host cues at the initiation of infection. The production of cuticle-degrading enzymes, chitinases, lipases and proteases, has long been recognized as important determinant of the infection process in various fungi, facilitating penetration as well as providing nourishment for further development (Charnley, 1984; Dean & Domnas, 1983; Hussain et al, 2010b; Samsináková et al, 1971;). Among the proteases found in entomopathogenic fungi, the spore bound Pr1 has been well characterized and its role in cuticle invasion has been established (Hussain et al, 2010b; St. Leger, 1994). Ultrastructural studies of *M. anisopliae* penetration sites on *Manduca sexta* larvae have shown high levels of Pr1 coincident with hydrolysis of cuticular proteins (Goettel et al, 1989; St. Leger et al, 1989). Pr1 inhibition studies also showed delayed mortality in *M. sexta* larvae, resulting from delayed penetration of the cuticle (St. Leger et al, 1988). Furthermore, construction of a *M. anisopliae* strain with multiple copies of the gene encoding Pr1 and over-expressing the protease resulted in 25% reduction of time to death among *M. sexta* compared to those infected by the wild-type strain (St. Leger et al, 1996). Furthermore, it has also been reported that successive *in vivo* passage enhanced the capacity of the fungus to cause infection (Daoust et al, 1982; Hussain et al, 2010b), which ultimately increased the activity of spore bound Pr1 (Shah et al, 2007).

After penetration through the cuticle, the conidia invade into the hemocoel to form a dense mycelial growth (Zimmerman, 1993). Along with penetration, fungi also produce secondary metabolites, derivatives from various intermediates, some of which have insecticidal activities (Vey et al, 2001). It has been experimentally proved that the entomopathogens producing these toxins, infection has been shown to result in more rapid host death (McCauley et al, 1968), compared to strains that do not produce these metabolites (Kershaw et al, 1999; Samuels et al, 1988). The insecticidal properties of destruxins, cyclic depsipeptide toxins from *Metarhizium* spp, described by Kodaira, (1961) are shown to be produced in wax moth and silkworm larvae by Roberts, (1966) and Suzuki et al, (1971), Furthermore, these toxins have been tested against various insects (Roberts, 1981). Currently, over 28 different destruxins have been described, mostly from *Metarhizium* spp, with varying levels of activities against different insects (Vey et al, 2001). The level of destruxin has been correlated with virulence (Al-Aïdroos & Roberts, 1978) and host specificity (Amiri-Besheli et al, 2000). Studies on the activities of destruxins have also shown modulation of the host cellular immune system, including prevention of nodule formation (Huxham et al, 1989; Vey et al, 2001) and inhibition of phagocytosis (Vilcinskas et al, 1977) among infected insects. Destruxins are produced as the mycelium grows inside the insect. Other representative toxins produced by entomopathogenic fungi include oosporein, beauvericin, and bassianolide from *Beauveria* spp. (Eyal et al, 1994; Gupta et al, 1994; Suzuki et al, 1977), efrapeptins (Dtolypin) from *Tolypocladium* spp. (Weiser & Matha, 1988), and hirsutellin from *Hirsutella thompsonii* (Mazet & Vey, 1995). Inside the insect haemocoel, the fungus switches from filamentous hyphal growth to yeast-like hyphal bodies that circulate in the hemolymph. The proliferation of these hyphal bodies occurs through budding (Boucias & Pendland, 1982). Later the fungus switches back to a filamentous phase and invades internal tissues and organs (Mohamed et al, 1978; Prasertphon & Tanada, 1968). The fungus later erupts through the cuticle and an external mycelium covers all parts of the host and formed infective spores under appropriate environmental conditions (Boucias & Pendland, 1982; McCauley et al, 1968;). Under suboptimal conditions, some fungi form resting structures inside the cadaver as in the case of *Nomuraea_rileyi* under conditions of low relative humidity and temperature (Pendland, 1982). The life cycle of the fungus is completed when the hyphal bodies sporulate on the cadaver of the host. The external hyphae produce conidia that ripen and are released into the environment. This allows horizontal transmission of the disease within the insects (Khetan, 2001).

Among 85 genera of entomopathogenic fungi only six species are commercially available for field application (Table 1). However, comparatively few have been investigated as potential mycoinsecticides. Fungal pathogens particularly *B. bassiana, I. fumosorosea* and *M. anisopliae* are being evaluated against numerous agricultural and urban insect pests. Several species belonging to order Ilsoptera (Hussain et al., 2010a ; Hussain et al., 2011), Lepidoptera (Hussain et al., 2009), Coleoptera (Ansari et al., 2006), Hemiptera (Leite et al., 2005) and Diptera (St. Leger et al., 1987) are susceptible to various fungal infections. This has led to a number of attempts to use entomopathogenic fungi for pest control with varying degrees of success.

Fungus	Product and Company	Formulation
Aeschersorzia aleyrodis	Koppert / Holland	Wettable powder
Beauveria bassiana	Naturalis™, Troy Bio-Science, USA.	Liquid formulation
B. bassiana	Conidia, AgrEvo, germany, Columbia	Suspendible granules
B. bassiana	Brocani™, Laverlam, Columbia	Wettable powder
B. bassiana	Boverol / Czeck Republic	Wettable powder and dry pellets
B. bassiana	Mycontrol-WP / Mycotech. Corp. USA	Wettable powder
B. bassiana	Ostrinil / natural Plant Protection / France	Micrograpules of mycelium
B. brongniarti	Betel / Natural Plant Protection / France	Microgranules of mycelium
B. brongniarti	Engerlingspilz / Andermatt - Biocontrol / Switerzarland	Barley kernels colonized with the fungus
M. anisopliae	Bio -path™ / Eco Science / OSA	Conidia on a medium placed in trap / chamber
M. anisopliae	Biogreen / Biocare Technology Pvt. Ltd / Australia	Conidia produced on grains
M. anisopliae	Biologic Bio 1020 / Bayer AG Germany	Granules of mycelium
Paecilomyces fumosoroseus	Pfr 21 / WR Grace USA	Wettable powder
Verticillium lecanii	Mycotal / Koppert / Netherlands	Wettable powder
V. lecanii	Vertalec / koppart / Netherlands	Wettable powder

(Bhattacharyya et al., 2004)

Table 1. Commercial formulations of entomopathogenic fungal pesticides

The majority of fungal production systems consist of two stages system in which fungal inoculum of hyphal bodies is produced in liquid culture and then transferred to a solid substrate for production of aerial spores (Devi, 1994). For practical use of entomopathogenic fungi as bio-insecticides at each stage, it is necessary to develop culture medium and method that produce high concentrations of viable and virulent propagules at low cost (Jackson, 1997). These goals can be achieved by using the most favorable inexpensive components for fungal growth at the lowest concentration that afford high yield. Most common compounds for fungal entomopathogens include agro-industrial by-products such as corn steep liquor (Zhao et al., 2010) and sugarcane molasses (Hussain et al., 2011). Our previous investigations showed that both the by-products stimulate the growth of the propagules of entomopathogenic fungi. Corn steep solid and cotton seed flour with yeast extract and KCl, NaCl etc., optimized blastospore production under water stress conditions (Ypsilos & Magan, 2005).

Entomopathogenic fungi infect insects in an aggressive manner by secreting cuticle degrading enzymes such as esterases, lipases, N-acetylglucosaminidases and chitinases (St. Leger et al., 1986). However, the production of extracellular protease Pr1, a major virulence determinant, plays an important role in the success of entomopathogenic fungi in insect penetration, which leads to the subsequent pathogenicity in the target host (Hussain et al., 2010b; Shah and Butt 2005). Previously, agro-industrial by-products such as corn-steep liquor and molasses have been used as alternative growth substrates to produce exopolysaccharides (Fusconi et al., 2008; Sutherland, 1996). Sugarcane molasses (SM), an industrial by-product rich in fermentable sugars, was proposed as a nutritious medium to produce bacterial cellulose by *Zoogloea* sp. (Paterson-Beedle et al., 2000). While, the dried powder of corn steep liquor was used as an inexpensive substitute for beef extract in the medium, which enhanced the lipase production of the strain of *Serratia marcescens* (Zhao et al., 2010). In the past, there is no report on the activity of extracellular protease Pr1 enzyme from the spores of entomopathogenic fungi cultivated from rice, previously grown on media with different composition. The current study is initiated in order to evaluate the effects of three different sources of nitrogen and two sugar sources in different combinations and concentrations in order to determine i) production in liquid media ii) activity of extracellular protease Pr1 of the locally isolated strains of entomopathogenic fungi.

2. Materials and methods

2.1 Culturing of entomopathogenic fungi

The entomopathogenic fungi *M. anisopliae* (EBCL 02049), *B. bassiana* (EBCL 03005) and *I. fumosorosea* (EBCL 03011) were originally isolated from *C. formosanus* in China. The strains were deposited at European Biological Control Laboratory, France. The strains were successively sub-cultured on Potato Dextrose Agar (PDA, Difco Laboratories, Detroit, MI, US) at 26 ± 0.5 °C, in complete darkness. Fungal strains maintenance was identical to our previous study, where it was extensively described (Hussain et al., 2009). In brief, 24-day-old spores of studied strains cultivated on PDA were used as inoculum in all the growth media.

2.2 Influence of liquid media composition on the *in vitro* growth of entomopathogenic fungi

The three sources of nitrogen: peptone (Sigma), yeast extract (Sigma) and corn steep liquor (Shanghai Xiwang Starch Sugar Co., Ltd.), and two sources of sugar: glucose (Sigma) and sugar molasses, were used in different combinations as shown in Table 2. In all the treatments, the following salts were used at the concentration, $CaCl_2 . 2H_2O$ (0.06%), KCl (0.28%), $MgCl_2 . 6H_20$ (0.16%), $MgSO_4 . 7H_2O$ (0.2%), $NaHCO_3$ (0.03%) and $NaH_2PO_4 . H_2O$ (0.1%). To avoid reactions among the salts, they were prepared in compatible pairs: $CaCl_2$. $2H_2O$ with KCl; $MgCl_2$. $6H_20$ with NaH_2PO_4. H_2O; $MgSO_4$. $7H_2O$ with $NaHCO_3$. The other precautionary measures to avoid precipitation were adopted as described by Leite et al. (2005). Media preparations were finalized by adjusting the pH to 6.2, with filter-sterilized HCl (0.1%) and NaOH (10%).

GROWTH MEDIA	SUGAR AND NITROGEN SOURCES				
	Glucose (G) %	Sugar molasses (SM) %	Corn steep liquor (CSL) %	Peptone (PE) %	Yeast extract (YE) %
G + CSL	2.66		1		
G + PE	2.66			1	
G + YE	2.66				1
G + CSL + PE	2.66		0.50	0.50	
G + CSL + YE	2.66		0.50		0.50
G + PE + YE	2.66			0.50	0.50
G + CSL + PE + YE	2.66		0.333	0.333	0.333
SM + CSL		2.66	1		
SM + PE		2.66		1	
SM + YE		2.66			1
SM + CSL + PE		2.66	0.50	0.50	
SM + CSL + YE		2.66	0.50		0.50
SM + PE + YE		2.66		0.50	0.50
SM + CSL + PE + YE		2.66	0.333	0.333	0.333
G+SM+CSL+PE+YE	1.33	1.33	0.333	0.333	0.333

Table 2. Composition of the media of shake flask cultures in agitated liquid cultures of entomopathogenic fungi

The media were poured into 250 ml Erlenmeyer flasks and autoclaved. After cooling, all the flasks were inoculated with one milliliter of spore suspension (1×10^6 spores/ml) in 0.03 % Tween 80 (Sigma-Aldrich, St Louis, MO, US) from 24-day-old cultures of *B. bassiana, M. anisopliae* and *I. fumosorosea* grown on PDA. Four replicates were used for each growth medium. After inoculation, cultures (100 ml) were grown in 250 ml shake flasks at 120 rpm on a rotary shaker, 25 ± 0.5 °C and 16 h fluorescent light photophase. After 120 h of growth, fungal biomass of each flask was evaluated separately. The fungal biomass was filtered through Whatman No. 1 filter paper. After filtration, the filtrates were dried for 24 h at 70 °C and weighed.

2.3 Influence of liquid media composition on the growth of blastospores of entomopathogenic fungi

The optical densities of the inocula of all the studied fungi grown on different media after different time intervals such as 20 h, 40 h and 60 h were measured at 600 nm (OD_{600}) by using a spectrophotometer (Shimadzu UV-1800). The media without inoculation were used as control.

2.4 Production on solid substrates

The entomopathogenic fungi were grown on solid substrate (rice grains) as described in our previous study (Hussain et al., 2011). In brief, 50 % diluted 10 ml inoculum obtained from above media was added into a self-aerating bag containing 20 g of sterilized par boiled rice, separately. Inoculated rice granules were mixed thoroughly from the outside of the bag. Inoculated rice bags were then incubated for 18 days at ambient conditions (24 ± 2 °C, 75-85% RH). The sporulating rice in bags from each growth medium was then allowed to dry for 10 d at 30 °C. The spores were separated from the rice by sieving through a 300 µm mesh. A collecting vessel, such as a bucket was fitted to the plastic sheeting at the bottom of the sieve to create a funnel into the collecting vessel. The sieve was shaken until all the loose spores had been removed from the rice and had collected in the vessel below. The spores were then further sieved using a 106 µm sieve to separate the larger rice dust particles from the spores. The spores as powder were kept at 4 °C for subsequent analysis of the activity of extracellular protease Pr1.

2.5 Influence of artificial media on the activity of extracellular protease Pr1

The activity of Pr1 protease bound to 10 mg spores harvested from the rice of above mentioned media were assayed using a modified method of St. Leger et al. (1987). Briefly, 10 mg spores were washed once in 0.3% aq Tween 80 solution and twice in distilled water, were then incubated in 1 ml of 0.1M Tris HCl (pH 7.95), containing 1mM Succinyl-ala-ala-pro-phe-*p*-nitroanilide (Sigma) for 5 min at room temperature. The spores were pelleted by centrifugation for 5 min at 12000 g (Fastwin Bio-Tech Company Limited). A 200 µL of supernatant was transferred to quartz cuvette. The absorbance was measured at 405 nm by using a spectrophotometer (Shimadzu UV-1800). Buffered substrate was used as a reference. The amount of spore bound extracellular protease Pr1 is expressed as micromoles of nitroanilide (NA) released per minute.

2.6 Statistical analyses

All experiments were repeated four times except the activity of spore bound Pr1, repeated three times. Data were analyzed by analysis of variance using the ANOVA procedure of SAS (SAS Institute, 2000) for a completely randomized design. When the effect was significant ($P < 0.05$), means were separated using Duncan's Multiple Range Test.

3. Results

3.1 Effect of different sources of sugar and nitrogen on the dried fungal biomass of entomopathogenic fungi

Significant differences in dried biomass of *M. anisopliae* ($F = 10.632$; df = 14, 45; $P < 0.001$), *B. bassiana* ($F = 9.286$; df = 14, 45; $P < 0.001$) and *I. fumosorosea* ($F = 9.596$; df = 14, 45; $P < 0.001$) were observed when fungi were grown on media containing different sources of nitrogen and sugar in different combinations. The media supplemented with sugarcane molasses (SM) afforded comparatively higher growth of *M. anisopliae* and *I. fumosorosea* than glucose (Fig. 1a, c). The growth medium contained G + CSL + YE, showed the lowest growth (12.95 mg/ml) of *M. anisopliae*. The glucose in combination with CSL, PE and YE showed higher growth; while their combination among them did not afford higher fungal growth (Fig. 1a). The growth medium such as G + CSL + PE, G + CSL + YE and G + PE + YE, afforded the lowest growth of *I. fumosorosea* and non significant differences were observed among them (Fig. 1c). *B. bassiana* grown on media supplemented with SM in combination with CSL, PE and YE showed significantly higher growth compared to the media containing glucose as sugar source (Fig. 1b). In contrast, their combinations (CSL, PE and YE) with one another in the presence of glucose as sugar source produced higher growth compared with SM (Fig. 1b). The medium SM + CSL afforded higher growth ranges from 31.46-35.28 mg/ml in all the studied fungal strains (Fig. 1a-c).

(a)

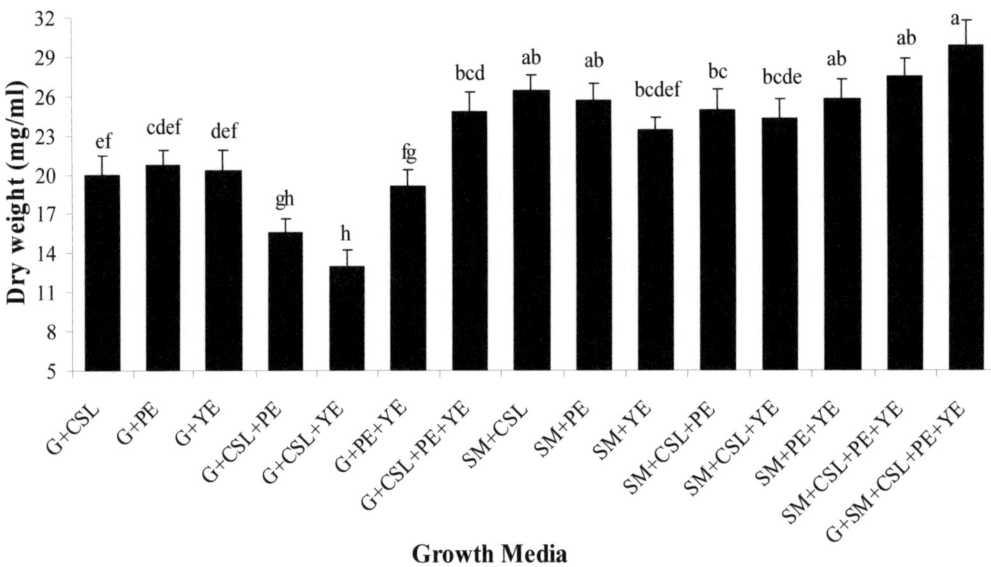

(b)

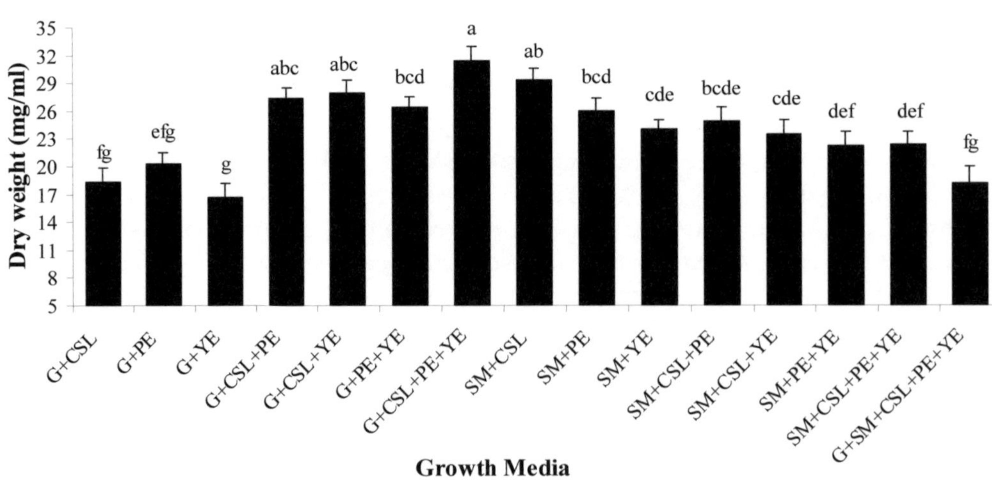

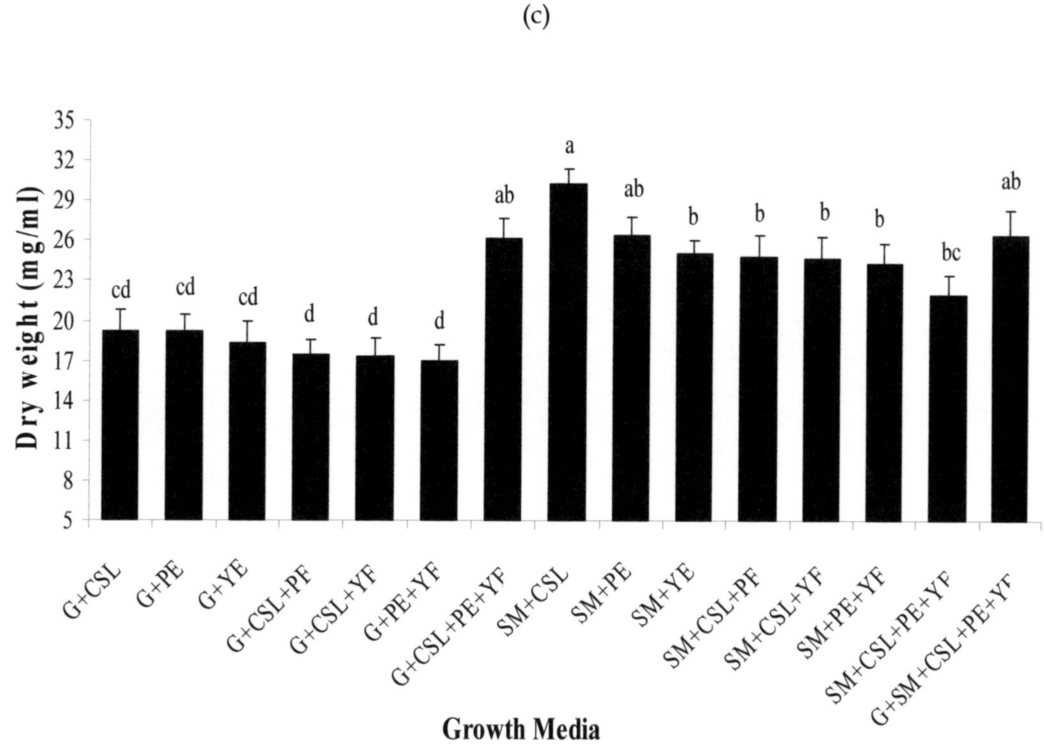

Fig. 1. Influence of media composition on the dried biomass (mg/ml) of (a) *M. anisopliae* (b) *B. bassiana* and (c) *I. fumosorosea*. Mean ± SE values with the same letter(s) along the bars of different growth media are not significantly different ($P < 0.05$). For detail of treatments see table 1.

3.2 Effect of different sources of sugar and nitrogen on the blastospores growth of entomopathogenic fungi

M. anisopliae growth observed from the shake flask cultures supported with different media differed significantly after 20 h ($F = 96.535$; df = 14, 45; $P < 0.001$), 40 h ($F = 77.536$; df = 14, 45; $P < 0.001$) and 60 h ($F = 67.381$; df = 14, 45; $P < 0.001$). As the incubation time elapsed, the concentration of the blastospores increased. After 20 h complete growth media (G + SM + CSL + PE + YE), afforded the highest growth of the yeast like hyphal bodies of the *M. anisopliae*.

While, G + PE, G + YE and G + PE + YE showed the lowest optical density of the growth media. After 40 and 60 h of incubation, the medium SM + CSL produced higher concentration of the blastospores (1.995) and remained significantly at higher level than any medium. While G + PE + YE media, showed the lowest values at both the studied time intervals (40 h and 60 h). The complete medium also afforded the growth of the fungi, which ultimately showed higher OD values of the blastospores but significantly at lower level than SM + CSL medium after 40 h and 60 h (Fig. 2a).

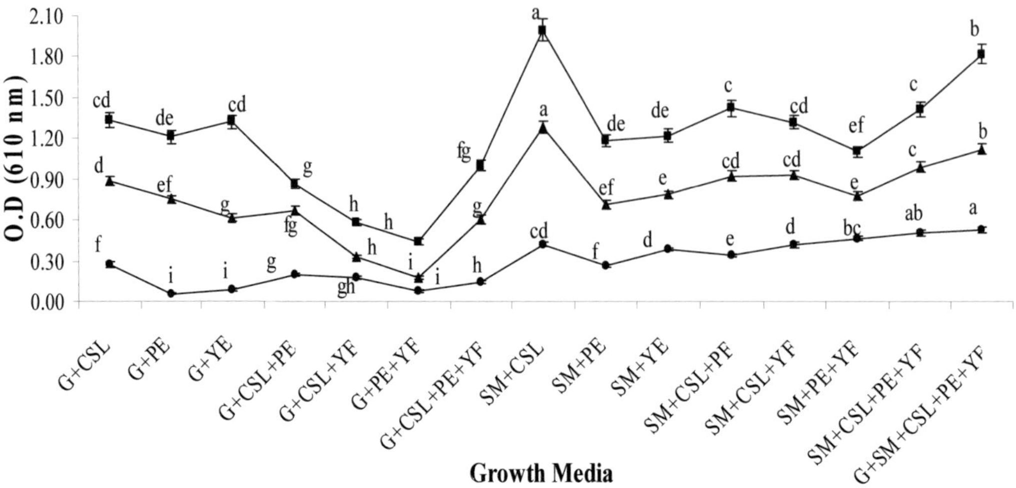

Fig. 2a. Blastospores concentration (OD) of *M. anisopliae* grown on different media supplemented with different sources of nitrogen and sugar after —●— 20 h, —▲— 40 h and —■— 60 h incubation. Mean ± SE values with the same letter along the bars of different growth media are not significantly different ($P < 0.05$). For detail of treatments see table 1.

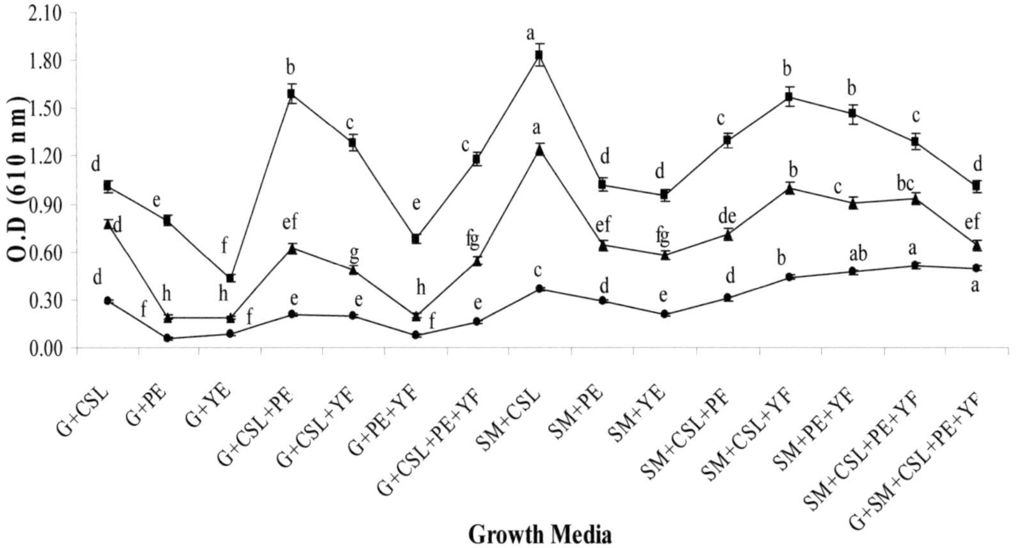

Fig. 2b. Blastospores concentration (OD) of *B. bassiana* grown on different media supplemented with different sources of nitrogen and sugar after —●— 20 h, —▲— 40 h and —■— 60 h incubation. Mean ± SE values with the same letter along the bars of different growth media are not significantly different ($P < 0.05$). For detail of treatments see table 1.

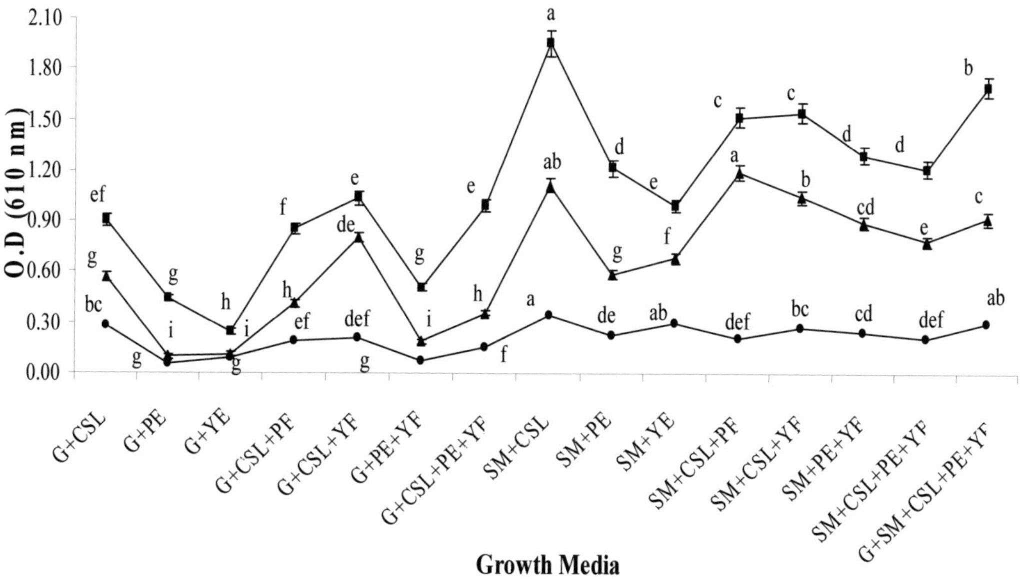

Fig. 2c. Blastospores concentration (OD) of *I. fumosorosea* grown on different media supplemented with different sources of nitrogen and sugar after —●— 20 h, —▲— 40 h and —■— 60 h incubation. Mean ± SE values with the same letter along the bars of different growth media are not significantly different ($P < 0.05$). For detail of treatments see table 1.

Blastospores growth was highly variable and significant differences among the growth of *B. bassiana* on all the growth media after 20 h (F = 172.619; df = 14, 45; P <0.001), 40 h (F = 103.584; df = 14, 45; P <0.001) and 60 h (F = 62.153; df = 14, 45; P <0.001) were observed. After 20 h of incubation, media supplemented with SM showed higher growth compared to all the media containing glucose as sugar source in all the combinations with CSL, PE and YE. On the whole, the media SM + CSL after 40 h and 60 h showed the highest growth of blastospores. CSL in combination with PE and YE in the presence of G as sugar source enhanced the growth of the fungi. Comparatively, the media supplemented with SM found to cause stimulant effect on the growth of the blastospores (Fig. 2b).

There was a significant difference in the blastospores concentration of *I. fumosorosea* after 20 h (F = 36.819; df = 14, 45; P <0.001), 40 h (F = 145.915; df = 14, 45; P <0.001) and 60 h (F = 110.554; df = 14, 45; P <0.001) grown on different media supplemented with different sources of nitrogen and sugar in all possible combinations. *I. fumosorosea* grown on SM + CSL exhibited the highest growth not only after 20 h of incubation but also after 60 h (Fig. 2c). While, SM + CSL also promoted the growth resulting higher blastospores but remained significantly lower than SM + CSL + PE, which showed the highest growth after 40 h of incubation. On the whole, media supplemented with SM as sugar source showed higher growth of the blastospores compared with the media containing glucose as sugar source (Fig. 2c).

(a)

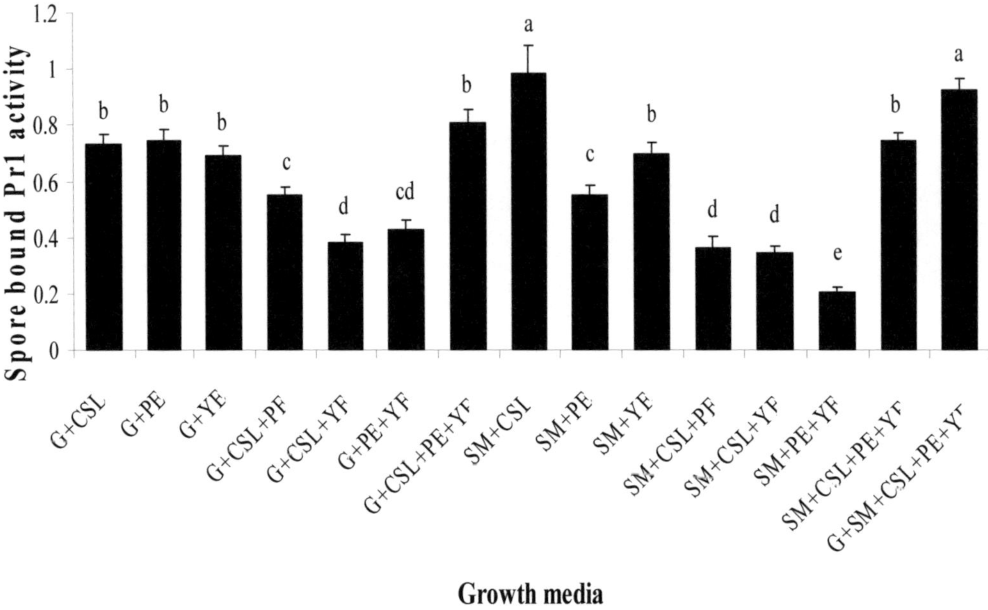

Spore bound Pr1 activity
1.2
1
0.8
0.6
0.4
0.2
0
b
b
b
c
d
cd
b
a
c
b
d
d
e
b
a
G+CSL
G+PE
G+YE
G+CSL+PE
G+CSL+YE
G+PE+YE
G+CSL+PE+YE
SM+CSL
SM+PE
SM+YE
SM+CSL+PE
SM+CSL+YE
SM+PE+YE
SM+CSL+PE+YE
G+SM+CSL+PE+YE
Growth media

(b)

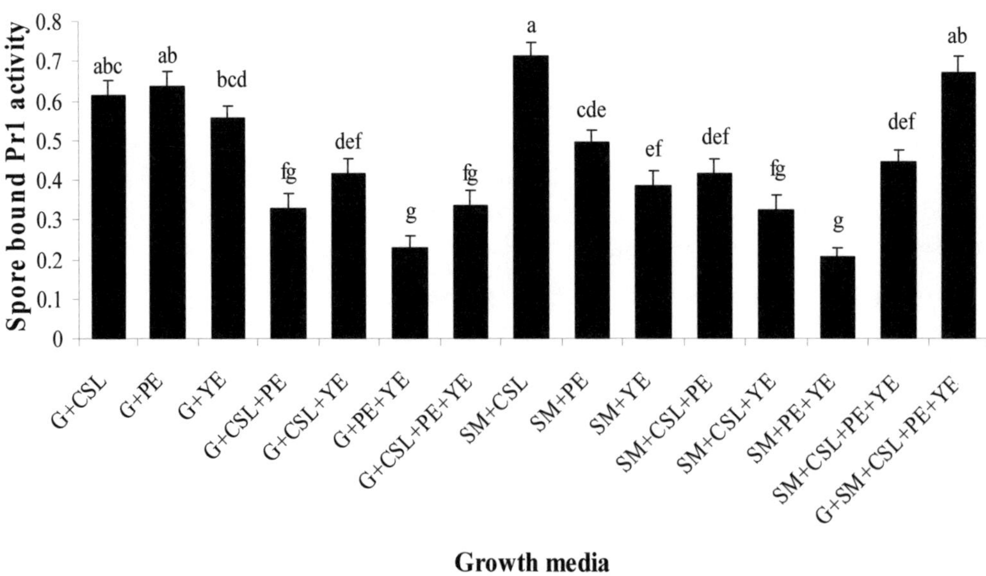

Spore bound Pr1 activity
0.8
0.7
0.6
0.5
0.4
0.3
0.2
0.1
0
abc
ab
bcd
fg
def
g
fg
a
cde
ef
def
fg
g
def
ab
G+CSL
G+PE
G+YE
G+CSL+PE
G+CSL+YE
G+PE+YE
G+CSL+PE+YE
SM+CSL
SM+PE
SM+YE
SM+CSL+PE
SM+CSL+YE
SM+PE+YE
SM+CSL+PE+YE
G+SM+CSL+PE+YE
Growth media

(c)

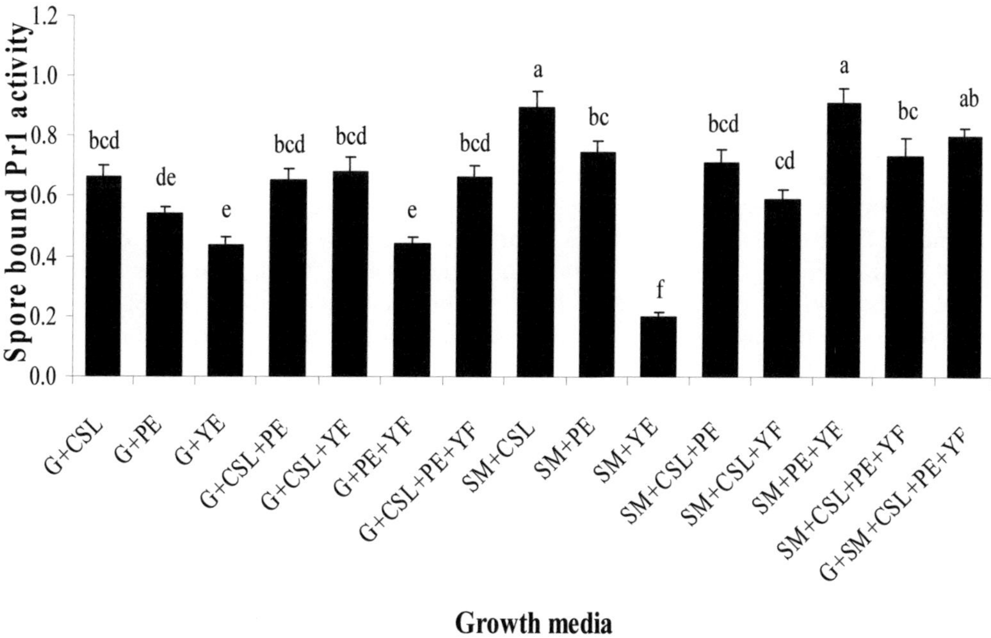

Fig. 3. Extracellular protease Pr1 activity (µmol NA ml^{-1} min^{-1}) of (a) *M. anisopliae* (b) *B. bassiana* and (c) *I. fumosorosea* grown on media supplemented with different sources of nitrogen and sugar. Mean ± SE values with the same letter along the bars of different growth media are not significantly different ($P < 0.05$). For detail of treatments see table 1.

3.3 Effect of different sources of sugar and nitrogen on the enzymatic activity (µmol NA ml-1 min-1) of spore bound Pr1 of entomopathogenic fungi

Significant differences in enzymatic activity of spore bound Pr1 were observed when different sources of sugar and nitrogen in all possible combinations were used to grow *M. anisopliae* ($F = 28.945$; df = 14, 30; $P < 0.001$), *B. bassiana* ($F = 20.302$; df = 14, 30; $P < 0.001$) and *I. fumosorosea* ($F = 21.376$; df = 14, 30; $P < 0.001$). The spores of all the studied entomopathogenic fungi cultivated from rice inoculated with medium supplemented with SM + CSL showed the highest activity of Pr1 (Fig. 3a-c). *M. anisopliae* spores grown on complete media also showed higher enzymatic activity (Fig. 3a). The growth medium SM + PE + YE in case of *M. anisopliae,* while SM + PE + YE and G + PE + YE, in case of *B. bassiana* showed the lowest enzymatic activity (Fig. 3a-b). The addition of PE and YE solely in combination with glucose produced spores of *B. bassiana* with relatively higher enzymatic activity compared to SM (Fig. 3b). In case of *I. fumosorosea,* CSL supplemented media showed comparatively higher enzymatic activity except the medium supplemented with SM + YE (0.20 µmol NA ml^{-1} min^{-1}) (Fig. 3c).

4. Discussion

These studies demonstrated that sugar and nitrogen sources significantly effect the growth of blastospores produced by cultures of *B. bassiana, M. anisopliae* and *I. fumosorosea*. Higher blastospores growth and dried fungal biomass was produced in cultures grown on media supplemented with sugarcane molasses (Fig. 1a, c). Previous studies with *Paecilomyces farinosus* (Hotmskiold) and *Paecilomyces lilacinus* (Thom.) showed that media supplemented with SM supported the highest growth of the studied fungi (Leena et al, 2003).

Mass production technology is an important way for improving mycoinsecticides based on the blastospores. Accelerated blastospores growth rates in the current study from the media supplemented with SM and corn steep liquor (CSL) greatly improved the dried fungal biomass production of the studied entomopathogenic fungi. While, our results disagreed with the findings of Leite et al, (2003) who concluded that replacement of CSL as nitrogen source gave relatively low yield of the three studied fungal strains. Our findings revealed that the fungal organisms directly interact with the culture conditions and strongly influence the growth of blastospores.

All the studied fungi did not have the same characteristics when cultivated in media with different complex sources of nitrogen and sugar. Sugarcane molasses, a by-product from the sugar industry, supported higher growth of dried biomass of *M. anisopliae* and *I. fumosorosea,* compared to the media supplemented with glucose. Thus, it may be speculated that SM efficiently enhanced the growth of blastospores, which ultimately led to the production of higher fungal biomass of the fungi. The complete media SM + G + CSL + PE + YE, greatly enhanced the growth of *M. anisopliae* and *I. fumosorosea*. These components when evaluated separately afforded lower growth for *M. anisopliae*. This suggests that these components differ concerning types of nutrients, and therefore provide complete nourishment when offered together.

CSL, peptone and yeast extract afforded higher growth of *B. bassiana* in fungal cultures supplemented with glucose, while these nitrogen sources did not enhance the growth of fungi in the presence of SM. On the other hand, these nitrogen sources when used in combinations (SM + CSL + PE, SM + CSL + YE, SM + PE + YE, SM + CSL + PE + YE), did not increase fungal production of *B. bassiana*. These combinations in the presence of glucose greatly enhanced the dried fungal biomass. Even, the complete medium showed significantly lower growth. On the basis of above findings, we may suggest that nutrition greatly influenced the growth of *B. bassiana*. The result of our study corroborates similar research on the effect of nutrition and propagule production in *Metarhizium* spp. and other entomopathogenic hyphomycetes (Inch et al, 1986; Rombach, 1989; Kleespies & Zimmermann, 1992; Jackson et al, 1997; Vidal et al, 1998).

CSL, a by-product from the corn industry, supported higher growth of the blastospores (Fig. 3a-c), and also ultimately led to the production of spores with higher enzymatic activity of Pr1. Since, peptone and yeast extract are expensive sources of nitrogen; CSL was chosen because of its stimulatory effect and cost economics and could be used to replace the nitrogen sources used previously. Supplementation with 1% CSL with 2.66% SM enhanced not only the growth of all the studied fungi but also enzymatic activity of spore bound Pr1 of only *M. anisopliae*. This is in agreement with the results of McCoy et al, (1988), that nutrition is one of the several factors that may determine the specificity of a fungal pathogen. The results suggest that the growth of blastospores can be efficiently improved from inexpensive CSL and SM, making fermentation an economical and environmental friendly process.

The management of arthropod pests generally involves preventive measures and remedial control (Lewis, 1997; Su & Scheffrahn, 1998). Currently registered insecticides have undergone rigorous field-testing, efficacy results have been mixed. Some insecticides are expensive and less persistent, leading to reduced longevity and the failure of the chemical barrier (Su and Scheffrahn, 1998). In addition, large quantities of persistent insecticides are raising concerns about applicator safety, environmental contamination and possible deleterious effects on non-target animals. By keeping in mind the above mentioned drawbacks, it is the urgent requirement to standardise the microbe base products against insects for the safety of human beings, animals and environment.

5. Conclusion

In conclusion, the replacement of nitrogen and sugar sources with CSL and SM respectively, in the liquid production medium significantly improved the growth and the activity of spore bound extracellular protease Pr1 of *M. anisopliae* for the first time. A more rapid growth rate for blastospores production permitted us to select this appropriate media for large scale commercial development of entomopathogenic fungi for the safe management of insect pests, in order to avoid the deleterious effects of insecticides. Entomopathogenic fungi being component of an integrated approach can provide significant and selective insect control. In the near future, we expect to see synergistic combinations of microbial control agents with other technologies (in combination with semiochemicals, soft chemical pesticides, other natural enemies, resistant plants, chemigation, remote sensing, etc.) that will enhance the effectiveness and sustainability of integrated control strategies.

6. References

Al-Aïdroos, K. & Roberts, D.W. (1978). Mutants of *Metarhizium anisopliae* with increased virulence toward mosquito larvae. *Canadian Journal of Genetics and Cytology,* Vol.20, pp. 211-219, ISSN 0008-4093.

Amiri-Besheli, B., Khambay, B. Cameron, M. Deadman, M.L. & Butt, T.M. (2000). Interand intra-specific variation in destruxin production by the insect pathogenic *Metarhizium,* and its significance to pathogenesis. *Mycological Research,* Vol.104, pp. 447-452, ISSN 0953-7562.

Ansari, M.A., Shah, F.A. Tirry, L. & Moens, M. (2006). Field trials against *Hoplia philanthus* (Coleoptera: Scarabaeidae) with a combination of an entomopathogenic nematode and the fungus *Metarhizium anisopliae* CLO 53. *Biological Control,* Vol.39, No.3, pp. 453-459, ISSN 1049-9644.

Bhattacharyya, A., Samal, A.C. & Kar, S. (2004). Entomophagous fungus in pest management. News Letter, Vol.5, pp. 1-4.

Boucias, D.G. & Pendland, J.C. (1991). Attachment of mycopathogens to cuticle: the initial event of mycoses in arthropod hosts. In: *The fungal spore and disease initiation in plants and animals.* G.T. Cole, H.C. Hoch. (Eds.), Plenum Press, New York, pp. 101-127.

Boucias, D.G. & Pendland, J.C. (1982). Ultrastructural studies on the fungus, *Nomuraea rileyi*, infecting the velvetbean caterpillar, *Anticarsia gemmatalis*. *Journal of Invertebrate Pathology*, Vol.39, pp. 338-345, ISSN 0022-2011.

Burges, H. D. (1981). Microbial Control of Pests and Plant Diseases 1970–1980. Academic Press, London, ISBN 9780121433604.

Chandler, D., Davidson, G. Pell, J.K. BALL, B.V. Shaw, K. & Sunderland K. D. (2000). Fungal biocontrol of Acari. *Biocontrol Science and Technology*, Vol.10, pp. 357-384, ISSN 0958-3157.

Charnley, A.K. (1984). Physiological aspects of destructive pathogenesis in insects by fungi: a speculative review. *British Mycological Society Symposium*, Vol.6, pp. 229-270.

Clarkson, J.M. & Charnley, A.K. (1996). New insights into the mechanisms of fungal pathogenesis in insects. *Trends in Microbiology*, Vol.4, pp. 197-203, ISSN 0966-842X.

Daoust, R.A., Ward, M.G. & Roberts, D.W. (1982). Effect of formulation on the virulence of *Metarhizium anisopliae* conidia against mosquito larvae. *Journal of Invertebrate Pathology*, Vol.40, pp. 228-236, ISSN 0022-2011.

Dean, D.D. & Domnas, A. (1983). The extracellular proteolytic enzymes of the mosquito–parasitizing fungus *Lagenidium giganteum*. *Experimental Mycology*, Vol.7, pp. 31-39, ISSN 0147-5975.

Devi, P.S.V. (1994). Conidia production of the entomopathogenic fungus *Nomuraea rileyi* and its evaluation for control of *Spodoptera litura* (Fab) on *Ricinus communis*. *Journal of Invertebrate Pathology*, Vol. 63, No. 2, pp. 145-150, ISSN 0022-2011.

Eyal, J., Mabud, M.D.A. Fischbein, K.L. Walter, J.F. Osborne, L.S. & Landa, Z. (1994). Assessment of *Beauveria bassiana* Nov. EO–1 strain, which produces a red pigment for microbial control. *Applied Biochemistry and Biotechnology*, Vol.44, pp. 65-80, ISSN 0273-2289.

Fargues, J. (1984). Adhesion of the fungal spore to the insect cuticle in relation to pathogenicity. In: *Infection processes of fungi*. D.W. Roberts, J.R. Aist (Eds.). The Rockefeller Foundation, New York, pp. 90-110, ISBN-13: 9780813300238

Ferron, P. (1978). Biological control of insect pests by entomogenous fungi. *Annual Review of Entomology*, 23, pp. 409-442, ISSN 0066-4170.

Fusconi, R., Godinho, M.J.L. & Bossolan, N.R.S. (2005). Culture and exopolysaccharide production from sugarcane molasses by *Gordonia polyisoprenivorans* CCT 7137, isolated from contaminated groundwater in Brazil. *World Journal of Microbiology & Biotechnology* Vol.24. No.7, pp. 937-943, ISSN 0959-3993.

Goettel, M.S., St. Leger, R.J. Rizzo, N.W. Staples, R.C. Roberts, D.W. (1989). Ultrastructural localization of a cuticle–degrading protease produced by entomopathogenic fungus *Metarhizium anisopliae* during penetration of the host (*Manduca sexta*) cuticle. *Journal of General Microbiology*, Vol.135, pp.2233-2239, ISSN 0022-1287.

Gupta, S., Montillot, C. & Hwang, Y.S. (1994). Isolation of novel beauvericin analogues from the fungus *Beauveria bassiana*. *Journal of Natural Products*, Vol.58, pp. 733-738, ISSN 0163-3864.

Hadley, N.F. (1981). Cuticular lipids of terrestrial plants and arthropods: a comparison of their structure, composition and waterproofing barrier. *Biology Reviews*, Vol.56, pp. 23-47, ISSN1469-185X.

Hajek, A.E. & St. Leger, R.J. (1994). Interaction between fungal pathogens and insect hosts. *Annual Review of Entomology*, Vol.39, pp. 293-322. ISSN 0066-4170.

Hussain, A., Ahmed, S. & Shahid, M. (2011) Laboratory and field evaluation of *Metarhizium anisopliae var. anisopliae* for controlling subterranean termites. *Neotropical Entomology,* Vol.40, No.2 pp. 244-250, ISSN 1519-566X.

Hussain, A., Tian, M.Y. He, Y.R. Bland, J.M. & Gu, W.X. (2010a). Behavioral and electrophysiological responses of *C. formosanus* towards entomopathogenic fungal volatiles. *Biological Control,* Vol.55, pp. 166-173, ISSN 1049-9644.

Hussain, A., Tian, M.Y. He, Y.R. & Lin, R. (2010b). *In vitro* and *in vivo* culturing impacts on the virulence characteristics of serially passed entomopathogenic fungi. *Journal of Food Agriculture & Environment,* Vol.8, No.3&4, pp. 481-487, ISSN 1459-0255.

Hussain, A., Tian, M.Y., He, Y.R. & Ahmed, S. (2009) Entomopathogenic Fungi disturbed the larval growth and feeding performance of *Ocinara varians* Walker (Lepidoptera: Bombycidae) Larvae. *Insect Science,* Vol.16, No.6, pp. 511–517, ISSN 1672-9609.

Huxham, I.M., Lackie, A.M. & McCorkindale, N.J. (1989). Inhibitory effects of cyclodepsipeptides, destruxins, from the fungus *Metarhizium anisopliae,* on cellular immunity in insects. *Journal of Insect Physiology,* Vol.35, pp. 97-105, ISSN 0022-1910.

Inch, J.M.M., Humphreys, A.M, Trinci, A.P.J. & Gillepsie, A.T. (1986) Growth and blastospore formation by *Paecilomyces flimosoroseus,* a pathogen of brown planthopper (*Nilaparvata lugens*). *Transactions of the British Mycological Society,* Vol.87, pp. 215–222, ISSN 0007-1536.

Jackson, M.A. (1997) Optimizing nutritional conditions for the liquid culture production of effective fungal biological control agents. *Journal of Industrial Microbiology and Biotechnology,* Vol.19, No.3, pp. 180–187, ISSN 1367-5435.

Jackson, M.A., McGuire, M.R, Lacey, L.A. & Wraight S.P. (1997) Liquid culture production of desiccation tolerant blastospores of the bioinsecticidal fungus *Paecilomyces fumosoroseus. Mycological Research,* Vol.101, pp. 35-41, ISSN 0953-7562.

Juárez, M.P. & Fernández, G.C. (2007). Cuticular hydrocarbons of triatomins. Comparative Biochemistry and Physiology - Part A: Molecular and Integrative Physiology. Vol.147, No.3, pp. 711-730, ISSN: 1095-6433.

Kershaw, M.J., Moorhouse, E.R. Bateman, R. Reynolds, S.E. & Charnley, A.K. (1999). The role of destruxins in the pathogenicity of *Metarhizium anisopliae* for three species of insect. *Journal of Invertebrate Pathology,* Vol.74, pp. 213-223, ISSN 0022-2011.

Khetan, S.L. (2001). Microbial pest control. Marcel Dekker, Inc, New York, pp. 211-256, ISBN 9780824704452.

Kleespies, R.G. & Zimmermann, G. (1992) Production of blastospores by three strains of *Metarhizium anisopliae* (Metch.) Sorokin in submerged culture. *Biocontrol Science and Technology,* Vol.2, No.2, pp. 127-135, ISSN 0958-3157.

Kodaira, Y. (1961). Biochemical studies on the muscardine fungi in silkworms, *Bombyx mori.* In: J. Fac. (Ed.), Text. Sci. Tech. 5. Sinshu. University Sericulture. pp. 1-68.

Krasnoff, S.B., Gupta, S. St. Leger, R.J., Renwick, J.A.A. & Roberts, D.W. (1991). Antifungal and insecticidal properties of efrapeptins: metabolites of the fungus *Tolypocladium niveum. Journal of Invertebrate Pathology,* Vol.58, pp. 180-188, ISSN 0022-2011.

Lacey, L.A. & Kaya, H.K. (Eds.). (2000). Field Manual of Techniques in Invertebrate Pathology: Application and Evaluation of Pathogens for Control of Insects and Other Invertebrate Pests. Kluwer Academic, Dordrecht, ISBN 9781402059322.

Leena, M.D., Easwaramoorthy S. & Nirmala R. (2003) *In vitro* production of entomopathogenic fungi *Paecilomyces farinosus* (hotmskiold) and *Paecilomyces lilacinus* (Thom.) Samson using byproducts of sugar industry and other agro-industrial byproducts and wastes. *Sugar Tech,* Vol.5, pp. 231-236, ISSN 0972-1525.

Leite L.G., Alves S.B, Filho A.B. & Roberts D.W. (2005) Simple, inexpensive media for mass production of three entomophthoralean fungi. *Mycological Research,* Vol.109, No.3, pp. 326-334, ISSN 0953-7562.

Leite, L.G., Alves, S.B, Filho, A.B. & Roberts, D.W. (2003) Effect of salts, vitamins, sugars and nitrogen sources on the growth of three genera of Entomophthorales: *Batkoa, Furia,* and *Neozygites. Mycological Research,* Vol.107, No.7, pp. 872-878, ISSN 0953-7562.

Lewis, V.R. (1997). Alternative control strategies for termites. *Journal of Agricultural Entomology,* Vol.14, pp. 291- 307, ISSN 0735-939X.

Madelin, M.F., Robinson, R.F. & Williams, R.S. (1967). Appressorium like structures in insect parasitizing deuteromycetes. *Journal of Invertebrate Pathology,* Vol.9, pp. 404–412, ISSN 0022-2011.

Mazet, I. & Vey, A. (1995). Hirsutellin A, a toxic protein produced *in vitro* by *Hirsutella thompsonii. Microbiology,* Vol.141, pp. 1343-1348, ISSN 1350-0872.

McCauley, V.J.E., Zacharuk, R.Y. & Tinline, R.D. (1968). Histopathology of the green muscardine in larvae of four species of Elateridae (Coleoptera). *Journal of Invertebrate Pathology,* Vol.12, pp. 444–459, ISSN 0022-2011.

McCoy, C.W., Samson, R.A. & Boucias, D.G. (1988). Entomogenous fungi. In: *Handbook of Natural Pesticides vol V. Microbial insecticides, part A. Entomogenous protozoa and fungi.* C.M. Ignoffo. (Ed.), 151-236, ISBN CRC Press, Boca Raton, FL, ISBN 9780849336607.

Mohamed, A.K.A., Sikorowski, P.P. & Bell, J.V. (1978). Histopathology of *Nomuraea rileyi* in larvae of *Heliothis zea* and *in vitro* enzymatic activities. *Journal of Invertebrate Pathology,* Vol.31, pp. 345-352, ISSN 0022-2011.

Paterson-Beedle, M., Kennedy, J.F, Melo, F.A.D, Lloyd, L.L. & Medeiros, V. (2000) A cellulosic exopolysaccharide produced from sugarcane molasses by a *Zoogloea* sp. Carbohydrate Polymers, Vol.42, No.4, pp. 375-383, ISSN 0144-8617.

Pendland, J.C. (1982). Resistant structures in the entomogenous hypomycete, *Nomuraea rileyi*: an ultrastructural study. *Canadian Journal of Botany,* Vol.60, pp. 1569-1576, ISSN 1480-3305.

Prasertphon, S. & Tanada, Y. (1968). The formation and circulation, in *Galleria,* of hyphal bodies of entomophthoraceous fungi. *Journal of Invertebrate Pathology,* Vol.11, pp. 260-280, ISSN 0022-2011.

Roberts, D.W. & Hajek, A.E. (1992). Entomopathogenic fungi as bioinsecticides. In: *Frontiers in Industrial Mycology.* G. Leathan (Ed.). Chapman and Hall, New York, pp. 144-159.

Roberts, D.W. (1966). Toxins from the entomogenous fungus *Metarhizium anisopliae.* II. Symptoms and detection in moribund hosts. *Journal of Invertebrate Pathology,* Vol.8, pp. 222-227, ISSN 0022-2011.

Roberts, D.W. (1981). Toxins of entomopathogenic fungi. In: Microbial control of pests. H.D. Burges (Ed.). Academic Press, London, pp. 441-464, ISBN 9780121433604.

Rombach, M.C. (1989) Production of *Beauveria bassiana* [*Deuteromycotina, Hyphomycetes*] sympodulo-conidia in submerged culture. *BioControl,* Vol.34, No.1, pp.45-52, ISSN 1386-6141.

Samsináková, A., Misiková, S. & Leopold, J. (1971). Action of enzymatic systems of *Beauveria bassiana* on the cuticle of the greater wax moth larvae (*Galleria mellonella*). *Journal of Invertebrate Pathology,* Vol.18, pp. 322-330, ISSN 0022-2011.

Samuels, R.J., Charnley, A.K. & Reynolds, S.E. (1988). The role of destruxins on the pathogenicity of 3 strains of *Metarhizium anisopliae* for the tobacco hornworm *Manduca sexta*. *Mycopathologia,* Vol.104, pp. 51-58. ISSN 0301-486X.

SAS Institute. 2000. SAS user's guide: statistics, SAS Institute, Cary, North Carolina.

Shah, F.A. Allen, N. Wright, C.J. & Butt, T.M. (2007). Repeated *in vitro* subculturing alters spore surface properties and virulence of *Metarhizium anisopliae*. *FEMS Microbiology Letters,* Vol.276, pp. 60-66, ISSN 0378-1097.

Shah, F.A. & Butt, T.M. (2005) Influence of nutrition on the production and physiology of sectors produced by the insect pathogenic fungus *Metarhizium anisopliae*. *FEMS Microbiology Letters,* Vol.250, No.2, pp. 201-207, ISSN 0378-1097.

St. Leger, R.J., Allee, L.L. May, B. Staples, R.C. & Roberts, D.W. (1992). World-wide distribution of genetic variation among isolates of *Beauveria* spp. *Mycological Research,* Vol.96, pp. 1007-1015, ISSN 0953-7562.

St. Leger, R.J., Butt, T.M. Staples, R.C. & Roberts, D.W. (1989). Synthesis of proteins including a cuticle-degrading protease during differentiation of the entomopathogenic fungus *Metarhizium anisopliae*. *Experimental Mycology,* Vol.13, pp. 253-262, ISSN 0147-5975.

St. Leger, R.J., Durrands, P.K. Charnley, A.K. & Cooper, R.M. (1988). Role of extra-cellular chymoelastase in the virulence of *Metarhizium anisopliae* for *Manduca sexta*. *Journal of Invertebrate Pathology*. Vol.52, pp. 285-293, ISSN 0022-2011.

St. Leger, R.J., Joshi, L. Bidochka. & M. Roberts, D.W. (1996), Construction of an improved mycoinsecticide over expressing a toxic protease. *Proceedings of National Academy of Sciences USA*, Vol.93, pp. 6349–6354, ISSN 0027-8424.

St. Leger, R.J. (1994). The role of cuticle degrading proteases in fungal pathogenesis of insects. *Canadian Journal of Botany*. Vol.73 (Suppl. 1), pp. 1119-1125, ISSN 1480-3305.

St. Leger, R.J., Charnley A.K. & Cooper R.M. (1986) Cuticle degrading enzymes of entomopathogenic fungi: Synthesis in culture on cuticle. *Journal of Invertebrate Pathology,* Vol.48, No.1, pp. 85-95, ISSN 0022-2011.

St. Leger, R.J., Cooper, R.M. & Charnley, A.K. (1987). Production of cuticle-degrading enzymes by the entomopathogen *Meterhizium anisopliae* during infection of cuticles from *Calliphora vomitoria* and *Manduca sexta*. *Microbiology,* Vol.133, No.5, pp. 1371-1382, ISSN 1350-0872.

Steinhaus, E.A. (1956). Microbial control: The emergence of an idea. *Hilgardia,* Vol.26, pp. 107-160. ISSN 0073-2230.

Steinhaus, E.A. (1975). Disease in a Minor Cord. Ohio State Univ. Press, Columbus, OH, ISBN 9780814202180.

Su, N.Y. & Scheffrahn, R.H. (1998). A review of subterranean termite control practices and prospects for integrated pest management programs. *Integrated Pest Management Reviews,* Vol.3, pp. 1-13. ISSN 1572-9745.

Sutherland, I.W. (1996) Microbial biopolymers from agricultural products: production and potential. *International Biodeterioration & Biodegradation* 38: 249–261, ISSN 0964-8305.

Suzuki, A., Kanaoka, M. Isogai, A. Murakoshi, S. Ichinoe, M. & Tamura, S. (1977). Bassianolide, a new insecticidal cyclodepsipeptide from *Beauveria bassiana* and *Verticillium lecanii*. *Tetrahedron Letters,* Vol.18, pp. 2167-2170, ISSN 0040-4039.

Suzuki, A., Kawakami, K. & Tamura, S. (1971). Detection of destruxins in silkworm larvae infected with *Metarhizium anisopliae*. *Agricultural and Biological Chemistry,* Vol.35, pp. 1641-1643. ISSN 0002-1369.

Tanada, Y. & Kaya, H.K. (1993). Insect Pathology. Academic Press, New York, ISBN 0-12-683255-2.

Vey, A., Hoagland, R.E. & Butt, T.M. (2001). Toxic metabolites of fungal control agents. In: *Fungi as Biocontrol Agents.* T.M. Butt, C. Jackson, N. Magan, (Eds.), CAB International, New York, pp. 311-346, ISBN 0-85199-356-7.

Vidal C., Fargues, J. Lacey, L.A. & Jackson, M.A. (1998) Effect of various liquid culture media on morphology, growth, propagule production, and pathogenic activity to *Bemisia argentifolii* of the entomopathogenic hyphomycete, *Paecilomyces fumosoroseus. Mycopathologia,* Vol. 143, No. 1, pp. 33-46, ISSN 0301-486X.

Vilcinskas, A., Matha, V. & Gotz, P. (1977). Inhibition of phagocytic activity of plasmatocytes isolated from *Galleria mellonella* by entomogenous fungi and their secondary metabolites. *Journal of Insect Physiology,* Vol.43, pp. 475-483, ISSN 0022-1910.

Weiser, J. & Matha, V. (1988). Tolypin: A new insecticidal metabolite of fungi of the genus *Tolypocladium. Journal of Invertebrate Pathology,* Vol.51, pp. 94-96, ISSN 0022-2011.

Wraight, S.P., Inglis, G.D. & Goettel, M.S. (2007). Fungi, In: *Field manual of techniques in invertebrate pathology,* L.A. Lacey & H.K. Kaya, (Eds.), 223-248, 2nd edition, Springer, Dordrecht, ISBN 978-1-4020-5931-5..

Ypsilos, I.K. & Magan, N. (2005). Characterization of optimum cultural environmental conditions for the production of high numbers of *Metarhizium anisopliae* blastospores with enhanced ecological fitness. *Biocontrol Science and Technology,* Vol. 15, No. 7, pp. 683-699, ISSN 0958-3157.

Zacharuk, R.Y. (1970a). Fine structure of the fungus *Metarhizium anisopliae* infecting three species of larval Elateridae (Coleoptera). II. Conidial germ tubes and appressoria. *Journal of Invertebrate Pathology,* Vol.15, pp. 81-91, ISSN 0022-2011.

Zacharuk, R.Y. (1970b). Fine structure of the fungus *Metarhizium anisopliae* infecting three species of larval Elateridae (Coleoptera): III. Penetration of the host integument. *Journal of Invertebrate Pathology,* Vol.15, 372-396, ISSN 0022-2011.

Zhao, L.L., Chen, X.X. & Xu, J.H. (2010). Strain improvement of *Serratia marcescens* ECU1010 and medium cost reduction for economic production of lipase. *World Journal of Microbiology and Biotechnology,* Vol. 26, No. 3, pp. 537-543, ISSN 0959-3993.

Zimmermann, G. (1993). The entomopathogenic fungus *Metarhizium anisopliae* and its potencial as a biocontrol agent. *Pesticide Science,* Vol.37, pp. 375-379, ISSN 1096-9063.

Neurophysiological Recording Techniques Applied to Insect Chemosensory Systems

Vonnie D.C. Shields[1] and Thomas Heinbockel[2]
[1]*Department of Biological Sciences, Towson University, Towson, MD,*
[2]*Department of Anatomy, Howard University College of Medicine, Washington, DC,*
USA

1. Introduction

The aim of this chapter is to discuss current electrophysiological recording techniques used to study the processing of olfactory and gustatory information in insects. More specifically, we will describe methods employed (a) to determine the physiological properties of gustatory (GRCs) and olfactory receptor cells (ORCs) in the peripheral nervous systems and (b) to physiologically characterize identifiable olfactory neurons in the central nervous system.

In studying the structural and functional organization of the nervous system, it is at times advantageous to use animal models, such as insects, for experimentation. Key factors that make insect nervous systems excellent models for analyzing gustatory and olfactory mechanisms are that they possess a relatively simple peripheral and central nervous system compared with their vertebrate counterparts. They also bear easily accessible sensory organs (sensilla) and individually identifiable neural structures. The number of receptor cells mediating mechanisms involved in olfaction and gustation is relatively small. Interestingly, some brain regions in invertebrates and vertebrates show remarkable morphological and physiological similarity and, therefore, the insect nervous system provides insights into general principles underlying taste and odor coding that occur in higher vertebrates.

Gustatory and olfactory cues play vital roles in shaping insect behavior. Insects rely on these senses in the sampling and selection of food sources, avoidance of noxious or toxic compounds, mating, and locating egg-laying sites. These chemical cues are detected by GRCs and ORCs housed in sensilla located mainly on the mouthparts, legs, and antennae and are converted into a neural code of action potentials, which is sent to the central nervous system for processing. GRCs and ORCs constitute sensory filters for environmental taste and smell signals in insects. They form the first layer of a decision making process and transfer information directly to centers in the brain.

2. Insect chemosensory organs

Insects are ideal models for both olfactory and gustatory studies. They bear numerous sensory organs or sensilla, which allow them to constantly monitor and respond to changes in their internal and external environments so as to maintain themselves under the most favorable conditions for survival. Sensilla are used in the sensory perception for smell, taste, sound, touch, vision, proprioception, and geo-, thermo-, and hygroreception. These specialized cuticular structures vary in size and shape and act as their first level of environmental perception. Receptor cells that innervate these sensilla are designed to detect environmental status and change and transmit information regarding the nature of the change to the central nervous system. In insects, it is specifically taste (gustatory) and smell (olfactory) stimuli in the environment that control the behavior of these animals.

Gustatory and olfactory sensilla are typically innervated by more than one bipolar sensory neuron. These neurons bear dendrites which are wrapped by accessory or sheath cells. Sensory information is transduced by receptor cells into an electrical signal, resulting in the generation of nerve signals comprised of action potentials. The absolute frequency and temporal distribution of action potentials in a spike train contain information about the stimulus. This information is transmitted by axons of sensory neurons to modality-specific brain centers. Unraveling the sensory code can be achieved by stimulating specific sensilla and electrophysiologically quantifying the trains of action potentials (input), as well as quantifying the behavior (output).

2.1 Peripheral olfactory processing

Olfactory stimuli play an important role in the orientation of many animals in their environment. Moths detect odor cues with their main olfactory organs, paired antennae. The antennae detect diverse mixtures of volatiles by means of ORCs residing in various types of sensilla. These sensilla are the crucial interface between the outer world and the central nervous system of the moth. The olfactory systems of both invertebrates and vertebrates share many similarities (see Hildebrand and Shepherd, 1997) and are capable of detecting and discriminating among a large number of odorants that differ in size, shape, and complexity. The olfactory organs of invertebrates (e.g., paired antennae of insects) and vertebrates (e.g., nose in mammals) are adapted to detect a vast array of odorants by means of receptors that are located on ORCs. These ORCs are associated with various types of sensilla in invertebrates (e.g., insects) or the olfactory epithelium lining a portion of the nasal cavity of vertebrates (e.g., mammals). Clyne et al. (1999) and Vosshall et al. (1999) identified a novel family of seven-transmembrane-domain proteins, which are encoded by 100-200 genes and are likely to function as *Drosophila melanogaster* olfactory receptors. An individual ORC in the antenna of *D. melanogaster* is thought to express one or a few of the candidate olfactory receptor genes and therefore, each ORC is functionally distinct (Vosshall, 2001).

Because of the fundamental morphological and physiological similarities between invertebrate and vertebrate olfactory systems and since current evidence indicates that basic olfactory processing is similar across all phyla (Hildebrand and Shepherd, 1997), it is feasible to use insects, such as the sphinx moth *Manduca sexta*, as a model system and thereby gain insights into the neural mechanisms of odor recognition and discrimination common to insects and to other animals (Hildebrand, 1995; 1996; Hildebrand and Shepherd, 1997). Insects are ideal experimental models because they possess readily accessible olfactory sensilla, have relatively simple peripheral and central nervous systems, possess individually identifiable neural structures, and have a relatively small number of sensory cells mediating olfactory mechanisms. In moths, the olfactory system comprises two parallel subsystems. One system processes information about plant-associated volatiles odors and is very similar to the main olfactory pathway in vertebrates. The other system is narrowly specialized to detect and respond to information regarding the sex pheromone emitted by conspecific females. Many researchers have classified ORCs based on their individual response profiles and have used terms such as "specialist" and "generalist" to describe those ORCs. Specialist ORCs have been described as those responding with a high sensitivity and selectivity to a single type of odor molecule, such as a pheromone component or a narrow range of related odorants, whereas generalist ORCs have been classified as those responding with relatively lower sensitivity to a broad range of odorants, such as plant-associated volatiles (Schneider et al., 1964). Interestingly, several researchers have found, that some insect ORCs respond to plant-associated volatiles and exhibit relatively high selectivity and sensitivity for effective stimulus molecules (e.g., Priesner, 1979; Dickens, 1990; Anderson et al., 1995; Heinbockel and Kaissling, 1996; Pophof, 1997; Hansson et al., 1999; Shields and Hildebrand, 2001a).

2.2 Antennal morphology

The antennae of both sexes of *M. sexta* comprise three segments, two small, basal segments (scape and pedicel) and a long distal segment (flagellum) (Fig. 1A, B). In adult *M. sexta*, the antennal flagellum is about 2 cm long, comprises approximately 80 subsegments called annuli (or flagellomeres) (Fig. 1C), and is sexually dimorphic (Sanes and Hildebrand, 1976). Each male and female antenna has approximately 4 x 10^5 ORCs (Sanes and Hildebrand, 1976) and 3.0 x 10^5 - 3.4 x 10^5 ORCs (Oland and Tolbert, 1988), respectively. The antenna of each sex is associated with about 10^5 sensilla (Sanes and Hildebrand, 1976; Keil, 1989; Lee and Strausfeld, 1990; Shields and Hildebrand, 1999a; 1999b) and each annulus may bear approximately 2100-2200 sensilla (Fig. 1C, D) (Lee and Strausfeld, 1990; Shields and Hildebrand, 1999b). Male flagella possess long and hair-like (trichoid) male-specific sensilla that house ORCs specialized to detect components of the conspecific female's sex pheromone (Sanes and Hildebrand, 1976; Kaissling et al., 1989). Females also possess trichoid sensilla (i.e., trichoid type-A) (Fig. 1C, D). They are much shorter than in males and respond to plant volatiles. In addition, both male and female antennae carry several other types of olfactory sensilla, some of which resemble short pegs, thought to contribute to the detection of plant-associated odorants. In total, there are five types present in males (Sanes and Hildebrand, 1976; Lee and Strausfeld, 1990) and six types present in females (Shields and Hildebrand, 1999a; 1999b). In the case of trichoid type-A sensilla, circumferential cuticular ridges are present which form a helical pattern (Fig. 1E). The cuticular shaft of all olfactory sensilla is pierced by a multitude of pores (Fig. 1E). These pores extend through the entire thickness of the shaft. It is generally believed that olfactory molecules gain entry through these pores and, in order to reach the dendritic receptor sites of an ORC, the small volatile lipid-soluble odor molecules must traverse an aqueous phase, perhaps with the aid of odorant-binding proteins. The binding of these odorant ligands to odor-specific receptors coupled to G-proteins initiates a cascade of intracellular second messengers (e.g. cyclic AMP, IP_3) that ultimately activate cyclic nucleotide-gated cation-permeable channels (e.g. Ca^{2+}, Cl^-, K^+) (reviewed by e.g. Firestein, 1992; Shepherd, 1994). These events result in the generation of action potentials in temporal patterns (in each ORC axon) and spatial patterns (across the array of ORC axons) that represent features of the stimulus and travel along the ORC axons to the primary olfactory center in the brain (i.e., the antennal lobe, AL, of insects or the olfactory bulb of vertebrates). The axons of antennal ORCs project to and terminate in compartments of condensed synaptic neuropil (glomeruli) in the AL (Hildebrand and Shepherd, 1997). Mounting evidence indicates that glomeruli are discrete anatomical and functional units, each dedicated to collecting and processing olfactory information about a subset of odor molecules (see below). Information about the odor stimulus is conveyed to a particular glomerulus by the axons of ORCs that express a particular olfactory receptor protein (Buck, 1996; Hildebrand and Shepherd, 1997; Mombaerts, 1996).

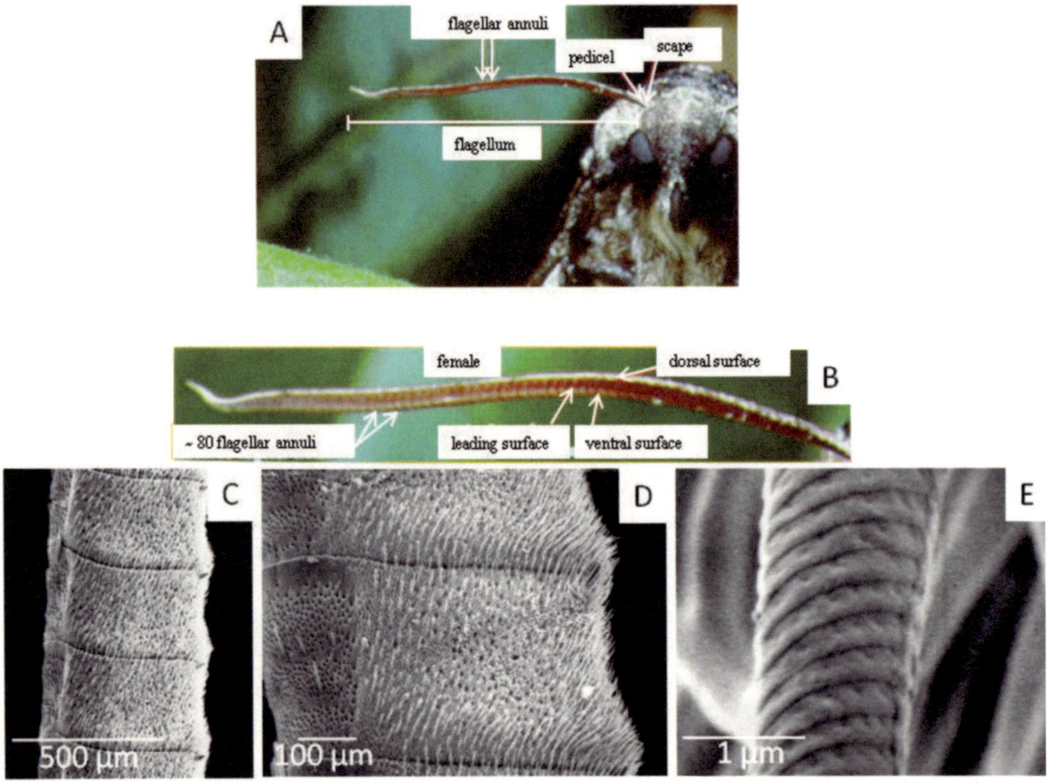

Fig. 1. Female *Manduca sexta* antenna. (A) Light micrographs showing a female *M. sexta* moth and its antenna. The antenna comprises three segments, two small, basal segments (scape and pedicel) and a long distal segment (flagellum). The antennal flagellum is about 2 cm long and comprises approximately 80 subsegments called annuli (or flagellomeres). (B) Higher magnification view of (A) showing the multitude of annuli. (C) Scanning electron micrograph of a portion of an adult female antennal flagellum showing a higher magnification of three annuli. (D) Higher magnification view of a single annulus. Long, hair-like sensory organs (sensilla) called trichoid sensilla are abundant on the surface of an annulus. (E) Higher magnification view of a single trichoid type-A sensillum showing the cuticular shaft and pores that extend through its entire thickness. The shaft bears circumferential cuticular ridges which form a helical pattern. Odorant molecules diffuse through these pores and interact with the underlying dendrites.

2.3 Peripheral gustatory processing

Insect larvae depend largely on their sense of taste and smell to find food. These larvae possess elaborate sensory organs (i.e., sensilla) located on the antennae and mouthparts that serve to gather olfactory and gustatory information on the chemical composition of the food plant. The feeding response in lepidopterous larvae is controlled by input from gustatory sensilla located on the mouthparts (Schoonhoven and Dethier, 1966; Shields, 1994). Food plant recognition is thought to be primarily mediated by the input from a bilateral pair of styloconic sensilla (Schoonhoven and Dethier, 1966; de Boer et al., 1977; de Boer and Hanson, 1987) located on the mouthparts. Each sensillum houses four GRCs (Figs. 2, 6) that are thought to play a primary role in hostplant discrimination. They have been referred to as the salt- sugar-, inositol-, and deterrent-sensitive cells (Schoonhoven, 1972; Schoonhoven et al., 1992), since they typically respond to salt, sweet, inositol, and bitter compounds, respectively (e.g. Schoonhoven, 1972; Frazier, 1986; Shields and Mitchell, 1995; Bernays et al., 1998; Glendinning et al., 1999). During feeding, the sensilla are in continuous contact with the sap liberated from the plant leaf and are capable of detecting different chemicals (i.e., phytochemicals) present in the plant. This gustatory sensory input is encoded as patterns of nerve impulses by GRCs and this information is then transferred to taste centers in the brain of the insect. Therefore, GRCs form the first layer of the decision-making process that ultimately determines whether food is acceptable or should be rejected. Thus, the insect faces the task to decipher individual tastants in a complex multimolecular mixture and to make appropriate feeding choices.

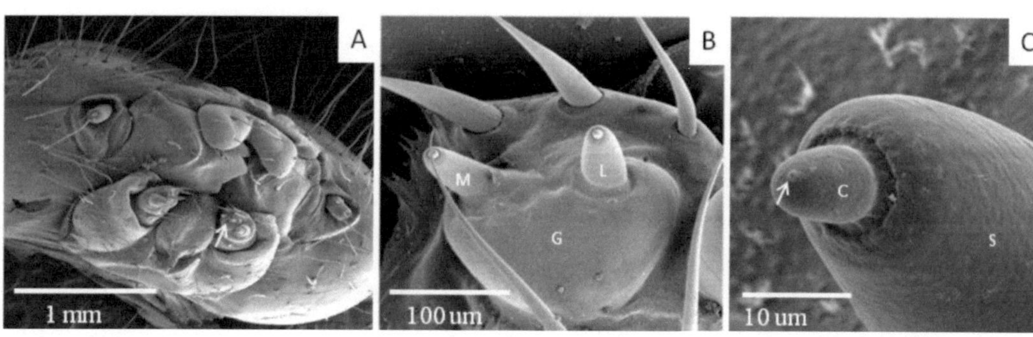

Fig. 2. Scanning electron micrographs showing the (A) whole head of a gypsy moth, *Lymantria dispar*, larva. The arrow denotes the location of the styloconic sensilla. (B) Higher magnification view of the lateral (L) and medial (M) styloconic sensilla located on the galea (G). (C) Higher magnification view of a lateral styloconic sensillum. The sensillum is comprised of a cone (C) or peg inserted into a style (S) or column. A terminal pore (arrow) is visible at the apex of the cone.

Insect GRCs transduce the quality and quantity of the complex plant chemistry into a neural code of action potentials. Complex stimuli resulting from e.g. plant saps often evoke spike trains in several receptor cells innervating one or more sensilla. The frequency of action potentials and the temporal distribution of action potentials in a spike train contain information about the stimulus. The axons of GRCs travel to, and converge in, the first relay station, the subesophageal ganglion (SOG), without intermittent synapses. Unraveling the sensory code occurs by analyzing "input-output" relationships (Schoonhoven and van Loon, 2002) and can be achieved by stimulating specific sensilla and quantifying electrophysiologically the trains of action potentials (input), as well as quantifying the behavior (output) on the basis of how much food is consumed (Bernays and Chapman, 1994).

To better understand the neural communication between the chemosensory organs and the central nervous system that results in acceptance or rejection behavior, sensory responses have been categorized as (1) labeled line system, (2) across-fiber patterning, and (3) temporal patterning. The first theory suggests that the more important a single compound is for controlling or modifying behavior, the more likely its detection will be coded by a single cell (Stadler, 1984). This "labeled line" (i.e., line or axon along which information is transferred to the brain) to the central nervous system would only carry information from cells with a narrow and well defined sensitivity spectrum of a specific chemical (or family of chemicals) and would be directly linked to a specific behavioral response (Schoonhoven and Blom, 1988). Such chemosensory cells seem to be quite unique for specialized herbivorous insects and have not been documented for other animal groups, such as vertebrates. The second theory suggests that the nervous system bases its decision for behavioral output by evaluating the responses from many individual sensory cells with different but overlapping response spectra and the central nervous system extracts meaningful information by reading and processing the simultaneous inputs across all afferent sensory fibers (axons) (across-fiber patterning) (Dethier and Crnjar, 1982), also known to occur in vertebrates (Dethier, 1982). The third theory suggests that temporal patterning may be superimposed on across-fiber patterning, suggesting that the ratios of firing across different cells changes with time and can modify a particular message (Schoonhoven, 1982). Most importantly, it should be noted that all three theories (code types) are not mutually exclusive and can be amalgamated into one model (Schoonhoven et al., 1992).

Sensory codes mediating acceptance can: (i) stimulate specific sugar cells coding for acceptance profile; (ii) stimulate broad spectrum sugar cells that the CNS recognizes as an acceptance profile (Schoonhoven, 1982; 1987), and (iii) inhibit specific phagodeterrent receptors; this contributes to the neural coding of acceptance (Schoonhoven et al., 1998). Feeding deterrents may alter sensory input by: (i) stimulating specific deterrent receptors; (ii) stimulating broad spectrum receptors; (iii) stimulating some cells and inhibiting others, thereby changing complex and subtle codes; (iv) inhibiting specific phagostimulant receptors; this contributes to the neural coding of deterrence, and (v) evoking highly unnatural impulse patterns, often at high frequency (Schoonhoven et al., 1998). The ability of a deterrent neuron to respond to a wide range of chemicals is due to it having a diverse range of receptor sites, each with its own structure-function specificity, or due to the active chemicals having common features making them able to interact with a single receptor site (Blaney et al., 1988).

Deterrent cells possess a number of unique characteristics: (i) they generally adapt more slowly than cells which respond to phagostimulatory compounds; (ii) the tonic activity of the deterrent receptor stabilizes at a higher level than in other cell types; (iii) there may be a relatively long latency period prior to the tonic response; (iv) there may be a slow increase in spike frequency following stimulus application, and (v) there may be an increase in spike amplitude with stimulus concentration (Schoonhoven, 1982); Hanson and Peterson, 1990). Schoonhoven (1982) used differential adaptation rates to explain that the sensory code changes with time, with the result that the deterrent receptor activity gradually becomes more pronounced in the sensory message sent to the brain. Food, which at the beginning of a meal may be acceptable, soon becomes unacceptable because of the more prominent share of the deterrent in the total sensory impression.

Recently, work by Wanner and Robertson (2008) revealed a family of 65 gustatory receptor (Gr) genes from the silkworm moth, *Bombyx mori*, genome. These authors revealed Gr genes for sugar, as well as those for cuticular hydrocarbons and carbon dioxide. Interestingly, they also found 55 Gr genes that are predominantly bitter receptors involved in the detection of a large variety of secondary plant chemicals and suggested that these Gr genes mediate food choice and avoidance, as well as oviposition site preference. This finding may provide new tools for controlling pest damage and lead to better understanding of the peripheral taste system and is noteworthy since about 99% of the 150,000 described species of lepidopterous insects are phytophagous feeders (Grimaldi and Engel, 2005). Functional characterization of GRCs may also provide a better understanding of the molecular and cellular basis of taste coding. Interestingly, Clyne et al. (2000) found a large Gr gene family and characterized the GRCs expressing divergent GR genes in the fruit fly, *D. melanogaster*. In the sequenced *D. melanogaster* genome, 68 receptors, encoded by 60 genes, were identified and were predicted to encode G protein-coupled receptors (Robertson et al., 2003). Expression and behavioral studies of two *D. melanogaster* Gr genes, *Gr5a* and *Gr66a*, revealed that inactivation of *Gr5a*-positive neurons resulted in a diminished behavioral responses to sugars and low concentrations of salt, whereas inactivation of Gr66a-positive neurons lowered behavioral responses to some bitter compounds (Thorne et al., 2004; Wang et al., 2004). Interestingly, molecular studies revealed that these "sugar" and "bitter" neurons also project to distinct and non-overlapping regions within the SOG (Dunipace et al., 2001; Scott et al., 2001; Thorne et al., 2004; Wang et al., 2004; Dahanukar et al., 2007).

3. Olfactory extracellular recording methods

A powerful tool for studying olfactory ORCs lies in physiologically recording from individual sensilla using a recording technique termed single-unit or single-sensillum recording (Figs. 3, 4). This technique monitors the electrical events elicited by ORCs when stimulated by different odor stimuli. Despite the fact that multiple GRCs are present within a single sensillum, the distinct electrophysiological responses of each ORC can be distinguished using specialized computer-aided software for spike sorting by differing spike amplitudes. Knowledge of the number of ORCs present within a particular sensillum, as visualized with transmission electron microscopy, can greatly contribute to a better interpretation and verification of electrophysiological results.

To record from olfactory ORCs to test the effect of volatile compounds, female *M. sexta* moths reared on artificial diet (17h light: 7h dark; ca., 60% relative humidity), 1-2 days post-eclosion were used for these studies (Bell and Joachim, 1976; Sanes and Hildebrand, 1976). Recordings were carried out at ambient temperatures (24-26°C). To record extracellulary from ORCs, the cut-tip recording technique was used (Van der Pers and Den Otter, 1978; Kaissling, 1995) (Fig. 4B). This technique involved restraining each moth in a plastic tube (5.2 x 1.2 cm i.d.) so that the head protruded from one end (Fig. 3). One of the moth's antennae was stabilized with a minimal amount of low melting point paraffin wax to allow for easier manipulation of the sensilla. Alternatively, double-sided tape can be used. The tip of a trichoid type-A sensillum, positioned on the distal or proximal margins of the dorsal, ventral, and leading surfaces of an annulus, was cut between two sharpened glass knives (2-

mm diameter, Corning Inc., Horseheads, NY) positioned at 90 degrees with respect to each other (micro knife-beveling device; type EG-03, Syntech, Hilversum, The Netherlands) (Van der Pers and Den Otter, 1978; Kaissling, 1995). This method is typically feasible for sensilla less than 50 µm in length. For longer sensilla, an alternative method can be used, whereby the sensillum is cut by pinching off the tip using sharpened forceps (Fig. 4A), thus preventing the opening and subsequent damage of the ORC dendrites due their depolarization. The flat side of the forceps is coated with a thin film of Vaseline to prevent desiccation of the sensillum (i.e., loss of the sensillum lymph surrounding the dendrites) (Kaissling, 1995).

Recording and ground electrodes (tip diameters: 2.5-3.5 µm and 4-5 µm, respectively) were made from borosilicate glass tubing (1.5 mm o.d., 1.10 mm i.d., Sutter Instrument Co., Novato, CA) on a Flaming-Brown micropipette puller (model P-97, Sutter, Novato, CA). The recording electrode was filled with sensillum-lymph saline solution (Kaissling and Thorson, 1980) and positioned over the tip of a trichoid type-A sensillum with the aid of a micromanipulator (Leitz) and a Wild M3Z Kombistereo microscope (Leica, Germany; total magnification, 640X) (Figs. 3, 4). The ground electrode was filled with hemolymph saline solution (Kaissling and Thorson, 1980) and inserted into the moth's eye or body. Both ground and recording electrodes were held in place by mounting them in electrode holders (1.5 mm i.d., Syntech) containing 0.25 mm silver wire (>99.99 %, Aldrich Chemical Co., Milwaukee, WI) (Fig. 3). Difficulties in cutting the tip of olfactory sensilla can arise if they are very short and small. In this case, a modification of the tip-cutting technique can be used called sidewall recording. In this method, a sharpened tungsten electrode is plunged into the base of the sensillum to establish contact with the underlying olfactory ORCs. In either case, these electrophysiological recording techniques record the contribution of one or all of the ORCs present within a sensillum and preparations can remain stable for several hours. The electrophysiological components of the set-up were placed on a vibration-free air table (Figs. 3).

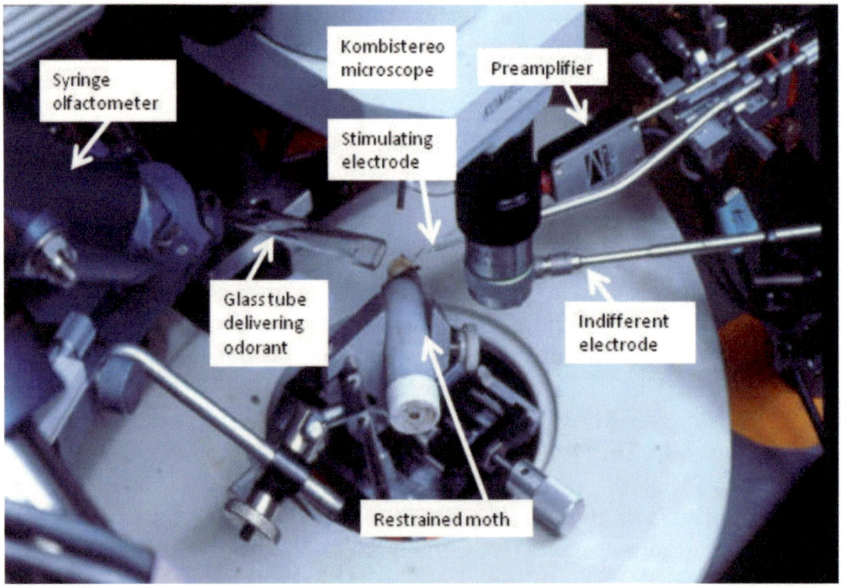

Fig. 3. Typical olfactory extracellular recording set-up.

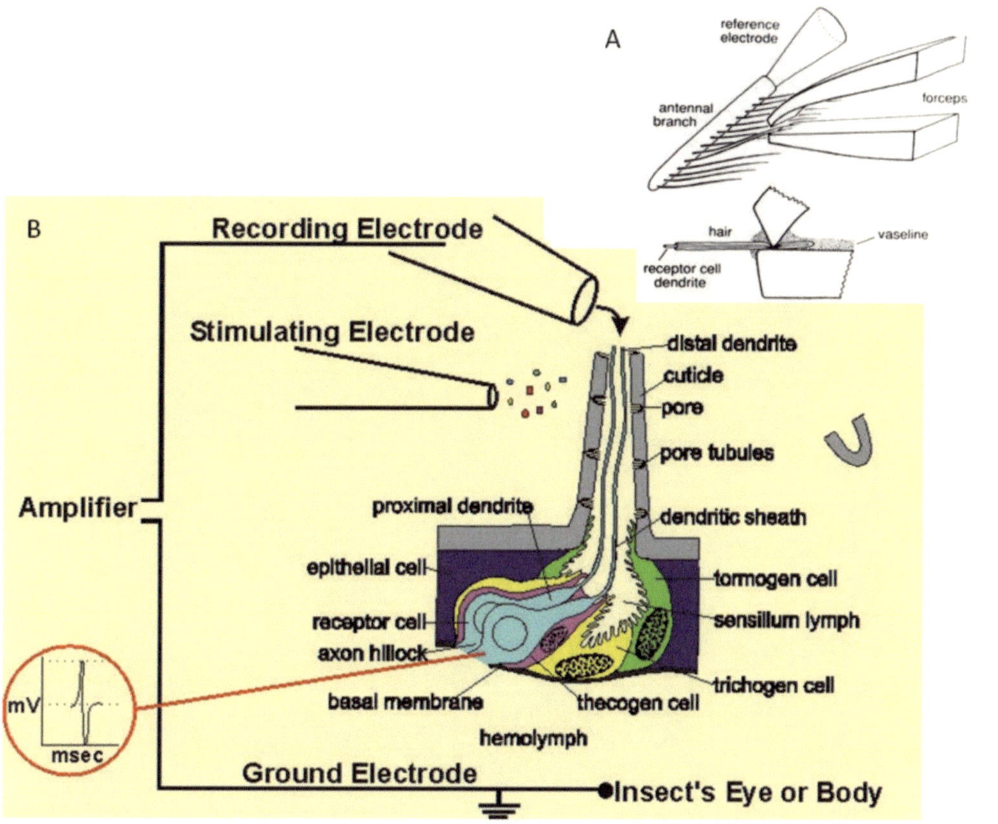

Fig. 4. Extracellular recording technique. (A) After the sensillum is cut, shown here using sharpened forceps, the recording electrode (B) is placed over the tip of the sensillum. The stimulating electrode, containing the odorant, is used to deliver the odorant to the sensillum. Odor stimuli are presented randomly to the sensillum. Action potentials are amplified, recorded, and relayed to a computer for further analyses. Fig. (A) was taken from Kaissling, 1995.

4. Gustatory extracellular recording methods

Randomly selected, fifth instar, 12-24 hours post-molt, *Lymantria dispar* larvae, reared on artificial diet (12h light: 12h dark; ca., 60% relative humidity) (Shields et al., 2006) were used for all experiments. The larvae were food deprived at least 12 hours prior to the experiments and were naive to the test compounds. Recordings were carried out at ambient temperatures (24-26°C). Each larva was transected just behind the head and a blunt-tipped saline-filled glass electrode was inserted into the head with sufficient force to cause the eversion of the lateral and medial styloconic sensilla. This pipette was filled with an electrically conductive solution (typically KCl or NaCl) and served as the ground or indifferent electrode (Figs. 5, 6). The cut end of the head was sealed with a minimal amount of melted bee's wax (Fig. 5A). The stimulating or recording electrode was filled with a test solution and was positioned over the tip of a single styloconic sensillum (Figs. 5A, 6) with the aid of a micromanipulator (Leitz) under visual control on a vibration-free air table (Fig. 5B). The preparation lasted, on average, one to two hours.

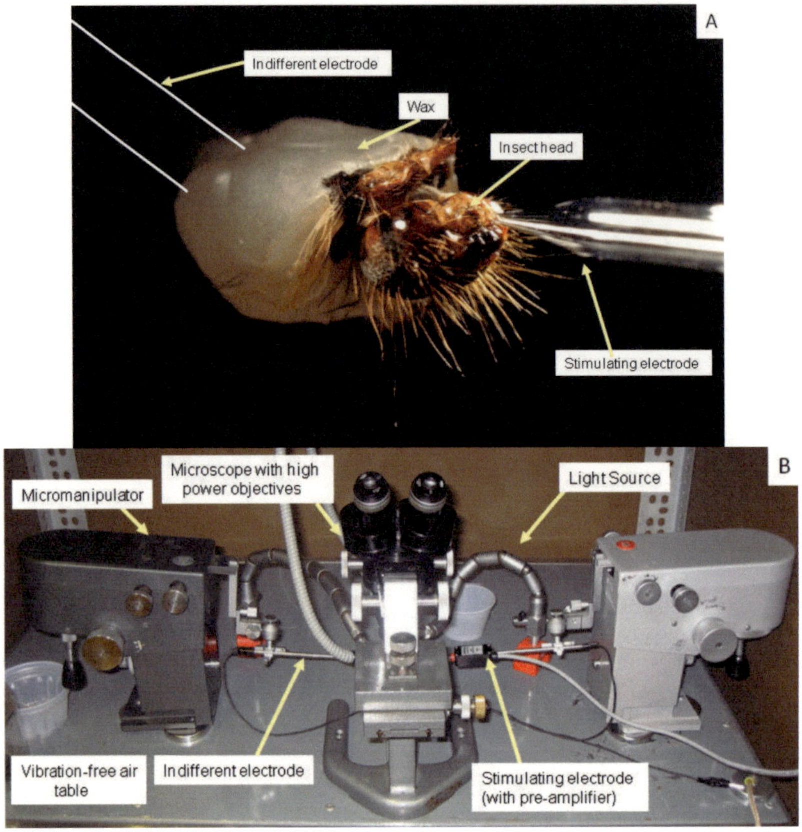

Fig. 5. (A) Higher magnification view of the insect preparation. (B) Gustatory extracellular recording set-up. Fig. (A) was modified from Shields and Martin, 2012.

5. Olfactory stimuli and odor delivery

Olfaction is the principal sensory modality through which insects locate their food sources, mates, and oviposition sites. Over the past three decades, or so, moth olfaction has focused primarily on mechanisms through which male moths detect, process sensory information about, and respond behaviorally to the sex pheromones emitted by conspecific females. In contrast, relatively little is known about similar mechanisms with respect to non-pheromonal odors, such as plant volatiles and the mechanisms by which female moths detect and discriminate plant-associated volatiles (odorants) for foraging and oviposition purposes and how this information is processed by the olfactory system.

Previous studies have indicated that scent emitted by sphinx moth-pollinated flowers (Knudsen and Tollsten, 1993; Raguso and Willis, 1997; Raguso and Light, 1998), as well as floral and vegetative volatiles of tobacco, tomato, and other hostplants (Andersen et al., 1988; Buttery et al., 1987a; 1987b) attract adult *M. sexta* for feeding (Morgan and Lyon, 1928; Yamamoto et al., 1969; Raguso et al., 1996; Raguso and Light, 1998). Flowers having a "white floral" scent, such as that associated with many night-blooming moth-pollinated flowers, possess acyclic terpene alcohols (e.g. linalool, nerolidol, and farnesol), as well as the corresponding hydrocarbons, aromatic alcohols, and esters derived from them, in addition to esters of salicylic acid (Knudsen and Tollsten, 1993). Also thought to play a significant role in the attraction of insects to, and their recognition of, their hostplants are various plant-associated odorants, such as green-leaf volatiles (mainly saturated and unsaturated C_6 alcohols and aldehydes), terpenoids, and benzenoid compounds (e.g. Boeckh, 1974; Visser and Avé, 1978; Renwick, 1989; Heath et al., 1992; Knudsen and Tollsten, 1993; Raguso and Willis, 1997).

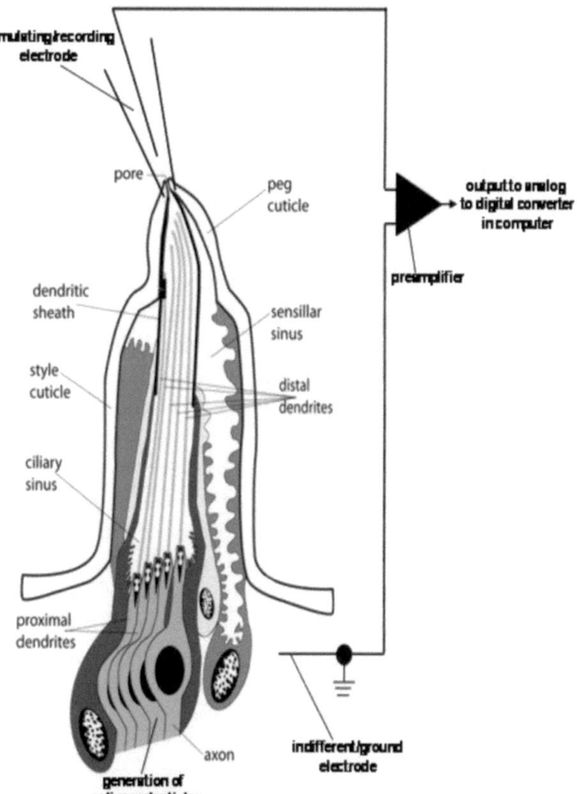

Fig. 6. Schematic diagram showing extracellular recording from a single gustatory styloconic sensillum showing the stimulating electrode containing the stimulus positioned over the tip of the sensillum. This figure was taken from Shields and Martin, 2010.

The majority of the selected, purified, and synthetic odorants tested (chemical purity of 95-99.9%) (Shields and Hildebrand, 2001a) represented floral headspace volatiles of native, night-blooming flowers, such as *Datura wrightii* (jimson weed), *Hymenocallis sonorensis* (spider lily), and *Oenothera caespitosa* (evening primrose), to which *M. sexta* and other sphinx moths are attracted for nectar-feeding and were available from Sigma (St. Louis, MO) or Aldrich (Milwaukee, WI). Complex blends of volatiles emitted by the headspace inflorescence from *D. wrightii* (i.e., jimsonweed) and the foliage of two solanaceous plants (i.e., *Lycopersicon* (tomato) and *Nicotiana* (tobacco) species were also tested. This was carried out by placing the entire inflorescence or two to three undamaged leaves in a 20 ml capped syringe, which was allowed to equilibrate at room temperature for 24 h prior to use. Tomato and tobacco foliage are preferred by female *M. sexta* for oviposition (Yamamoto et al., 1969), while jimsonweed provides a nectar feeding source.

For testing individual compounds, each odorant was dissolved in odorless mineral oil (light white oil, Sigma, St. Louis, MO) at a dilution of 1:10 (v:v). A 30-µl aliquot was pipetted onto a 1.5-cm² piece of grade-1 filter paper (Whatman, Kent, ME) and inserted into a disposable 20 ml syringe. Control syringes, loaded with 30 µl of mineral oil, were also prepared. All syringes were capped and allowed to equilibrate, similar to that described above. After each experimental session, the capped odorant syringes were stored at 4°C. Some odorants eliciting the strongest responses were tested in dose-response experiments. For these experiments, dilutions were prepared in decadic steps (v:v) over at least four log units from stock solutions. Stimulus loads ranged from 3 x 10^{-4} to 3 µl of the test compound (approximately 0.3 µg to 3 mg). Odorants were tested in order of increasing concentration to prevent adaptation of the ORCs. Also included among the odorants, were two key components of the female moth's sex pheromone, E10,E12-hexadecadienal (bombykal) and E11,Z13-pentadecadienal ("C15"), a relatively stable mimic of E10,E12,Z14-hexadecatrienal (see below) (Kaissling et al., 1989). Each compound was dissolved in *n*-hexane and applied directly (without mineral oil) at 1 ng·μl^{-1} and 10 ng·μl^{-1} to the filter paper.

The odor was delivered to the female antenna using a stimulating device (type CS-01, Syntech) (Shields and Hildebrand, 2001a). The tip of the syringe containing the odorant was introduced into a 20 cm long, L-shaped glass tube (1 cm i.d.) whose open end was positioned 5-7 mm from the test antenna (Fig. 3). The end of the tube was flattened to 0.5 cm to correspond with the length of the antenna. The antenna was continuously flushed with a 5 cm·s^{-1} stream of charcoal-filtered, humidified air. To deliver a stimulus, 2 ml of odor-laden air was injected at a constant rate over 200 ms from the syringe into the airflow using a motor-driven syringe olfactometer, resulting in increased airflow of 20 cm s^{-1} (measured by a thermo-anemometer) (Fig. 3). A glass funnel (3.5 cm i.d.) attached to an air evacuation line was positioned near the preparation to draw away any odor-bearing air following stimulus delivery. Odor stimuli were presented randomly to the sensilla to be tested. Blank controls were tested repeatedly, but never elicited a response from the ORCs. An interstimulus interval of at least 60 s, or longer, if necessary for spontaneous activity of the ORC to return to their initial value, was allowed between stimulations.

6. Olfactory responses from trichoid type-a sensilla

Alternating current (AC) signals were recorded for 6 s, starting 1 s before stimulation (olfactory experiments) and for 10 s, beginning 0 s before stimulation. In either case, these signals were pre-amplified 10X using a Syntech Probe run through a 16-bit analog-to-digital interface (IDAC-02, olfactory, or IDAC-4, gustatory) (Syntech), and then analyzed off-line with Autospike software (Syntech). This software allowed the user to more easily interpret how many cells were responding to specific compounds, as determined by their size, shape, and firing frequency of the biphasic action potentials. The net number of spikes within the first 500 ms after stimulus delivery was considered to be a measure of the strength of excitation and was calculated as the mean number of spikes within 500 ms after the termination of stimulus delivery minus the mean number of spikes within 500 ms before stimulus onset. The responses of ORCs were classified as excitatory (Figs. 7A-F, 7H-M, 8A-N, 8P, 8Q) or inhibitory (Figs. 7G, 8O) if there was cessation in spiking activity for more than 500 ms after stimulation. Dose-response experiments were conducted after the general screening of the odorants was completed and only if the response of an ORC remained constant throughout the experimental session. In order to increase the accuracy of the data analyses, vapor pressure of a particular odorant was taken into consideration when analyzing the sensitivities of selected ORCs.

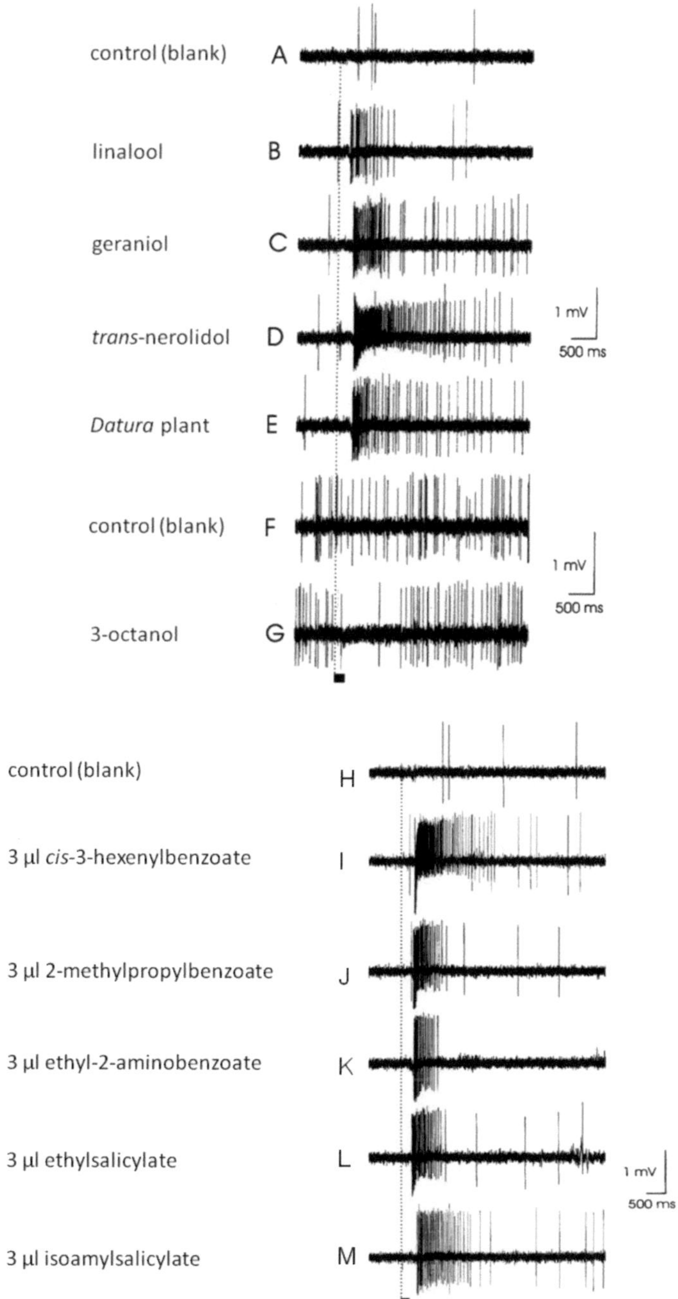

Fig. 7. Representative extracellular olfactory electrophysiological recordings of from three different olfactory receptor cells (ORCs) from female *Manduca sexta* trichoid type-A sensilla showing excitatory (A-F, H-M) and inhibitory (G) responses. The responses in (F) and (G) are from a different ORC than those in (A-E) and (H-M). The stimulus bar is represented by a filled rectangle and represents 200 ms and represents the onset of odor delivery. This figure from modified from Shields and Hildebrand, 2001b.

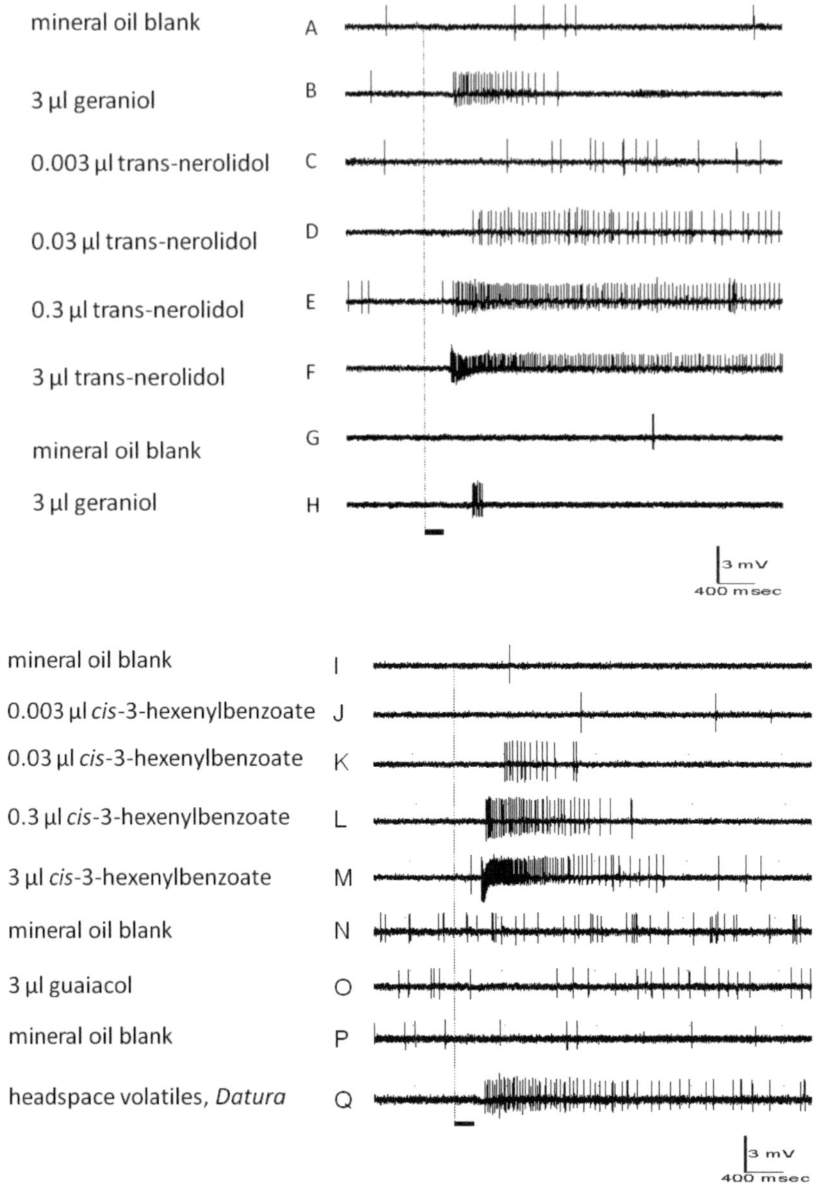

Fig. 8. Representative extracellular olfactory electrophysiological recordings from five different olfactory receptor cells (ORCs) from female *Manduca sexta* trichoid type-A sensilla showing excitatory responses (A-N, P, Q) and inhibitory (O) responses. Note the increasing excitatory activity of the ORC to increasing concentrations of trans-nerolidol (C-F) and *cis*-3-hexenylbenzoate (J-M). Note the two different types of excitatory responses to geraniol (B and H). The stimulus bar is represented by a filled rectangle and represents 200 ms and represents the onset of odor delivery. This figure from modified from Shields and Hildebrand, 2001a.

7. Staining and mapping of ORC axonal projections

The axons of ORCs project to and terminate in compartments of condensed synaptic neuropil (i.e., glomeruli) in the primary olfactory centers in the CNS of invertebrates and vertebrates (Hildebrand and Shepherd, 1997). Each olfactory glomerulus is a discrete anatomical and functional unit and represents a specific anatomical "address" dedicated to collecting and processing specific molecular features about the olfactory environment, conveyed to it by ORC axons expressing specific olfactory receptor proteins (Buck, 1996; Buonviso and Chaput, 1990; Christensen et al., 1996; Hildebrand and Shepherd, 1997; Mombaerts, 1996). Over approximately the past two decades, there has been mounting evidence that the arrays of glomeruli in the ALs of insects and the olfactory bulbs of vertebrates are organized chemotopically (e.g., Sharp et al., 1975; Rodrigues and Buchner, 1984; Hansson et al., 1992; Mombaerts, 1996; Friedrich and Korsching, 1997; 1998; Galizia et al., 1999), analogous to visuotopy, in visual systems, and tonotopy, of auditory systems.

The axons of antennal ORCs project via the antennal nerve to the ipsilateral AL in *M. sexta*. In the ALs, they form synapses with processes of a subset of the approximately 1,200 central neurons (Homberg et al., 1988; Christensen et al., 1995; Rössler et al., 1998; 1999). In both sexes of *M. sexta*, each AL bears 60 ordinary, sexually isomorphic glomeruli and three sexually dimorphic glomeruli (Rospars and Hildebrand, 2000). Male *M. sexta* moths bear three prominent glomeruli (cumulus, toroid 1, toroid 2), which constitutes the male-specific macroglomerular complex (MGC). This complex processes information about the conspecific female's sex pheromone (Hansson et al., 1991; Heinbockel et al., 1999; Rospars and Hildebrand, 2000) (discussed in more detail, below). Interestingly, females also bear two homologous, sexually dimorphic glomeruli (i.e., large female glomeruli or LFG, lateral and medial) (Rössler et al., 1998; 1999; Rospars and Hildebrand, 2000). Central neurons with arborizations leading to the lateral LFG were found to display a preferential response to linalool and certain other monoterpenoids (King et al., 2000). LFGs have been implicated in being involved in olfactory information attributed to the interactions of females with hostplants or with courting males.

To determine where the axons of ORCs of trichoid type-A sensilla project in the AL of female *M. sexta*, we performed anterograde labeling using dextrantetramethylrhodamine of ORCs in groups of 5-10 sensilla on various surfaces of a single annulus in the middle of an antennal flagellum (Fig. 1C, D). This was carried out by restraining an adult female moth in a plastic tube (Fig. 4) and cutting the tips of 5-10 trichoid type-A sensilla (Fig 4); (see Shields and Hildebrand, 1999a; Shields and Hildebrand, 2001) using the method outlined, above. In order to stain the ORCs associated with these cut sensilla, we created a small well from melted paraffin wax around the selected annulus and filled it with a small amount of 1% solution of dextran-tetramethylrhodamine (3000 MW, anionic, lysine-fixable, D-3308, Molecular Probes, Eugene, OR) until the annulus became completely submerged. The area was then covered with a small amount of petroleum jelly. We transferred the preparation to a humid chamber and kept it in darkness for 2-3 days. Following this period, we excised the brain and fixed it for 24-48 h in 2.5% formaldehyde solution in 0.1 M sodium phosphate buffer (pH 7.4) containing 3% sucrose. Following this, the tissue was dehydrated in a graded ethanol series, cleared in methyl salicylate, and viewed as a temporary whole mount in a laser-scanning confocal microscope.

We found that a majority of axonal projections from these ORCs terminated in the two LFG located in the dorsolateral region of the ipsilateral (AL) (Fig. 9A), more specifically, near the entrance of the antennal nerve into the AL. We also found that in addition to the LFGs, a subset of the other 60 spheroidal, ordinary, sexually isomorphic glomeruli received sparse projections of a subset of ORC afferent axons (Fig. 9). The results of these anatomical studies and our electrophysiological results that some trichoid type-A sensilla are tuned mainly to terpenoids and aromatic esters, provide a basis to hypothesize that information about odorants belonging to those chemical classes is processed in the LFGs (King et al., 2000; Shields and Hildebrand, 2001a; b). To improve visualization (i.e., resolution) of successfully labeled preparations, samples were embedded in Spurr's low-viscosity embedding medium (Electron Microscopy Sciences, Fort Washington, PA) and sectioned at 48 μm. The preparations were then viewed in a laser scanning confocal microscope (Bio-Rad MRC-600; Cambridge, MA) equipped with a Nikon Optiphot-2 microscope and both 15-mW krypton-argon and 100-mW argon laser light sources and YHS filter cube (excitation wavelength 568 nm). Serial optical sections were collected at 3-μm intervals through the whole mount or 2 μm intervals from the embedded sectioned preparations (Fig. 9). Image processing and analysis were performed using Confocal Assistant 4.02 (copyrighted by Todd Brelje, distributed by Bio Rad, Cambridge, MA), Corel Photopaint 8, and Corel Draw 8 (Corel Corporation, Ottawa, Ontario, Canada).

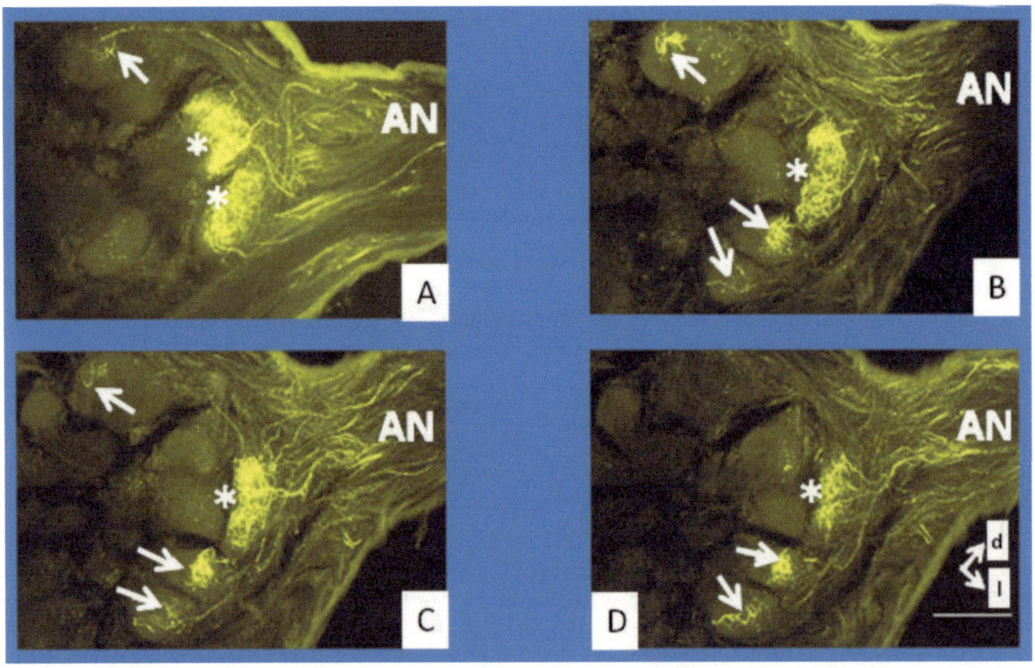

Fig. 9. Confocal microscopic images (A-D) showing serial optical sections of female *Manduca sexta* specimens embedded in plastic and sectioned to improve resolution taken at different depths through the antennal lobe. Images show the central projections of axons from olfactory receptor cells from trichoid type-A sensilla stained with the fluorescent dye dextran-tetramethylrhodamine. Askterisks indicate the sexually dimorphic large female glomeruli located in the dorsolateral region of the antennal lobe, near the site of entry of the antennal nerve (AN). The arrows indicate ordinary sexually isomorphic glomeruli. d, dorsal; l, lateral. Scale bars = 100 μm. Figs. (A) and (B) were modified from Shields and Hildebrand, 2001a.

8. Gustatory stimuli and responses from the medial styloconic sensilla

For all gustatory experiments, stimulus compounds were dissolved in 30 mM potassium chloride (KCl) (control) (Fisher Scientific, Fair Lawn, New Jersey) in distilled water to enhance the electrical conduction of the recording electrode and to improve the signal-to-noise ratio. This inorganic salt was chosen since the hemolymph of plant-feeding (i.e., phytophagous) feeders typically shows high K^+ and low Na^+ concentrations (Kaissling, 1995). KCl was also used to fill the indifferent electrode (Figs. 5, 6). Selected carbohydrates (i.e., sucrose and inositol), as well as alkaloids (Fig. 10) were tested in this study to observe

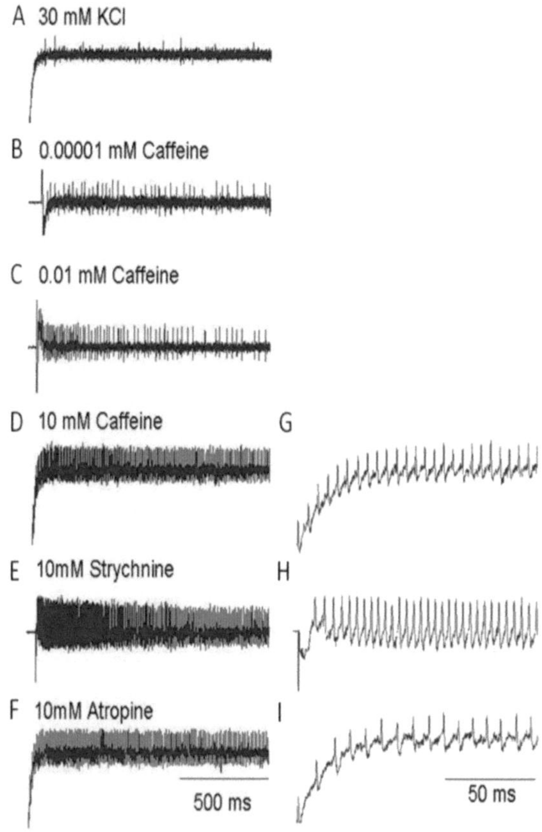

Fig. 10. Representative extracellular gustatory electrophysiological recordings from the medial styloconic sensillum of *Lymantria dispar*. The deterrent-sensitive cell responds to (A) potassium chloride (control) and to the alkaloids, (B-D) caffeine, (E) strychnine, and (F) atropine. The recordings in G, H, and I are higher magnifications of the action potentials from the deterrent-sensitive cell to (D) caffeine, (E) strychnine, and (F), atropine, respectively. The deterrent-sensitive cell displays a typical phasic-tonic response (i.e., a transient pattern of firing changing to that of a more sustained pattern of firing) to these alkaloids. This feature is more apparent at higher concentrations (compare (C) and (D) with (B)) and is shown for caffeine. A weaker concentration of caffeine (B) elicits fewer spikes from the deterrent-sensitive cell. This figure was modified from Shields and Martin, 2010.

the responses of the deterrent-sensitive, inositol-sensitive, sucrose-sensitive, and salt-sensitive cells (see also Shields and Martin, 2010; 2012). The alkaloids were diluted in 10% ethanol solution, which was added to the electrolyte solution. Ethanol, at this concentration, was found to have no discernible effect on the activity of cell(s) (referred to as KCl-sensitive cell(s)) (see below) responding to the control (i.e., KCl) alone. Each stimulus to be tested was applied to the tip of the styloconic sensillum by way of the stimulating electrode (Figs. 5, 6).

Electrical activity from individual styloconic sensilla was amplified and conditioned (bandpass filter set at 100-1200 Hz) prior to being digitized by a 16-bit analog-to-digital interface (IDAC-4, Syntech). For each electrophysiological recording, we stimulated a sensillum for total of 10 sec. Action potentials generated 50 ms after contact with the sensillum were analyzed off-line using a computer equipped with Autospike (Syntech).A pause of at least 2 minutes was allowed between successive stimulations to minimize any carry-over effects. For dose-response experiments, only those phytochemicals that elicited a robust response were tested. Test stimuli were presented to the sensillum from lowest to highest concentration. To ensure a reproducible response to a particular concentration, 1-2 replications of each were made. The number of action potentials generated 0.05 s after the contact artifact was quantified in 100 ms increments.

We found two cells (i.e., one large spike amplitude and one small spike amplitude) in the medial styloconic sensillum that responded to KCl (Fig. 10A). Another cell in the same sensillum responded robustly to various alkaloids (i.e., deterrent-sensitive cell) (Fig. 10B-F). This cell displayed a typical phasic-tonic response (i.e., the initial high firing frequency of the cell changed rapidly to a more sustained pattern of firing). The ability of the deterrent-sensitive cell to respond to more than one alkaloid may be due to either the presence of multiple receptor sites, each with their own structure-function specificity, or a broadly tuned single receptor site (Blaney et al., 1988). These results correlated well with feeding behavioral studies using gypsy moth larvae (Shields et al., 2006), where it was shown that an increase in feeding deterrency (decrease in consumption) occurred with increasing alkaloid concentration.

9. Intracellular recording methods

ORCs in the antenna send olfactory information as trains of action potentials to the ipsilateral ALs of the insect brain where the axon terminals of ORCs synapse onto central neurons in neural structures known as olfactory glomeruli. Central neurons in the ALs, such as projection neurons and local interneurons, can be characterized by intracellular recordings with sharp electrodes. These glass electrodes are filled with a physiological solution that mimics the intracellular fluid of the recorded neuron. In addition, the electrodes can contain intracellular markers such as fluorescent dyes. The development of new intracellular markers provides the basis for rapid and complete reconstruction of individual neurons with little or no toxicity to the neuron. For examples, central neurons are injected iontophoretically with Lucifer yellow, neurobiotin, or biocytin. Brains are then dissected and fixed overnight in formaldehyde with sucrose in phosphate buffer. To visualize biocytin-injected neurons, brains are incubated with, e.g., Cy3-conjugated streptavidin. After subsequent histological processing, neurons are further investigated by laser-scanning confocal microscopy. This approach allows the study of both the physiological as well as morphological properties of the recorded neuron in a relatively undisturbed *in vivo* preparation as described below.

9.1 Olfactory neurobiology

Female moths release sex pheromones that attract conspecific males over long distances (Kaissling, 1987; Hildebrand, 1996). An outstanding challenge in olfactory neurobiology is to understand how a male moth is able to locate a mate, namely, a conspecific female releasing sex-pheromone. The olfactory brain of a male moth must integrate information about qualitative, quantitative, and spatiotemporal features of an attractive blend of volatile compounds, the sex pheromone, released by a conspecific female (Hildebrand, 1995; 1996; Christensen et al., 1996). Central processing of this information occurs in a sexually dimorphic cluster of olfactory glomeruli, the MGC, in the male moth's AL. Since information about the sex-pheromone is primarily processed in the MGC, it serves as a model for studies of the functional architecture of glomeruli, as well as the physiological relationships between glomeruli in the olfactory system (Hansson and Christensen, 1999; Christensen and White, 2000; Christensen and Hildebrand, 2002; Reisenman et al., 2008; Lei et al., 2010).

Several species of moths use the same chemical compounds in their sex-pheromone blend, but the attractant signal produced by each species is nevertheless unique because each blend has characteristic proportions of the components (Arn et al., 1992; Kaissling, 1996). Chemical studies revealed that in the sphinx moth *M. sexta*, the sex-pheromone blend comprises eight components. Behavioral data shows that two of the eight components are required to evoke olfaction-modulated flight in males (Tumlinson et al., 1989). The antennae of male *M. sexta* are covered by different types of sensilla (see above). The long trichoid sensilla on the male's antennae house highly selective and sensitive ORCs that detect one or the other of the two key pheromone components (Kaissling et al., 1989). Each of these ORC populations sends their axons to a different glomerulus in the MGC (Christensen et al., 1995). Projection neurons (PNs) that arborize in one of these glomeruli, the toroid 1 [T1] (Strausfeld, 1989; Homberg et al., 1995) respond to antennal stimulation with *E*,*Z*-10,12-hexadecadienal (bombykal or BAL, one of the two key components of the female's sex pheromone). PNs that arborize in a neighboring glomerulus (the cumulus) respond selectively to *E*,*E*,*Z*-10,12,14-hexadecatrienal (the second key component) or its more stable mimic, *E*,*Z*-11,13-pentadecadienal (C15; Kaissling et al., 1989; Hansson et al., 1991). An MGC-PN with arborizations in both glomeruli is activated by both BAL and C15 (Hansson et al., 1991; Heinbockel et al., 2004).

For a number of moth species it has been demonstrated in behavioral studies in the field and in laboratory wind tunnels that pheromone-modulated upwind flight is regulated in a dose-dependent manner (Baker, 1989). Similarly, electrophysiological studies of AL neurons in male *M. sexta* have shown how the responses of MGC-PNs to pheromone components and extracts of the female's pheromone gland are influenced by stimulus intensity (Christensen and Hildebrand, 1990). In a technological advance, neural-ensemble recordings have revealed concentration-dependent patterns of synchronous firing between MGC-PNs responding to the two key pheromone components (Christensen et al., 2000). A growing body of evidence suggests that glomeruli can receive converging excitatory and inhibitory input. Subpopulations of MGC-PNs receive excitatory input driven by BAL and inhibitory input driven by C15 (or vice versa) (Christensen and Hildebrand, 1987; Hansson et al., 1991; Heinbockel et al., 1999; Lei et al., 2002). Input from both excitatory and inhibitory pathways enhances the ability of MGC-PNs to resolve multiple pulses of pheromone such as intermittent odor stimuli found in natural stimulus situations (Christensen and Hildebrand,

1997; Christensen et al., 1998; Heinbockel et al., 1999). Here, we describe how this processing function is optimized at particular stimulus concentrations or ratios in different PNs (Heinbockel et al., 2004). By using intracellular recording and staining methods, we examined the effect of changing these quantitative attributes of the pheromonal stimulus on the responses of identified PNs innervating different glomeruli in the MGC.

9.2 Experimental preparation

Male *M. sexta* were reared on an artificial diet (modified from that of Bell and Joachim, 1976) at 25°C and 50-60% relative humidity under a long-day photoperiod regimen (17h light : 7h dark) as described previously (Sanes and Hildebrand, 1976; Prescott et al., 1977). Adult moths (1-3 days post-eclosion) were immobilized and prepared by standard methods (Christensen and Hildebrand, 1987). With the antennae intact, the head was separated from the rest of the body and pinned in a Sylgard-coated (Dow Corning, Midland, Michigan) recording chamber (volume <0.5 ml) (Heinbockel and Hildebrand, 1998; Heinbockel et al., 1998). Isolation of the head from the rest of the thorax had no detectable adverse effect on neural responses (Christensen and Hildebrand, 1987). Part of the AL was then desheathed with fine forceps to facilitate insertion of the recording electrode. The brain was superfused constantly with physiological saline solution (modified from that of Pichon et al., 1972; ca. 2 ml/min) containing 150 mM NaCl, 3 mM KCl, 3 mM $CaCl_2$, 10 mM TES buffer (pH 6.9), and 25 mM sucrose to balance osmolarity with that of the extracellular fluid

9.3 Electrophysiological recordings and data analysis

Sharp glass microelectrodes for intracellular recording were produced from borosilicate tubing (o.d. 1.0 mm, i.d. 0.5 mm, World Precision Instruments, Sarasota, Florida) with a Flaming-Brown Puller (P-2000, Sutter Instrument Co., Novato, California). The tip of each electrode was filled with a solution of an intracellular stain (see below), and the shaft was filled with filtered (0.2 µm pore size) 2.5 M KCl. Electrode resistance ranged from 60-100 MΩ, measured in the tissue. Movements of microelectrodes were controlled with a Burleigh Inchworm (Model 6000/ULN; Burleigh Instruments Inc., Fishers, New York) attached to a Leitz micromanipulator (Leitz, Wetzlar, Germany). Recordings were made from neurites in the synaptic neuropil of the MGC. Because the site of the electrode impalement in a neuron can affect the amplitude of postsynaptic potentials, impalements targeted the same area of neuropil in all preparations. Typically, one neuron was recorded per animal, except when a second recorded neuron had different pheromone response patterns (excited by BAL vs. excited by C15). Intracellular recording and current injection were carried out in bridge mode with an Axoclamp-2A amplifier (Axon Instruments - Molecular Devices, Sunnyvale, California), and data were initially stored on magnetic tape (Hewlett Packard Instrumentation Tape Recorder 3968A, Palo Alto, California) and subsequently transferred to a computer and analyzed with Experimenter's Workbench (Datawave Technologies Co., Longmont, Colorado) or Autospike (Syntech, Kirchzarten, Germany).

The data presented below were analyzed statistically for differences using one-way analysis of variance (Kruskal-Wallis One-Way Analysis of Variance by Ranks, ANOVA) (Zar, 1984). A multiple-comparison procedure was used to isolate groups from each other (All Pairwise Multiple-Comparison Procedures, Dunn's Method, or Student-Newman-Keuls Method; $P<0.05$) (Sigma Stat, Jandel Scientific Software, Richmond, California, Version 1.0, Statistical Software).

9.4 Intracellular staining and confocal microscopy

Neurons were injected iontophoretically with either Neurobiotin (Vector Laboratories, Burlingame, California; 3-5% in 2 M KCl with 0.05 M Tris buffer, pH 7.4) or biocytin (Sigma, St. Louis, Missouri; 3-5% in 2M KCl with 0.05 M Tris buffer, pH 7.4). Alternating hyperpolarizing and depolarizing current pulses (30 sec, 1 nA) for about 10 min were used to inject either tracer. Brains were then dissected and fixed overnight in 2.5% formaldehyde with 3% sucrose in 0.1 M sodium phosphate buffer. To visualize injected neurons, brains were incubated with Cy3-conjugated streptavidin (Jackson Immuno Research Laboratories, West Grove, Pennsylvania; diluted 1:100 with 0.2 M sodium phosphate buffer containing 0.3% Triton X-100) for 3 days on a shaker at 4°C. The brains were then dehydrated with increasing concentrations of ethanol and cleared in methyl salicylate. Neurons were further investigated by laser-scanning confocal microscopy (Fig. 11) using a BioRad MRC 600 (Bio-Rad, Cambridge, Massachusetts) with a Nikon Optiphot-2 microscope, a Krypton/Argon (15 mW) laser light source, and appropriate dichromatic filter cubes (Sun et al., 1993). Serial 2-µm optical sections were imaged through whole mounts and saved as series of images on disks.

9.5 Olfactory stimulation

We used synthetic pheromone compounds kindly provided by Dr. J.H. Tumlinson (USDA, Gainesville, FL). The following olfactory stimuli were applied: (1) *E,Z*-10,12-hexadecadienal (bombykal or BAL); (2) *E,Z*-11,13-pentadecadienal (C15), a synthetic mimic of the second pheromone component, *E,E,Z*-10,12,14-hexadecatrienal, and hereinafter referred to as the second component; and (3) mixtures of BAL and C15 with various blend ratios. These compounds were dissolved in *n*-hexane and applied to a piece of filter paper (1 x 2 cm), which was inserted into a glass cartridge (acid-cleaned glass syringe barrel), as described previously (Christensen and Hildebrand, 1987). The stimulus load on the filter paper was reported as the number of ng of compound applied in hexane solution. A pulse of charcoal-filtered, humidified air (1,000 ml/min) moving through the cartridge carried the stimulus, roughly proportional to the loading on the filter paper, to the proximal half of the antenna ipsilateral to the impaled AL. The antenna typically was stimulated with five 50-ms pulses from a cartridge at 5 pulses sec^{-1}.

The two key pheromone components used for sexual communication in *M. sexta* are found in a 2:1 ratio in solvent rinses of intact pheromone glands of calling virgin females (Tumlinson et al., 1989). Because the mimic (C15) of the second key component is less potent than the natural component itself, a 1:1 BAL-to-C15 ratio was used to elicit physiological responses comparable to those recorded in response to the pheromone-gland extract, as found in earlier electrophysiological studies (Christensen and Hildebrand, 1987; Kaissling et al., 1989). In our experiments, different blend ratios above and below 1:1 were tested in sequence at intervals of 1 min. For one series of blend ratios, the BAL stimulus load was held constant (1 ng) while the C15 stimulus load was varied (0.01, 0.1, 1.0, and 10 ng). For the second series of blend ratios, the C15 stimulus load was held constant while the BAL stimulus load was varied in a similar fashion. In addition to these stimulus series, an elevated stimulus level (10 ng) of either BAL or C15 and a blend of BAL and C15 (10 ng of each) were also tested.

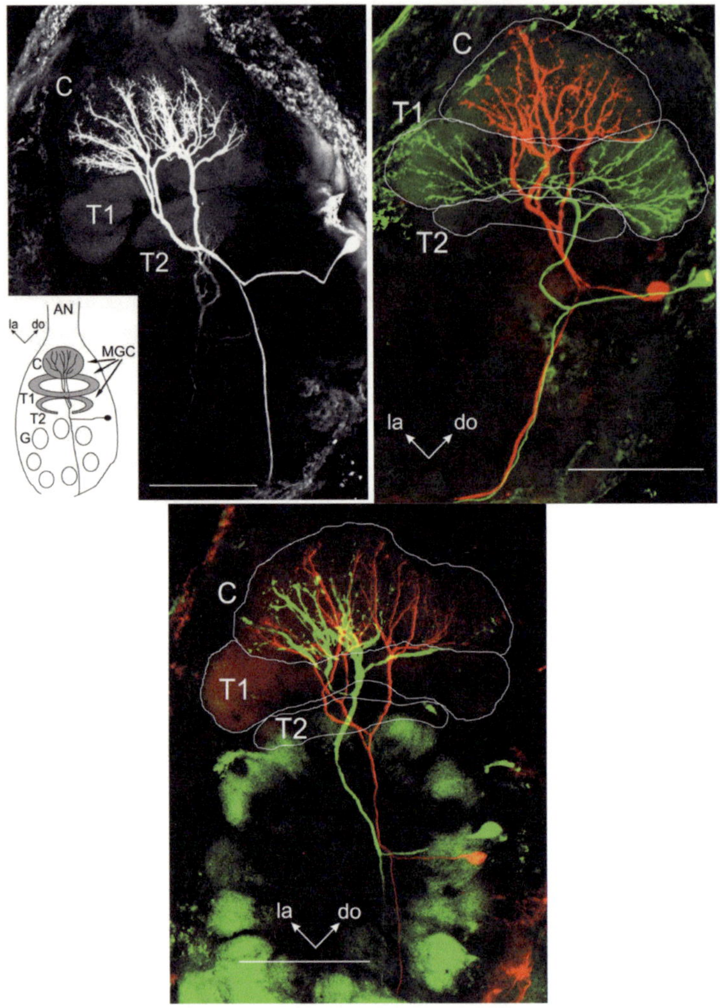

Fig. 11. Laser-scanning confocal micrographs of antennal lobe projection neurons in the moth antennal lobe of male *Manduca sexta*. Left panel: Image of a C15-specialist MGC-PN with arborizations confined to the cumulus. The inset illustrates the organization of the antennal lobe with the macroglomerular complex (MGC) and other glomeruli (G). Right panel: Two specialist MGC-PNs, one neuron, stained with Lucifer Yellow (colored red here), had arborizations confined to the cumulus (C), and the other neuron, stained with biocytin (colored green here) had arborizations confined to the toroid (T1). Areas of apparent overlap between the 2 neurons are shown in yellow and are possible sites of synaptic contact. Bottom panel: Morphological diversity in cumulus neurons. Image of two C15-specialist MGC-PNs with arborizations confined to the cumulus. While the branches of the two neurons apparently overlapped in certain parts of the cumulus (indicated in yellow), other parts were innervated by just one of the two neurons. The green neuron was stained with Lucifer Yellow and the red neuron with biocytin. C - cumulus, T1 - toroid 1, T2 - toroid 2; do, dorsal; la, lateral. Scale bar = 100 μm. Modified from Heinbockel and Hildebrand, 1998 (left panel); Heinbockel et al., 1999 (right and bottom panels).

9.6 Calculating stimulus "tracking" in MGC-PNs

We applied several criteria to assess the ability of PNs to register each odor pulse of a series of pulses with a discrete burst of action potentials. In most cases, a stimulus pulse was scored as "tracked" if the evoked burst comprised at least five action potentials and was separated from the next burst by a period of inactivity of at least 50 ms, which corresponded to the inter-stimulus interval. Alternatively, in cases where PNs showed a higher level of spontaneous activity, the instantaneous spike frequency (ISF) was calculated for each spike train. A given stimulus pulse was scored as "tracked" if (a) the ISF showed a two-fold increase in response to the odor pulse and (b) there was a clear decrease in ISF before the next odor pulse (<50% of the maximum ISF observed during the previous pulse). The "tracking index" thus ranged from 1 to 5 and was defined as the number of pulses tracked in response to the 5-pulse stimulus train. Essentially identical results were obtained with both methods.

9.7 Olfactory processing

Changes in stimulus intensity can modulate the patterns of glomerular activity and output from the olfactory bulb and AL as has been shown for olfactory systems of vertebrates and invertebrates alike (Anton et al., 1997; Hildebrand and Shepherd, 1997; Christensen and White, 2000; Keller and Vosshall, 2003; Leon and Johnson, 2003; Sachse and Galizia, 2003; Stopfer et al., 2003). Until recently, little was known about how the responses of uniglomerular PNs can be influenced by specific stimulus blends (Wu et al., 1996). Below, we present direct evidence for PNs in the AL of *M. sexta* that the temporal responses of some PNs are optimized for a particular ratio of stimulus compounds in a blend and that altering this ratio dramatically changes the central representation of the blend at the first stage of processing in the brain.

9.7.1 MGC-PNs have different dynamic ranges and response thresholds

When we changed the intensity of the pheromone stimulus blend, we observed pronounced effects on the strength of the responses (IPSPs, EPSPs, number of impulses) in two major types of MGC-PNs: the BAL-selective (*BAL⁺ PNs*) and C15-selective cells (*C15⁺ PNs*) (Fig. 12). However, not all of these neurons exhibited classical concentration-dependent responses in the range of concentrations used in this study. A subset of *BAL⁺ PNs* innervating the T1 glomerulus and *C15⁺ PNs* innervating the cumulus were extremely sensitive at all concentrations tested (Fig. 12, left panel), while others with similar morphology failed to respond except at the highest stimulus concentration (Fig. 12, right panel). Such functional diversity is not an exclusive property of the male MGC as shown in a study in female *M. sexta* (Reisenman et al., 2004). Similar diversity exists in female *M. sexta* among PNs that innervate a single glomerulus and respond selectively to one enantiomer of linalool, a common plant volatile. The results present us with evidence from both male and female moths to clearly demonstrate functional heterogeneity with respect to threshold sensitivity and concentration-response characteristics among PNs innervating a single glomerulus.

The observed diversity in the PN population of a single glomerulus could reflect distinct functional roles for different PNs under diverse environmental conditions, as previously proposed (Christensen et al., 2000). In this scenario, PNs that are little affected by a change in pheromone concentration might signal the presence of the pheromone without regard to its intensity, whereas the PNs with higher response thresholds, recruited into the coding ensemble only at relatively high pheromone concentrations, might therefore function in source location (Murlis, 1997).

Concentration changes may be encoded in two distinct ways by PNs: (1) by individual neurons that give incremental responses to increasing concentrations and/or (2) by recruitment of different populations of PNs at different concentrations. Our data presented evidence for such a dual coding strategy in MGC-PNs (*BAL⁺ PNs* and *C15⁺ PNs*) in that some displayed a monotonic response across all concentrations, whereas others showed concentration-dependent responses over several orders of magnitude of stimulus concentration. The representation of sex-pheromonal information in the AL of *M. sexta* is sparse at low stimulus intensities because only a subset of MGC-PNs is active under these conditions, whereas the representation becomes increasingly combinatorial and complex as pheromone concentration increases because it involves a greater number and functional variety of types of MGC-PNs. This is similar to the neural coding of general olfactory stimuli in AL glomeruli in fruit flies and honey bees (Galizia et al., 1999, Ng et al., 2002; Wang et al., 2003).

A large number of pheromone-responsive afferent neurons converges onto many fewer central neurons. Therefore, threshold values are expected to be lower for MGC-PNs than for ORCs. Kaissling et al. (1989) determined a 1-ng threshold stimulus load in the stimulus-delivery cartridge for BAL-specific ORCs and a 100-ng load for C15-specific ORCs for *M. sexta*. The threshold stimulus loads found in our experiments are 0.01 ng for *BAL⁺ PNs* and 0.1 ng for *C15⁺ PNs*. These >100-fold lower corresponding thresholds of MGC-PNs are thought to overcome the critical signal-to-noise ratio, when pheromone molecules activate only a few ORCs. Similar response threshold relationships between ORCs and AL neurons have been observed in other insect species (Boeckh, 1984).

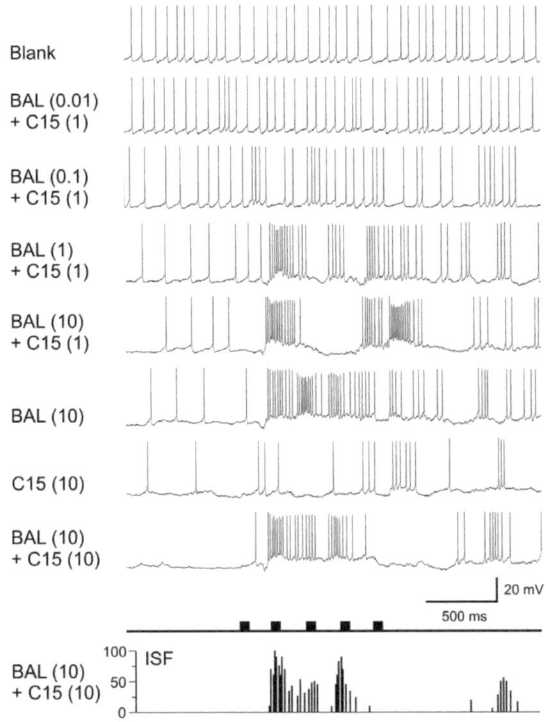

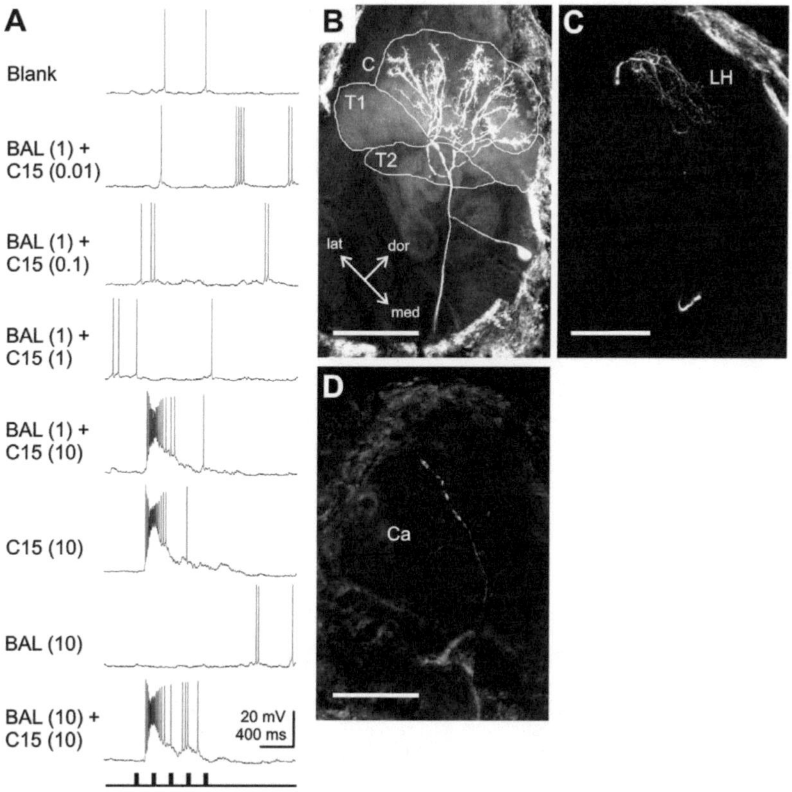

Fig. 12. Panel on previous page: Intracellular recordings from male *Manduca sexta* showing one class of *BAL⁺ PN* that was exclusively responsive to BAL and thus classified as a *BAL⁺-specialist PN*. This type of PN responded to BAL in a dose-dependent manner but is unresponsive to C15, and thus did not show a blend effect. Small BAL stimulus loads evoked only a membrane depolarization (EPSP) accompanied by spiking, but elevated dosages also triggered a brief IPSP preceding the EPSP. In each case, five identical stimulus pulses were delivered to the ipsilateral antenna at a frequency of 5 sec^{-1} (stimulus markers for the 50-ms pulses are shown beneath the records). A plot of instantaneous spike frequency (ISF) calculated from the last record illustrates that the firing dynamics of this type of MGC-PN could not accurately track all pulses of a stimulus train, thus leading to a low tracking index as defined in the text. Panels on this page: (A-D): Olfactory responses and morphology (frontal view) from one class of *C15⁺ PN* of male *M. sexta* that was solely responsive to C15 (*C15⁺-specialist PN)*, and therefore did not exhibit a blend effect. (A) This neuron was strongly depolarized by the first pulse of any stimulus that contained 10 ng of C15, but it did not respond to lower stimulus intensities. The neuron showed no clear response to BAL even at the elevated 10-ng load. In each trace, the ipsilateral antenna received five 50-ms stimulus pulses at 5 sec^{-1} (stimulus markers are shown beneath the records). (B) The branches of this PN were confined to the cumulus (C) of the MGC. *dor*, dorsal; *lat*, lateral; *med*, medial; *T1*, toroid 1; *T2*, toroid 2. (C) Frontal view of the axon's ramifications in the lateral horn (*LH*) of the protocerebrum. (D) Collateral branches innervating the calyces (*Ca*) of the mushroom body in the posterior protocerebrum. Scale bar = 100 μm. From Heinbockel et al., 2004.

9.7.2 Effects of blend ratio on temporal response dynamics in PNs

In subpopulations of *BAL⁺ PNs* and *C15⁺ PNs*, a change in the ratio of BAL and C15 clearly affected pulse tracking of these PNs (Figs. 13, 14).

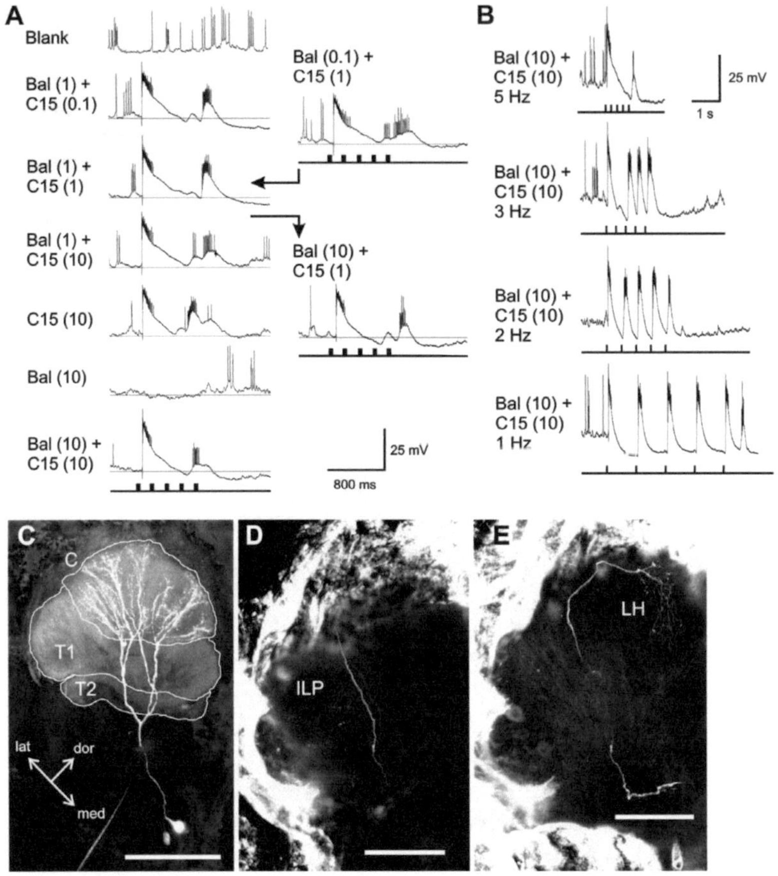

Fig. 13. Responses and anatomy of a *C15⁺ PN* from male *Manduca sexta* that gave depolarizing responses to C15 and inhibitory responses to BAL (*C15⁺ / BAL⁻ PN*). (A) Responses from left to right are to blends containing 1 ng C15 plus increasing amounts of BAL. The ipsilateral antenna received five 50-ms stimulus pulses at 5 pulses sec^{-1} (stimulus markers are shown beneath the records). This neuron had a low threshold for excitation and gave strong responses to stimuli containing only 0.1 ng of C15. The horizontal lines in the records indicate the membrane potential observed prior to stimulation. Notice that BAL alone hyperpolarized the PN and suppressed spiking activity (*C15⁺ / BAL⁻ PN*). (B) Responses of the same PN to different pulse rates of antennal stimulation (BAL, C15, 10 ng each) ranging from 1 to 5 sec^{-1}. At higher frequencies, the PN was unable to repolarize sufficiently to track each of the five stimulus pulses. The tracking index improved dramatically at lower pulse frequencies. (C) The neuron in (A) branched in the cumulus and not in the toroid 1 of the MGC (frontal view). *C*, cumulus; *do*, dorsal; *la*, lateral; *me*, medial; *T1*, toroid; *T2*, toroid 2. (D) More anterior view of the ramifications of the axon in the protocerebrum. *ILP*, inferior lateral protocerebrum. (E) More posterior aspect of protocerebral ramifications of the axon. *LH*, lateral horn. Scale bar = 100 µm. From Heinbockel et al., 2004.

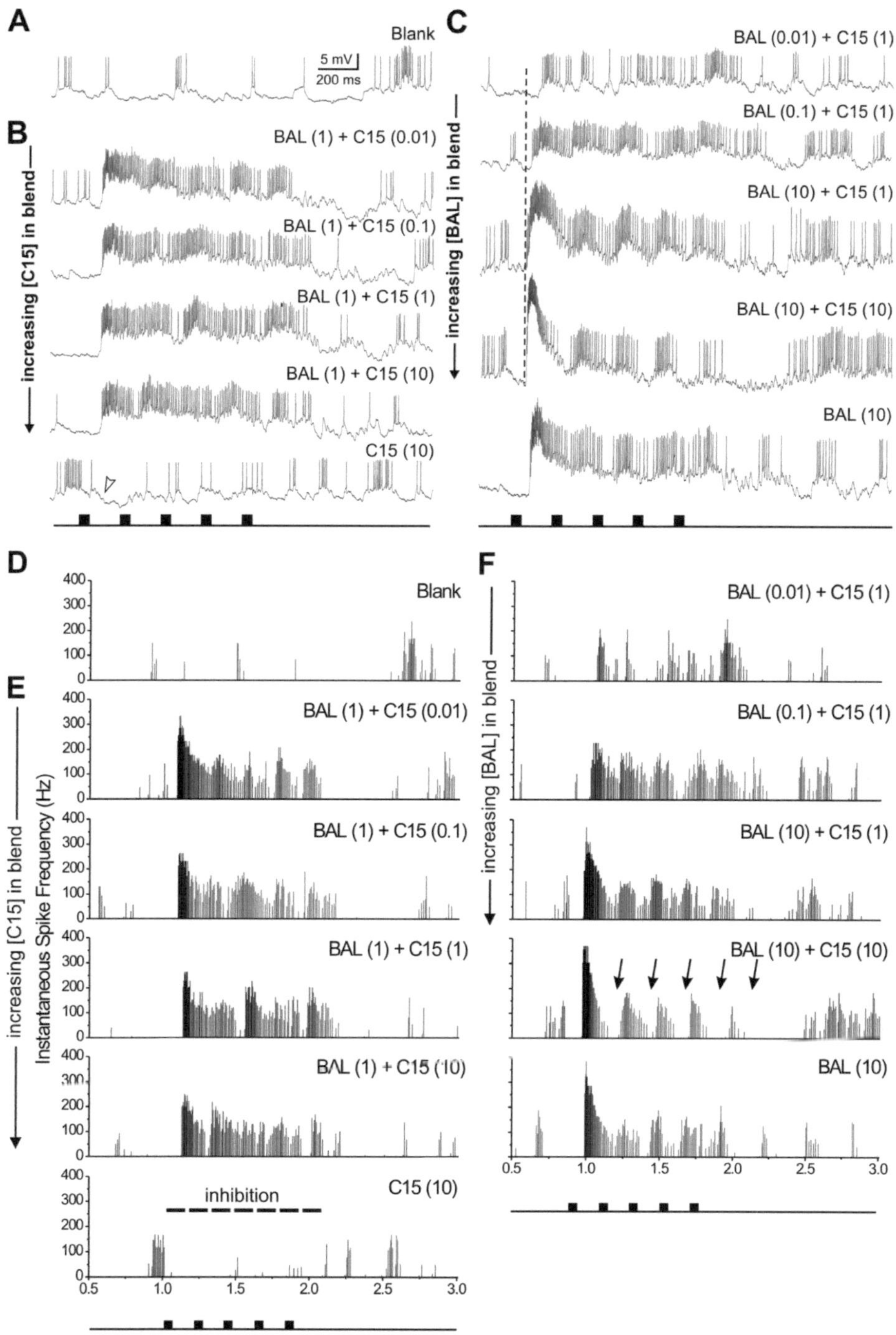

Fig. 14. Intracellular recordings and ISF plots from one *BAL⁺ PN* from male *Manduca sexta* that gave depolarizing responses to BAL and an inhibitory response to C15 (*BAL⁺ / C15⁻ PN*). Responses to blanks are shown in (A) and (D). (B), (E) Intracellularly recorded responses and corresponding ISF plots to blends containing 1 ng BAL plus increasing

amounts of C15. Repetitive stimulus pulses were delivered to the ipsilateral antenna at a frequency of 5 pulses sec^{-1} (stimulus markers are shown beneath the records). Note the modulations of PN firing with each stimulus pulse. C15 at 10 ng led to a hyperpolarization and suppression of firing in the PN (*dashed line*). (C, F) Responses (intracellular records and ISF plots) of the same neuron to blends containing 1 ng C15 plus increasing amounts of BAL. The tracking index of this PN improved at a ratio of 10 ng BAL to 1 ng C15, but was optimized at a 1:1 ratio with 10 ng of each component. Note the distinct periods of inactivity (*arrows*) between spike bursts evoked by consecutive stimulus pulses. The concentration-dependency of this neuron was also indicated by a marked reduction in response latency as the amount of BAL in the blend was increased. From Heinbockel et al., 2004.

These MGC-PNs integrated convergent excitatory and inhibitory afferent input and better resolved intermittent olfactory signals. This processing function was optimized at particular blend ratios for different PNs. *C15⁺ / BAL⁻ PNs* and, particularly, *BAL⁺ / C15⁻ PNs* were best suited to convey the temporal structure of the stimulus to higher brain centers, whereas responses of *BAL⁺*- and *C15-specialist PNs* often lacked the temporal precision to represent rapid changes in pheromone concentration. These results show that information about individual components and stimulus timing is transmitted from the MGC in different output channels.

We also observed blend effects in *BAL⁺ / C15⁺ PNs* and the *C15⁺ / BAL⁺ PNs* (Fig. 15). These blend effects were clearly ratio-dependent. These MGC-PNs are excited by stimulation with either of the two key pheromone components and, therefore, they are considered to be pheromone "generalists" (Christensen and Hildebrand, 1987; 1990; Christensen et al., 1996).

Principally, the responses of these MGC-PNs to stimulation with either pheromone component were primarily excitatory. However, the cells often received different amounts of excitatory and inhibitory input when stimulated with blends of the two components. Several PNs responded with a strong, long-lasting excitatory response to stimulation with one component and a less-intense and brief response to stimulation with the other component. Stimulation with the blend evoked a "mixed" response of intermediate character depending on the concentrations of the components. The excitatory phase of this blend response was not simply the sum of the excitatory responses elicited by each of the two components alone but the response was typically weaker than expected from adding the two responses. MGC-PNs with temporally distinct blend responses could serve specific functions in olfactory information processing. They could function as the neural substrate in the AL for rapid behavioral changes in response to the appropriate component ratio of a pheromone blend (Heinbockel et al., 2004).

9.7.3 Phasis vs. tonic patterning of postsynaptic response

Both key pheromone components (represented in this study by BAL and C15) must be present to elicit upwind olfaction-modulated flight in a wind tunnel in *M. sexta* (Tumlinson et al., 1989). Individual MGC-PNs gave multiphasic responses (inhibition-excitation-afterhyperpolarization) (Figs. 12-15), which are typical of PNs in the AL of *M. sexta* (Christensen and Hildebrand, 1987; Christensen et al., 1998). However, *BAL⁺ PNs* had larger IPSPs, smaller EPSPs, weaker responses during 200 ms after stimulus onset, and stronger responses over 1000 ms after stimulus onset than did *C15⁺ PNs*. This suggests that BAL evokes sustained responses and C15 evokes shorter, more phasic responses. Corresponding response patterns were observed among ORCs in the antenna of *M. sexta* (Kaissling et al., 1989) such that responses of BAL-specific ORCs to antennal stimulation with BAL are typically more phasic-tonic whereas the phasic part of the response is more pronounced in C15-specific receptor cells. Different ORCs with selective responses to the same stimulus compound could nevertheless exhibit different response dynamics as has been shown for pheromone-specific ORCs of moths (Almaas et al., 1991) and ORCs in *D. melanogaster* (de Bruyne et al., 2000). It is tempting to speculate that MGC-PNs with phasic olfactory responses could be synaptic targets of phasic ORCs, whereas PNs giving tonic responses could receive input largely from ORCs that exhibit tonic firing patterns. In this scenario, at least some aspects of temporal coding and coding of stimulus concentration and mixture ratio are regulated through the activation of different populations of ORCs.

If MGC-PNs respond to pheromonal stimulation with sustained firing outlasting the stimulus duration they are likely to report the onset of stimulation rather than its termination and could serve as a substrate for more temporally complex spike codes (Heinbockel and Kloppenburg, 1999; Friedrich and Stopfer, 2001; Laurent et al., 2001; Christensen and Hildebrand, 2002). If MGC-PNs respond with a more phasic spike pattern and show better pulse-following, they could encode both the beginning and the end of each stimulus pulse. These MGC-PNs are able to convey information about rapidly changing signals and by doing so transmit information about the physical dynamics of the stimulus to higher olfactory centers in addition to cues about stimulus identity (Heinbockel et al., 1999; Vickers et al., 2001).

Since each AL is innervated by fewer than 50 MGC-PNs (Homberg et al., 1989), each MGC-PN contributes significantly to information flow from the MGC to higher order olfactory centers. Different features of the olfactory stimulus are likely to be encoded by these different outputs arising from a single MGC glomerulus, i.e., functional heterogeneity exists in the population of olfactory output channels from the MGC. In the AL of male *M. sexta*, both component-specific and blend-specific MGC-PNs exist that are affected by changes in concentration and/or blend ratio, but the performance of different PNs is optimized under different ambient conditions.

Advances in intracellular recording, staining, confocal microscopy and data analysis allowed us to characterize the responses of individual MGC-PNs and their ability to encode features of the stimulus, e.g., to follow intermittent olfactory stimuli. PNs that integrated information about the two-component pheromone blend (i.e., they received excitatory input from one component and inhibitory input from the other) were particularly well suited to track a train of stimulus pulses. Stimulus-pulse tracking was furthermore optimized at a synthetic blend ratio that mimics the physiological response to an extract of the female's pheromone gland. Our results show that optimal responsiveness of a PN to repetitive stimulus pulses depends not only on stimulus intensity but also on the relative strength of the two opposing synaptic inputs that are integrated by MGC-PNs.

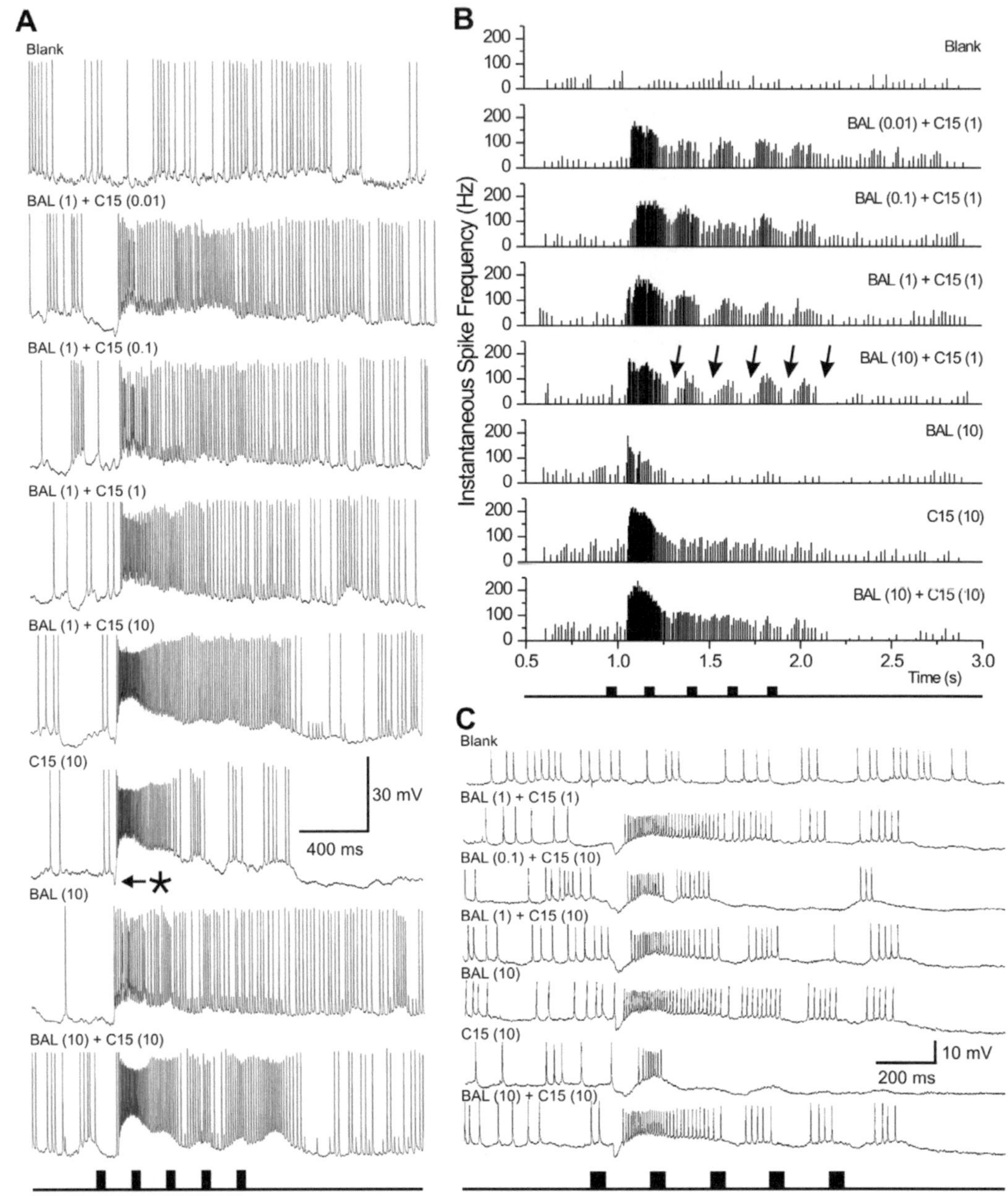

Fig. 15. Responses of *BAL⁺ / C15⁺ PNs* and *C15⁺/BAL⁺ PNs* from male *Manduca sexta.* These MGC-PNs were depolarized by antennal stimulation with either BAL or C15. (A) This neuron responded with a strong response to C15 and a more sustained response to BAL (*C15⁺/BAL⁺ PNs*). Stimulation with C15 also evoked a distinct IPSP (marked by asterisk) resulting in a mixed response (inhibition / excitation / inhibition). Varying the blend ratio changed the response character to more phasic or more tonic, depending on the load in the stimulus blend. (B) ISF response plots of a *C15⁺/BAL⁺ PN* to various BAL and C15 stimulus loads and ratios revealed a sustained phasic-tonic response to C15 and a brief excitatory

response to BAL followed by inhibition. Addition of increasing amounts of BAL to the blend resulted in improved stimulus-pulse tracking. Note the distinct periods of inactivity (*arrows*) between spike bursts evoked by consecutive stimulus pulses. (C) In this neuron, stimulation with C15 evoked a brief excitatory response preceded by an IPSP and followed by a strong and prolonged inhibitory response phase. Stimulation with BAL resulted in a mixed response (inhibition / excitation / inhibition) comprising a strong IPSP and a strong excitatory response phase (*BAL⁺/C15⁺ PN*). From Heinbockel et al., 2004.

10. Conclusions

This chapter details both extracellular and intracellular recording methods used with insect preparations in the fields of insect olfaction and/or gustation. Using one or both olfactory recording techniques contributes to a better understanding of (i) how and what olfactory information is processed in the insect brain, (ii) the chemical identification of important plant volatiles for insect-plant interactions, and (iii) which components of the female sex pheromone, as well as plant-associated volatiles, play important roles in male and female moth orientation, respectively. Complementing these recording techniques with the use of fluorescent markers allows axonal projections to be traced to the brain or central neurons to be labeled individually. Gustatory extracellular recording methods can ultimately (i) lead to a clearer understanding of the importance of gustatory cues involved in larval host-plant interactions, (ii) give us a better perception on how taste stimuli code for different behavioral responses, and (iii) provide ideas and strategies for crop protection from insect predation. Overall, either one or both of these methods (i.e., extracellular or intracellular) can contribute to an increased understanding of how and what chemosensory information is processed in the insect brain and shed more light on how nervous systems recognize, analyze, and respond to complex sensory information.

11. Acknowledgments

This work was supported by NIH grants 1R15DC007609-01 and 3R15DC007609-01S1 to V.D.S., NIH grant DC-02751, as well as grants from Towson University, the FCSM URG and TU URG. Grants from the Whitehall Foundation, NIH-NIGMS (S06GM08016) and NIH-NINDS (U54NS039407) to T.H. also supported this work. The authors would like to gratefully acknowledge J.G. Hildebrand, T.A. Christensen, T.L. Martin, C.E. Reisenman, R. Bennett, J. Klupt, R. Kuta, and M. Chen.

12. References

Andersen, R.A., Hamilton-Kemp, T.R., Loughrin, J.H., Hughes, C.G., Hildebrand, D.F., & Sutton, T.G. (1988). Green leaf headspace volatiles from *Nicotiana tabacum* lines of different trichome morphology. *Journal of Agricultural and Food Chemistry,* Vol. 36, No. 1, (January 1988), pp. 295-299. ISSN 0021-8561

Anderson, P., Hansson, B.S., & Löfqvist, J. (1995). Plant-odour-specific receptor neurones on the antennae of female and male *Spodoptera littoralis. Physiological Entomology,* Vol. 20, No. 3, (September 1995), pp. 189-198

Almaas, T.J., Christensen, T.A., & Mustaparta, H. (1991). Chemical communication in heliothine moths. I. Antennal receptor neurons encode several features of intra- and interspecific odorants in the male corn earworm moth *Helicoverpa zea*. *Journal of Comparative Physiology A*, Vol. 169, No. 3, (September 1991), pp. 259-274. ISSN 0340-7594

Anton, S., Löfstedt, C., & Hansson, B.S. (1997). Central nervous system processing of sex pheromones in two strains of the European Corn Borer *Ostrinia nubilalis* (Lepidoptera: Pyralidae). *Journal of Experimental Biology*, Vol. 200, No. 1, (April 1997), pp. 1073-1087. ISSN 0022- 0949

Arn H., Tóth, M., & Priesner, E. (1992). List of sex pheromones of Lepidoptera and related attractants, 2nd edition, International Organization for Biological Control, West Palearctic Regional Section, ISSN 92-9067-002-9, Montfavet, France

Baker, T.C. (1989). Sex pheromone communication in the Lepidoptera: New research progress. *Experientia*, Vol. 45, No. 3, (March 1989), pp. 248-262. ISSN 0014-4754

Bell, R.A. & Joachim, F.A. (1976). Techniques for rearing laboratory colonies of tobacco hornworms and pink bollworms. *Annals of the Entomological Society of America*, Vol. 69, No. 2, (March 1976), pp. 365-373. ISSN 0013-8746

Bernays, E.A. & Chapman, R.F. (1994). *Host-plant selection by phytophagous insects*. ISSN 0-412-03131-0, Chapman Hall, New York

Blaney, W.M., Simmonds, M.S.J., Ley, S.V., & Jones, P.S. (1988). Insect antifeedants: a behavioural and electrophysiological investigation of natural and synthetically derived clerodane ditepenoids. *Entomologia Experimentalis et Applicata*, Vol 46, No. 3, (March 1988), pp. 267-274. ISSN 0013-8703

Boeckh, J. (1974). Die Reaktionen olfaktorischer Neurone im Deutocerebrum von Insekten im Vergleich zu den Antwortmustern der Geruchssinneszellen. *Journal of Comparative Physiology*, Vol. 90, pp. 183-205. ISSN 0340-7594

Boeckh, J. (1984). Neurophysiological aspects of insect olfaction. In: *Insect Communication*, Lewis T (ed), pp 83-104, Academic Press, ISSN 0-12-447175-7, Orlando, FL

Buck, L.B. (1996). Information coding in the vertebrate olfactory system. *Annual Review of Neuroscience*, Vol., 19, (March 1996), pp. 517–544. ISSN 0147-006X

Buonviso, N. & Chaput, M.A. (1990) Response similarity of odors in olfactory bulb output cells presumed to be connected to the same glomerulus: electrophysiological study using simultaneous single-unit recordings. *Journal of Neurophysiology*, Vol., 63, No. 3, (March, 1990), pp. 447-454. ISSN 0022-3077

Buttery, R.G., Teranishi, R., & Ling, L.C. (1987a) Fresh tomato aroma volatiles: a quantitative study. *Journal of Agricultural and Food Chemistry*, Vol. 35, No. 4, (July 1987), pp. 540-544. ISSN 0021-8561

Buttery, R.G., Ling, L.C., & Light, D.M. (1987b) Tomato leaf volatile aroma odorants. *Journal of Agricultural and Food Chemistry*, Vol. 35, No. 6, (November 1987), pp. 1039-1042. ISSN 0021-8561

Christensen, T.A. & Hildebrand, J.G. (1987). Male-specific, sex-pheromone-selective projection neurons in the antennal lobes of the moth *Manduca sexta*. *Journal of Comparative* Physiology, *A* Vol. 160, No. 5, (September 1987), pp. 553-569. ISSN 0340-7594

Christensen, T.A. & Hildebrand, J.G. (1990). Representation of sex-pheromonal information in the insect brain. In: *Proceedings of the Tenth International Symposium on Olfaction and Taste, ISOT X*, Døving, K.B. (ed), pp (142-150), University of Oslo, Oslo, Norway

Christensen, T.A. & Hildebrand, J.G. (1997). Coincident stimulation with pheromone components improves temporal pattern resolution in central olfactory neurons. *Journal of Neurophysiology,* Vol. 77, No. 2, (February 1997), pp. 775-781. ISSN 0022-3077

Christensen, T.A. & White, J. (2000). Representation of olfactory information in the brain. In: *The neurobiology of taste and smell,* Vol 2, Finger, T.E., Silver, W.L., Restrepo, D. (eds), pp (201-232), Wiley-Liss, ISBN 0-47125-721-5, New York

Christensen, T.A. & Hildebrand, J.G. (2002). Pheromonal and host-odor processing in the insect antennal lobe: how different? *Current Opinions in Neurobiology,* Vol. 12, No. 4, (August 2001), pp. 393-399. ISSN 0959-4388

Christensen T.A., Mustaparta, H., & Hildebrand, J.G. (1989). Discrimination of sex pheromone blends in the olfactory system of the moth. *Chemical Senses,* Vol. 14, No. 3, (June 1989), pp. 463-477. ISSN 0379-864X

Christensen T.A., Harrow, I.D., Cuzzocrea, C., Randolph, P.W., & Hildebrand, J.G. (1995). Distinct projections of two populations of olfactory receptor neurons in the antennal lobe of the sphinx moth *Manduca sexta. Chemical Senses,* Vol. 20, No. 3, (June 1995), 313-323. ISSN 0379-864X

Christensen, T.A., Heinbockel, T., & Hildebrand, J.G. (1996). Olfactory information processing in the brain: encoding chemical and temporal features of odors. *Journal of Neurobiology,* Vol. 30, No. 1, (May 1996), pp. 82-91. ISSN 0022-3034

Christensen, T.A., Waldrop, B.R., & Hildebrand, J.G. (1998). Multitasking in the olfactory system: context-dependent responses to odors reveal dual GABA-regulated coding mechanisms in single olfactory projection neurons. *Journal of Neuroscience,* Vol. 18, No. 15, (August 1998), pp. 5999-6008. ISSN 0270-6474

Christensen, T.A., Pawlowski, V.M., Lei, H., & Hildebrand, J.G. (2000). Multi-unit recordings reveal context-dependent modulation of synchrony in odor-specific neural ensembles. *Nature Neuroscience,* Vol. 3, No. 9, (September 2000), pp. 927-931. ISSN 1097-6256

Clyne, P.J., Warr, C.G., Freeman, M.R., Lessing, D., Kim, J., & Carlson, J.R. (1999). A novel family of divergent seven-transmembrane proteins: candidate odorant receptors in *Drosophila. Neuron,* Vol. 22, No. 2 (February 1999), pp. 327-338. ISSN 0896-6273

Clyne, P. J., Warr, C. G., & Carlson, J. R. (2000). Candidate taste receptors in *Drosophila. Science,* 287, No. 5497, (December 2000), pp. 1830–1834. ISSN 0036-8075

Dahanukar, A., Lei, Y. T., Kwon, J. Y., & Carlson, J. R. (2007). Two Gr genes underlie sugar reception in *Drosophila. Neuron* Vol. 56, No. 3, (November 2007), pp. 503–516. ISSN 0896-6273

de Boer, G., Dethier, V.G., & Schoonhoven, L.M. (1977). Chemoreceptors in the preoral cavity of the tobacco hornworm, *Manduca sexta,* and their possible function in feeding behaviour. *Entomologia Experimatalis et Applicata,* Vol. 21, No. 3, (May 1977), pp. 287-298. ISSN 0013-8703

de Bruyne, M., Foster, K., & Carlson, J.R. (2000). Odor coding in the *Drosophila* antenna. *Neuron,* Vol. 30, No. 2, (November 2000), pp. 537-552. ISSN 0896-6273

Friedrich, R.W. & Stopfer, M. (2001). Recent dynamics in olfactory population coding. *Current Opinions in Neurobiology,* Vol. 11, No. 4, (August 2001), pp. 468-474. ISSN 0959-4388

Dethier, V.G. (1982). Mechanisms of host plant recognition. *Entomologia Experimentalis et Applicata* Vol. 31, No. 1, (January 1982), pp. 49-56. ISSN 0013-8703

Dethier, V.G. & Crnjar, R.M. (1982). Candidate codes in the gustatory system of caterpillars. *The Journal of General Physiology*, Vol. 79, No. 4 (April 1982), pp. 549-569. ISSN 0022-1295

Dickens, J.C. (1990). Specialized receptor neurons for pheromones and host plant odors in the boll weevil, *Anthonomus grandis* Boh. (Coleoptera: Curculionidae). *Chemical Senses*, Vol. 15, No. 3, pp. 311-331. ISSN 0379-864X

Dunipace, L., Meister, S., McNealy, C., & Amrein, H. (2001). Spatially restricted expression of candidate taste receptors in the *Drosophila* gustatory system. *Current Biology*, Vol. 11, No. 11, (June 2001), pp. 822–835. ISSN 0960-9822

Firestein, S. (1992). Electrical signals in olfactory transduction. *Current Opinions in Neurobiology*, Vol. 2, No. 4, (August 1992), pp. 444-448. ISSN 0959-4388

Frazier, J.L. (1986) The perception of plant allelochemicals that inhibit feeding. In: *Molecular Aspects of Insect-Plant Associations*, Brattsten, L.B. and Ahmad, S. (eds), pp. (1-42), Plenum Press, ISSN 03-064-25475, New York

Friedrich, R.W. & Korsching, S.I. (1997). Combinatorial and chemotopic odorant coding in the zebrafish olfactory bulb visualized by optical imaging. *Neuron*, Vol. 18, No. 5, (May 1997), pp. 737-752. ISSN 0896-6273

Galizia, C.G., Sachse, S., Rappert, A., & Menzel, R. (1999). The glomerular code for odor representation is species specific in the honeybee *Apis mellifera*. *Nature Neuroscience*, Vol. 2, No. 5 (May 1999), pp. 473-478. ISSN 1097-6256

Glendinning, J.I., Tarre, M., & Asaoka, K. (1999). Contribution of different bitter-sensitive taste cells to feeding inhibition in a caterpillar (*Manduca sexta*). *Behavioral Neuroscience*, Vol. 113, No. 4 (August 1999), pp. 840-854. ISSN 0735-7044

Grimaldi, D. & Engel, M.S. (2005). *Evolution of the Insects*. Cambridge University Press, New York, New York

Hansson, B.S. (1995). Olfaction in Lepidoptera. *Experientia* Vol. 51, No. 11, (November 1995), pp. 1003-1027. ISSN 0014-4754

Hansson, B.S. & Christensen, T.A. (1999). Functional characteristics of the antennal lobe. In: *Insect Olfaction*, Hansson, B.S. (ed), pp. (126-161), Springer, ISSN 3-540-65034-2, Berlin

Hansson B.S., Larsson M.C., & Leal, W.S. (1999). Green leaf volatile detecting olfactory receptor neurones display very high sensitivity and specificity in a scarab beetle. *Physiological Entomology*, Vol. 24, No. 2, (June 1999), pp. 121-126. ISSN 0307-6962

Hanson, F.E. & Peterson, S.C. (1990). Sensory coding in *Manduca sexta* for deterrence by a non-host plant, *Canna generalis*. *Insects-Plants, '89, Symposia Biologica Hungarica*, Vol. 39, pp. 29-37. ISSN 0082-0695

Hansson, B.S., Christensen, T.A., & Hildebrand, J.G. (1991). Functionally distinct subdivisons of the macroglomerular complex in the antennal lobe of the male sphinx moth *Manduca sexta*. *Journal of Comparative Neurology*, Vol. 312, No. 2, (October 1991), pp. 264-278. ISSN 1096-9861

Heath, R.R., Landolt, P.J., Dueben, B., & Lenczewski, B. (1992). Identification of floral compounds of night-blooming jessamine attractive to cabbage-looper moths. *Environmental Entomology*, Vol. 21, No. 4, (August 1992), pp. 854-859. ISSN 0046-225X

Heinbockel, T. & Kaissling, K.-E. (1996). Variability of olfactory receptor neuron responses of female silkmoths (*Bombyx mori* L.) to benzoic acid and (+)-linalool. *Journal of Insect Physiology*, Vol. 42, No. 6 (June 1996), pp. 565-578. ISSN 0022-1910

Heinbockel T. & Hildebrand, J.G. (1998). Antennal receptive fields of pheromone-responsive neurons in the antennal lobes of the sphinx moth *Manduca sexta*. *Journal of Comparative Physiology A*, Vol. 183, No. 2, (July 1998), pp. 121-133. ISSN 0340-7594

Heinbockel, T., Kloppenburg, P., & Hildebrand, J.G. (1998). Pheromone-evoked potentials and oscillations in the antennal lobes of the sphinx moth, *Manduca sexta*. *Journal of Comparative Physiology A*, Vol. 182, No. 6, (May 1998), pp. 703-714. ISSN 0340-7594

Heinbockel, T., Christensen, T.A., & Hildebrand, J.G. (1999). Temporal tuning of odor responses in pheromone-responsive projection neurons in the brain of the sphinx moth *Manduca sexta*. *Journal of Comparative Neurology*, Vol. 409, No. 1, (June 1999), pp. 1-12. ISSN 1096-9861

Heinbockel, T., Christensen, T.A., & Hildebrand, J.G. (2004). Representation of binary pheromone blends by glomerulus-specific olfactory projection neurons. *Journal of Comparative Physiology A*, Vol. 190, No. 12, (December 2004), pp. 1023-1037. ISSN 0340-7594

Hildebrand, J.G. (1995) Analysis of chemical signals by nervous systems. *Proceedings of the National Academy of Sciences* USA, Vol. 92, No. 1, (January 1995), pp. 67-74. ISSN 0027-8424

Hildebrand, J.G. (1996) Olfactory control of behavior in moths: central processing of odor information and the functional significance of olfactory glomeruli. *Journal of Comparative Physiology A*, Vol. 178, No. 1, (January 1996), pp. 5-19. ISSN 0340-7594

Hildebrand, J.G. & Shepherd, G. (1997). Mechanisms of olfactory discrimination: converging evidence for common principles across phyla. *Annual Review of Neuroscience*, Vol. 20, (March 1997), pp. 595-631. ISSN 0147-006X

Homberg, U., Montague, R.A., & Hildebrand, J.G. (1988). Anatomy of antenno-cerebral pathways in the brain of the sphinx moth *Manduca sexta*. *Cell and Tissue Research*, Vol. 254, No. 2, (November 1988), pp. 225-281. ISSN 0302-766X

Homberg, U., Christensen, T.A., & Hildebrand, T.A. (1989). Structure and function of the deutocerebrum in insects. *Annual Review of Entomology*, Vol. 34, pp. 477-501. ISSN 0066-4170

Homberg U., Hoskins S.G., & Hildebrand J.G. (1995) Distribution of acetylcholinesterase activity in the deutocerebrum of the sphinx moth *Manduca sexta*. *Cell and Tissue Research*, Vol. 279, No. 2, (February), pp. 249-259. ISSN 0302-766X

Kaissling K.-E. (1995) Single unit and electroantennogram recordings in insect olfactory organs. In: *Experimental Cell Biology of Taste and Olfaction: Current Techniques and Protocols*, Spielman A.I., Brand, J.G. (eds), pp. 361-377, CRC Press, ISSN 0-8493-7645-9, New York

Kaissling, K.-E. (1996). Peripheral mechanisms of pheromone reception in moths. *Chemical Senses* Vol. 21, No. 2, (April 1996), pp. 257-268. ISSN 0379-864X

Kaissling K.-E. & Thorson J (1980) Insect olfactory sensilla: structural, chemical and electrical components of the functional organization. In: *Receptors for neurotransmitters, hormones and pheromones in insects*, Sattelle, D.B., Hall, L.M., Hildebrand, J.G. (eds), pp (261-282), Elsevier, ISSN 0444802312, North-Holland, Amsterdam

Kaissling, K.-E., Hildebrand J.G., & Tumlinson, J.H. (1989). Pheromone receptor cells in the male moth *Manduca sexta*. *Archives of Insect Biochemistry and Physiology*, Vol. 10, No. 4, (April 1989), pp. 273-279. ISSN 0739-4462

Keil, T.A. (1989) Fine structure of the pheromone-sensitive sensilla on the antenna of the hawkmoth, *Manduca sexta*. *Tissue and Cell*, Vol. 21, No. 1, (February 1989), pp. 139–151. ISSN 0040-8166

Keller, A. & Vosshall, L.B. (2003). Decoding olfaction in *Drosophila*. *Current Opinions in Neurobiology*, Vol. 13, No. 1, (February 2003), pp. 103-110. ISSN 0959-4388

King, J. Roche, Christensen, T.A., & Hildebrand, J.G. (2000). Response characteristics of an identified, sexually dimorphic olfactory glomerulus. *Journal of Neuroscience*, Vol. 20, No. 6, (March 2000), pp. 2391-2399. ISSN 0270-6474

Kloppenburg, P. & Heinbockel, T. (2000). 5-hydroxytryptamine modulates pheromone-evoked local field potentials in the macroglomerular complex of the sphinx moth *Manduca sexta*. *Journal of Experimental Biology*, Vol. 203, No. 11, (June 2000), pp. 1701-1709. ISSN 1477-9145

Knudsen, J.T. & Tollsten, L. (1993). Trends in floral scent chemistry in pollination syndromes: floral scent composition in moth-pollinated taxa. *Botanical Journal of the Linnean Society*, Vol. 113, No. 3, (November 1993), pp. 263-284. ISSN 0024-4074

Laurent, G., Stopfer, M., Friedrich, R.W., Rabinovich, M.I., Volkovskii, A., & Abarbanel, H.D. (2001). Odor encoding as an active, dynamical process: experiments, computation, and theory. *Annual Review of Neuroscience*, Vol. 24, (March 2001), pp. 263-297. ISSN 0147-006X

Lee, J.K. & Strausfeld, N.J. (1990). Structure, distribution and number of surface sensilla and their receptor cells on the olfactory appendage of the male moth, *Manduca sexta*. *Journal of Neurocytology*, Vol. 19, No. 4 (August 1990), pp. 519–538. ISSN 0300-4864

Lei, H., Christensen, T.A., & Hildebrand, J.G. (2002). Local inhibition modulates odor-evoked synchronization of glomerulus-specific output neurons. *Nature Neuroscience*, Vol. 5, No., 6, (June 2002), pp. 557-565. ISSN 1097-6256

Lei, H., Oland, L.A., Riffell, J.A., Beyerlein, A., & Hildebrand, J.G. (2010). Microcircuits for olfactory information processing in the antennal lobe of *Manduca sexta*. In: *Handbook of brain microcircuits*, Shepherd, G.M., Grillner, S. (eds), pp. (417-426), Oxford University Press, ISBN 978-0-19-538988-3, New York

Leon M & Johnson BA (2003) Olfactory coding in the mammalian olfactory bulb. *Brain Research Reviews* Vol. 42, No. 1, (April, 2003), pp. 23-32. ISSN 0165-0173

Mombaerts, P. (1996). Targeting olfaction. *Current Opinions in Neurobiology*, Vol. 6, No. 4, (August 1996), pp. 481-486. ISSN 0959-4388

Morgan, A.C. & Lyon, S.C. (1928). Notes on amyl salicylate as an attractant to the tobacco hornworm moth. *Journal of Economic Entomology*, Vol. 21, pp. 189-191. ISSN 0022-0493

Murlis, J. (1997). Odor plumes and the signal they provide. In: *Insect pheromone research – new directions*. Cardé R.T., Minks, A.K. (eds), pp (221-231), Chapman & Hall, ISSN 0-412-99611-1, New York

Ng, M., Roorda, R.D., Lima, S.Q., Zemelman, B.V., Morcillo, P., & Miesenböck, G. (2002). Transmission of olfactory information between three populations of neurons in the antennal lobe of the fly. *Neuron*, Vol. 36, No. 3 (October 2002), pp. 463-474. ISSN 0896-6273

Oland, L.A. & Tolbert, L.P. (1988). Effects of hydroxyurea parallel the effects of radiation in developing olfactory glomeruli. *Journal of Comparative Neurology*, Vol. 278, No. 3, (December 1988), pp. 377–387. ISSN 1096-9861

Pichon, Y., Sattelle, D.B., & Lane, N.J. (1972). Conduction processes in the nerve cord of the moth *Manduca sexta* in relation to its ultrastructure and haemolymph ionic composition. *Journal of Experimental Biology*, Vol. 56, No. 3 (June 1972), pp. 717-734. ISSN 1477-9145

Pophof, B. (1997). Olfactory responses recorded from sensilla coeloconica of the silkmoth *Bombyx mori*. *Physiological Entomology*, Vol. 22, No. 3, (September 1997), pp. 239-248. ISSN 0307-6962

Priesner, E. (1979). Progress in the analysis of pheromone receptor systems. *Annals of Zoology, Ecology and Animals*, Vol. 11, pp. 533-546.

Prescott, D.J., Hildebrand, J.G., Sanes, J.R., & Jewett, S. (1977). Biochemical and developmental studies of acetylcholine metabolism in the central nervous system of the moth *Manduca sexta*. *Comparative Biochemistry and Physiology Part C: Comparative Pharmacology*, Vol. 56, No. 2, pp. 77-84. ISSN 1532-0456

Raguso, R.A. & Willis, M.A. (1997). Floral scent and its role(s) in hawkmoth attraction. *Chemical Senses*, Vol. 22, No. 6, (December 1997), pp. 774-775. ISSN 0379-864X

Raguso, R.A. & Light, D.M. (1998). Electroantennogram responses of male *Sphinx perelegans* hawkmoths to floral and "green-leaf volatiles." *Entomologia Experimentalis et Applicata*, Vol. 86, No. 3, (March 1998), pp. 287-293. ISSN 0013-8703

Reisenman, C.E., Christensen, T.A., & Hildebrand, J.G. (2004). Enantioselectivity of projection neurons innervating identified olfactory glomeruli. *Journal of Neuroscience*, Vol. 24, No. 11, (March 2004), pp. 2602-2611. ISSN 0270-6474

Reisenman, C.E., Heinbockel,T., & Hildebrand, J.G. (2008). Inhibitory interactions among olfactory glomeruli do not necessarily reflect spatial proximity. *Journal of Neurophysiology*, Vol. 100, No. 2, (August 2008), pp. 554-564. ISSN 0022-3077

Renwick, J. A. A. (1989). Chemical ecology of oviposition in phytophagous insects. *Experientia*, Vol. 45, No. 3, (March 1989), pp. 223-228. ISSN 1420-682X

Robertson, H. M., Warr, C. G., & Carlson, J. R. (2003). Molecular evolution of the insect chemoreceptor gene superfamily in *Drosophila melanogaster*. *Proceedings of the National Academy of Sciences U.S.A.*, Vol. 100, Suppl. 2, (November 2003), pp. 14537–14542. ISSN 0027-8424

Rodrigues, V. & Buchner, E. (1984). [^{3}H]2-deoxyglucose mapping of odor-induced neuronal activity in the antennal lobes of *Drosophila melanogaster*. *Brain Research*, Vol. 324, pp. 374-378. ISSN 0006-8993

Rospars, J.P. & Hildebrand, J.G. (2000). Sexually dimorphic and isomorphic glomeruli in the antennal lobes of the sphinx moth *Manduca sexta*. *Chemical Senses*, Vol. 25, No. 2, (April 2000), pp. 119-129. ISSN 0379-864X

Rössler, W., Tolbert, L.P., & Hildebrand, J.G. (1998). Early formation of sexually dimorphic glomeruli in the developing olfactory lobe of the brain of the moth *Manduca sexta*. *Journal of Comparative Neurology*, Vol. 396, No. 4, (July 1998), pp. 415-428. ISSN 1096-9861

Rössler, W., Randolph, P.W., Tolbert, L.P., & Hildebrand, J.G. (1999). Axons of olfactory receptor cells of transsexually grafted antennae induce development of sexually dimorphic glomeruli in *Manduca sexta*. *Journal of Neurobiology*, Vol. 38, No. 1, (January 1999), pp. 1-541. ISSN 0022-3034

Sachse, S. & Galizia, C.G. (2003). The coding of odour-intensity in the honeybee antennal lobe: local computation optimizes odour representation. *European Journal of Neuroscience*, Vol. 18, No. 8, (October 2003), pp. 2119-2132. ISSN 0953-816X

Sanes, J.R. & Hildebrand, J.G. (1976). Structure and development of antennae in a moth, *Manduca sexta*. *Developmental Biology* Vol. 51, No. 2, (July 1976), pp. 282-299. 0012-1606

Schneider, D., Lacher, V., & Kaissling, K.-E. (1964). Die Reaktionsweise und Reaktionsspektrum von Riechzellen bei *Antheraea pernyi* (Lepidoptera, Saturniidae). *Zeitschrift fuer Vergleichende Physiologie,* Vol. 48, No. 6, (November 1964), pp. 632-662. ISSN 0340-7594

Schoonhoven, L.M. (1973). Plant recognition by lepidopterous larvae. *Symposia of the Royal Entomological Society of London*, Vol. 6, pp. 87-99. ISSN 0080-4363

Schoonhoven, L.M. (1982). Biological aspects of antifeedants. *Entomologia Experimentalis et Applicata,* Vol. 31, No. 1, (January 1982), pp. 57-69. ISSN 0013-8703

Schoonhoven, L.M. (1987). What makes a caterpillar eat? The sensory coding underlying feeding behavior. In: *Perspectives in Chemoreception and Behavior*, Chapman, R.F., Bernays, E.A., and Stoffolano, J.G., (eds), pp. (69-97), Springer-Verlag, ISSN 9780387963747, New York

Schoonhoven, L.M. & Dethier, V.G. (1966). Sensory aspects of host-plant discrimination by lepidopterous larvae. *Archives Néerlandaises de Zoologie*, Vol. 16, No. 4, pp. 497-530. ISSN 0365-5164

Schoonhoven, L.M. & Blom, F. (1988). Chemoreception and feeding behaviour in a caterpillar: towards a model of brain functioning in insects. *Entomologia Experimentalis et Applicata,* Vol. 49, No. 1-2, (November 1988), pp. 123-129. ISSN 0013-8703

Schoonhoven L.M. & van Loon, J.J.A. (2002) An inventory of taste in caterpillars: each species its own key. *Acta Zoologica Academiae Scientiarum Hungaricae, Vol.* 48, No. 3, pp. 215-263. ISSN 1217-8837

Schoonhoven, L.M., Blaney, W.M., & Simmonds, M.S.J. (1992). Sensory coding of feeding deterrents in phytophagous insects. In: *Insect-Plant Interactions,* Vol. 4, Bernays, E.A. (ed), pp. (59-79), CRC Press, ISSN 0-8493-4124-8, Boca Raton, FL

Schoonhoven, L.M., Jermy, T., & van Loon, J.J.A. (1998) *Insect-Plant Biology. From Physiology to Evolution.* ISSN, 9780412587009, Chapman and Hall, London.

Scott, K., Brady, J. R., Cravchik, A., Morozov, P., Rzhetsky, A., Zuker, C., & Axel, R. (2001). A chemosensory gene family encoding candidate gustatory and olfactory receptors in *Drosophila*. *Cell,* Vol. 104, No. 5 (March 2001), pp. 661–673. ISSN 0092-8674

Sharp, F.R., Kauer, J.S., & Shepherd, G.M. (1975). Local sites of activity-related glucose metabolism in rat olfactory bulb during olfactory stimulation. *Brain Research,* Vol. 98, No. 3, (November 1975), pp. 596-600. ISSN 0006-8993

Shepherd, G.M. (1994). Discrimination of molecular signals by the olfactory receptor neuron. *Neuron* Vol. 13, No. 4 (October 1994), pp. 771-790. ISSN 0896-6273

Shields, V.D.C. (1994). Ultrastructure of the uniporous sensilla on the galea of larval *Mamestra configurata* (Walker) (Lepidoptera: Noctuidae). *Canadian Journal of Zoology,* Vol. 72, No. 11, (November 1994), pp. 2016-31. ISSN 0008-4301

Shields, V.D.C. & Mitchell, B.K. (1995). Responses of maxillary styloconic receptors to stimulation by sinigrin, sucrose and inositol in two crucifer-feeding, polyphagous lepidopterous species. *Philosphical Transactions of the Royal Society of London B.*, Vol. 347, No. 1322, (March 1995), pp. 447-457. ISSN 0962-8436

Shields V.D.C. & Hildebrand J.G. (1999a). Fine structure of antennal sensilla of the female sphinx moth, *Manduca sexta* (Lepidoptera: Sphngidae). I. trichoid and basiconic sensilla. *Canadian Journal of Zoology*, Vol. 77, No. 2, (August 1999), pp. 290–301. ISSN 0008-4301

Shields V.D.C. & Hildebrand J.G. (1999b). Fine structure of antennal sensilla of the female sphinx moth, *Manduca sexta* (Lepidoptera: Sphingidae). II. auriculate, coeloconic, and styliform complex sensilla. *Canadian Journal of Zoology*, Vol. 77, No. 2, (August 1999), pp. 302–313. ISSN 0008-4301

Shields V.D.C. & Hildebrand J.G. (2001a). Responses of a population of antennal olfactory receptor cells in the female moth *Manduca sexta* to plant-associated volatile organic compounds. *Journal of Comparative Physiology A*, Vol. 186, No. 12, (February 2001), pp. 1135–1151. ISSN 0340-7594

Shields V.D.C. & Hildebrand J.G. (2001b). Recent advances in insect olfaction, specifically regarding the morphology and sensory physiology of antennal sensilla of the female sphinx moth *Manduca sexta*. *Microscopy Research and Technique*, Vol. 55, No. 5, (December 2001), pp. 307–329. ISSN 1059-910X

Shields, V.D.C. & Martin, T.L. (2010). The effect of alkaloids on the feeding of lepidopteran larvae. In: *Alkaloids: Properties, Applications and Pharmacological Effects*, Chapter 6, Cassiano, Nicole M. (ed), pp. (109-138), Nova Science Publishers, Inc., ISSN 9781617611308, Hauppauge, New York

Shields, V.D.C. & Martin, T.L. (2012). The structure and function of taste organs in caterpillars. In: *The Sense of Taste*, Chapter 11, Lynch, E.J., Petrov, A.P. (eds), pp. (147-166), Nova Science Publishers, Inc., ISSN 978-1-61209-748-0, Hauppauge, New York

Shields, V.D.C., Rodgers, E.J., Arnold, N.S., & Williams, D. (2006). Feeding responses to selected alkaloids by gypsy moth larvae, *Lymantria dispar* (L.). *Naturwissenschaften*, Vol. 93, No. 3, (March 2006), pp. 127-130. ISSN 0028-1042

Städler, E. (1984) Contact chemoreception. In: *Chemical Ecology of Insects*, Bell, W. J., Cardé, R.T. (eds), pp. (3-35), Chapman and Hall, ISSN 9780878930692, New York

Stopfer, M., Jayaramann, V., & Laurent, G. (2003). Intensity versus identity coding in an olfactory system. *Neuron*, Vol. 39, No. 6 (September 2003), pp. 991-1004. ISSN 0896-6273

Strausfeld, N.J. (1989). Cellular organization in male-specific olfactory neuropil in the moth *Manduca sexta*. In: *Dynamics and Plasticity in Neuronal Systems:* Proceedings of the 17th Göttingen Neurobiology Conference, Elsner N, Singer W, eds), p (79), Thieme, ISSN 3137380014, Stuttgart, Germany,

Sun, X.J., Tolbert, L.P., & Hildebrand, J.G. (1993). Ramification pattern and ultrastructural characteristics of the serotonin immunoreactive neuron in the antennal lobe of the moth *Manduca sexta*: a laser scanning confocal and electron microscopic study. *Journal of Comparative Neurology*, Vol. 338, No. 1 (December 1993), pp. 5-16. 1096-9861. ISSN 1096-9861

Thorne, N., Chromey, C., Bray, S., & Amrein, H. (2004). Taste perception and coding in *Drosophila. Current Biology,* Vol. 14, No. 12, (June 2004), 1065–1079. ISSN 0960-9822

Tumlinson, J.H., Brennan, M.M., Doolittle, R.E., Mitchell, E.R., Brabham, A., Mazomenos, B.E., Baumhover, A.H., Jackson, D.M. (1989). Identification of a pheromone blend attractive to *Manduca sexta* (L.) males in a wind tunnel. *Archives of Insect Biochemistry and Physiology*, Vol. 10, No. 4, (April 1989), pp. 255-271. ISSN 1520-6327

Van der Pers, J.N.C. & Den Otter, C.J. (1978). Single cell responses from olfactory receptors of small ermine moths to sex attractants. *Journal of Insect Physiology* Vol. 24, No. 4, (April 1978), pp. 337-343. ISSN 0022-1910

Vickers, N.J., Christensen, T.A., Baker, T.C., & Hildebrand, J.G. (2001). Odour-plume dynamics influence the brain's olfactory code. *Nature*, Vol. 410, No. 6827, (March 2001), pp. 466-470. ISSN 0028-0836

Visser, J.H. & Avé, D.A. (1978). General green leaf volatiles in the olfactory orientation of the Colorado beetle, *Leptinotarsa decemlineata. Entomologia Experimentalis et Applicata,* Vol. 24, No. 3, (November 1978) pp. 538-549. ISSN 0013-8703

Vosshall, L.B. (2001). The molecular logic of olfaction in *Drosophila. Chemical Senses,* Vol. 26. No. 2, (February 2001), pp. 207-213. ISSN 0379-864X

Vosshall, L.B., Amrein, H., Morozov, P.S., Rzhetsky, A., & Axel, R. (1999). A spatial map of olfactory receptor expression in the *Drosophila* antenna. *Cell,* Vol. 96, No. 5, (March 1999), pp. 725-736. ISSN 0092-8674

Wang, J.W., Wong, A.M., Flores, J., Vosshall, L.B., & Axel, R. (2003). Two-photon calcium imaging reveals an odor-evoked map of activity in the fly brain. *Cell* Vol. 112, No. 2 (January 2003), pp. 271-282. ISSN 0092-8674

Wang, Z., Singhvi, A., Kong, P., & Scott, K. (2004). Taste representations in the *Drosophila* brain. *Cell,* Vol. 117, No. 7, (June 2004). pp. 981–991. ISSN 0092-8674

Wanner, K.W. & Robertson, H.M. (2008). The gustatory receptor family in the silkworm moth *Bombyx mori* is characterized by a large expansion of a single lineage of putative bitter receptors. *Insect Molecular Biology,* Vol. 17, No. 6, (December 2008), pp. 621-629. ISSN 1365-2583

Wu, W., Anton, S., Löfstedt, C., & Hansson, B.S. (1996). Discrimination among pheromone component blends by interneurons in male antennal lobes of two populations of the turnip moth, *Agrotis segetum. Proceedings of the National Academy of Sciences,* Vol. 93, No. 15, (July 1996), pp. 8022-8027. ISSN 0027-8424

Yamamoto, R.T., Jenkins, R.Y., & McClusky, R.K. (1969). Factors determining the selection of plants for oviposition by the tobacco hornworm, *Manduca sexta. Entomologia Experimentalis et Applicata,* Vol. 12, No. 5, (December 1969), pp. 504-508. ISSN 0013-8703

Zar, J.H. (1984). Biostatistical analysis, 2nd edition, Englewood Cliffs, ISSN 9780130779250, Prentice Hall, NJ.

8

Histopathological Alterations in some Body Organs of Adult *Clarias gariepinus* (Burchell, 1822) Exposed to 4-Nonylphenol

Alaa El-Din H. Sayed[1*], Imam A. Mekkawy[1,2] and Usama M. Mahmoud[1]

[1]*Zoology Department, Faculty of Science, Assiut University, Assiut,*
[2]*Biology Department, Faculty of Science, Taif University, Taif,*
[1]*Egypt*
[2]*Saudi Arabia*

1. Introduction

Endocrine-disrupting chemicals (EDCs) include synthetic and naturally occurring chemicals that affect the balance of normal functions in animals (Razia et al. 2006). It has been found that exposure to natural and synthetic estrogenic chemicals may adversely affect wildlife and human health (Colborn et al. 1993). In vitro exposures (Soto et al. 1992; Soto et al. 1994; Toomey et al. 1999) have confirmed the effects of EDCs on tissue structure and cellular processes. Nonylphenol ethoxylates (NPEs) are EDCs which are used globally in the production of plastics, pesticides, and cleaning products and are present in sewage effluents around the world (Talmage, 1994). It has been reported that NP is the most important degradation product of NPEs because of its enhanced resistance towards biodegradation, toxicity, ability to bioaccumulate in aquatic organisms, and estrogenicity (Ahel et al. 1994). NP is found in surface waters, aquatic sediments, and ground water (Bennie, 1999; Talmage, 1994) and it is estrogenic in various aquatic animals (Nimrod and Benson, 1996; Talmage, 1994; Servos, 1999).

The application of environmental toxicological studies on non-mammalian vertebrates is rapidly expanding; and for aquatic system, fish have become valuable indicator for the evaluation of the effects of noxious compounds (Khidr and Mekkawy, 2008). Histology and histopathology can be used as biomonitoring tools for health in toxicity studies (Meyers and Hendricks, 1985). Histoplathological alterations are biomarkers of effect exposure to environmental stressors, revealing alterations in physiological and biochemical function (Hinton et al. 1992). Histopathology, the study of lesions or abnormalities on cellular and tissue levels is useful tool for assessing the degree of pollution, particularly for sublethal and chronic effects (Bernet et al. 1999). More than one tissue may be studied for assessment of the biological effects of a toxicant on localized portions of certain organs and also for assessment of subsequent derangements (degradations) in tissues or cells in other locations and this allows for diagnoses of the observed changes (Adeyemo, 2008). NP has been shown

*Corresponding Author

to cause histopathological changes in the germ and Sertoli cells of the male eelpot (Christiansen et al. 1998). The skin of fish is continuously exposed to and in direct contact with the environment pollutants such as NP. Histological changes in skin of rainbow trout with mucosomes in goblet cells were recorded after exposure to 10 μg/l of 4-nonylphenol. Several studies demonstrated the high susceptibility of skin to environmental pollutant impacts (Burkhardt-Holm et al. 1997; Iger et al. 1995; Shephard, 1994). Burkhardt-Holm et al. (1997) hypothesized that in trout, xenobiotic estrogens might affect the skin, like natural estrogens, via the steroid receptor.

In trout species, nonylphenol was found to accumulate in the liver, gill, skin, gut, fat, and kidney tissue (Ahel et al. 1993; Coldham et al. 1998; Lewis and Lech, 1996). So that, 4-nonylphenol may affects those organs in corresponding with its impacts on reproductive ones. Most of NP studies revealed sever effects on the liver and gonads of fish tissues (Christiansen et al. 1998; Jobling et al. 1996; Lech et al. 1996) and the corresponding metabolism. The liver is important in many aspects of nutrition, including lipid and carbohydrates storage and alterations in liver structure may be useful as biomarker that indicate prior exposure to environmental stressors (Hinton and laurén, 1990). Stressors-associated alterations of hepatocytes may be found in the nucleus or cytoplasm or both (Marchand et al. 2008). Malik and Hodgson, (2002) reported that the liver plays a major role in complex enzymatic processes of thyroid hormones conversion. So, liver dysfunction and disease affects thyroid hormone metabolism. Although gills are not only the prime organs for gaseous exchange, they perform several other physiological functions including osmoregulation and excretion. Parashar and Banerjee, (2002) reported that changes in environmental parameters often damage this delicate vital organ which has direct contact with aquatic environment. Many studies demonstrated that increased concentrations of different pollutants including several heavy metals seriously damage the gills of teleostean fish (Dutta et al. 1996; Wendelaar Bonga, 1997)).

African catfish (*Clarias gariepinus*), an omnivore freshwater fish, is a popular delicacy relished throughout tropical Africa (Nguyen and Janssen, 2002) due to fast growth rate, high stocking-density capacities, high consumer acceptability and high resistance to poor water quality and oxygen depletion (Adewolu et al. 2008; Akinwole and Faturoti, 2007; Karami et al. 2010). Because it is a prominent culture species (Adeyemo, 2008), the African catfish has been used in many fundamental experimental researches (Mahmoud et al., 2009).

The present work is an extension of previous studies of the present authors (Mekkawy et al., 2011; Mahmoud et al., 2011; Sayed et al., 2011) to determine to what extent the histopathological variations in some organs of the adult catfish, *Clarias gareipinus* (Burchell, 1822) are simultaneously correlated with biochemical and physiological NP-induced changes especially in respect with endocrine disruption.

2. Materials and methods

2.1 Specimen collection

Specimens of adult catfish *C. gariepinus* were collected from the River Nile at Assiut and then were transported to Fish Biology Laboratory of Zoology Department, Faculty of Science, Assiut University. The fish (500–1200 g) were fed on a commercial pellet diet (3% of body weight per day) and kept together in 100 l rectangular tanks containing tap water (conductivity 2000 ls/cm; pH 7.5; oxygen 88–95% saturation; temperature 27-28 °C; photoperiod 12:12 light: dark). After 2 week acclimatization, fishes were used for the experimental setup.

2.2 4-nonylphenol

4-Nonylphenol was obtained from Sigma- Aldrich (Schnelldrof, Germany)

2.3 Experimental setup

The adapted adult fish classified into four groups (10 fish per each): control, 4-nonylphenol-treated group (for15 day/ for 0.05mg/l day), 4-nonylphenol-treated group (for 15 day/for 0.08mg/l day), and 4-nonylphenol-treated group (for 15 day/for 0.1 h/ day). In the present study, the range of NP exposures was 0.05-0.1 mg/l and the exposure concentrations are environmentally relevant. The conditions of the experiment were as that of acclimatization with changing all the tap water and concentrations of 4-nonylphenol every day.

2.4 Hematoxylin-Eosin (HE) and Masson's Trichrome (TRI) histopathological preparations

For microscopic preparations, after 15 days, 3 surviving fish of each group were removed and dissected. Small pieces of the liver, kidneys, gills, and skin were taken and immediately fixed in 10% neutral buffered formalin. Fixed tissues were processed routinely for paraffin embedding technique. Embedded tissues were sectioned at 5-7μ in thickness and then stained with Harris' hematoxylin and eosin stain (H & E) and Masson's Trichrome (TRI) stain according to Bancroft and Steven, (1982). Sections were visualized and studied using OLYMPUS microscope model BX50F4 (Olympus optical Co., LTP. Japan).

2.5 Transmission electron microscope (TEM)

Small pieces of liver of newly scarified fish were fixed in 2 % glutaraldehyde, washed in cacodylate buffer and post-fixed in 1% osmium tetroxide. Dehydration was carried out in ascending grads of alcohol and then embedded in epon-araldite mixture. Semithin sections of liver were cut at 1μm and stained with toluidine blue for examination under a light microscope. Ultrathin sections were stained with uranyl acetate followed by lead nitrate (Johannessen, 1978). Electron micrographs were obtained using a Jeol JEM 1200 EX Transmission Electron Microscope at Electron microscope center of Assiut University.

2.6 Ethical statement

All experiments were carried out in accordance with the Egyptian laws and University guidelines for the care of experimental animals. All procedures of the current experiment have been approved by the Committee of the Faculty of Science, Assiut University, Egypt.

3. Results

Throughout the duration of the experiment, the number of fish that died was 1.2, 3.75, 5.5, and 8 % for control, 0.05, 0.08 and 0.1 mg/l 4-nonylphenol respectively. Lesions were observed in the gills, skin, kidneys, and liver of sampled fish for all 4-nonylphenol at all exposure concentrations and durations. The occurrence and degree of alterations were positively related with the concentrations of 4-nonylphenol while samples taken from the control group remained normal for all the organs throughout the duration of the experiment.

3.1 Histopathological changes in the gills

Histologically, the gills of the adult catfish *Clarias gariepinus* are composed of primary lamellae (pl), secondary lamellae (sl), epithelial cell (epc), mucous cell (mc), and chloride cell (chc) (Fig. 1a). The initial lesions in the gills were manifested in groups exposed to 0.05, 0.08, and 0.1 mg/l of 4-nonylphenol for 15 days (Fig. 1b, c, d, e). The anomalies include epithelial lifting, edema, deformed secondary lamellae in fishes exposed to 0.05 mg/l 4-nonylphenol (Fig. 1b) while in fishes exposed to 0.08 mg/l 4-nonylphenol, desquamation and necrosis were recorded (Fig. 1c). As Fig. (1d, e) shows gills with degeneration of cartilaginous bar malformed secondary lamellae, increase in chloride cell size and number, epithelial hyperplasia, diffusion of secondary lamellae and increase number of mucous cells in fishes exposed to 0.1 mg/l 4-nonylphenol for 15 days.

3.2 Histopathological changes in the skin

Normal structure of skin of adult catfish, *Clarias gariepinus* was shown in fig. (2a), where it consists of alarm cell (ac), mucous cell (mc) and epithelium (ep) with pigment cell (p). The fishes exposed to 0.05 mg/l of 4-nonylphenol showed enlarged alarm cell (eac), with vacuoles (va) in their skin structure (Fig. 2f). As shown in fig. (2c) other changes such as ruptured epithelial cells (repc) and enlarged mucous cells (emc) in the skin of fishes exposed to 0.08 mg/l of 4-nonylphenol for 15 days. Severe damage was recorded in fishes exposed to 0.1 mg/l of 4-nonylphenol as in Fig. (2b, d, e) in which ruptured epithelial cells, necrotic cell (nc), granuled cells (gc), vacuoles (va) and fat cells (fc) were recorded.

3.3 Histopathological changes in the kidney

The functional units of the kidney of the control fish, *Clarias gareipinus* are nephrons which are composed of renal corpuscles (rc) and renal tubules (rt); these structures are surrounded by haemopoietic tissue (ht). The shape of the renal corpuscle is roughly spherical consisting of a double membraned capsule (Bowman's capsule) enclosing a tuft of blood capillaries (glomerulus) (g). Bowman's space; a space between the glomerulus and the capsule (Fig. 3a). Examination of kidney sections of fish exposed to 0.05 mg/l of 4-nonylphenol for 15 days revealed edema in the epithelium lining of some renal tubules (e) and some showed degeneration (d) and rupture of Bowman's capsule (r). Hypertrophy of the glomerulus (hyt) was observed with shrinkage (sh). Moreover, necrosis (n) and pyknosis (p) were observed in some renal tubules (fig. 3b). After 15 days of exposure to the 0.08 mg/l of 4-nonylphenol, similar histological changes were observed, however, proliferation in renal tubules and haemopoieatic tissue (pr) with dissociation in some tubules (di) were recorded. Dilated blood vessels (dbv) and mealnomacrophages (m) were also observed (fig. 3c). The sections of kidney in fishes exposed to 0.1 mg/l of 4-nonylphenol showed severe damage or complete degeneration with obliterated Bowman's space (obs). Masson's Trichrome stain indicated the degeneration of connective tissue and degeneration of renal tubule and glomerulus (fig. 3d).

3.4 Histopathological changes in the liver

The liver of the control fish *Clarias gariepinus* appears as a continuous mass of hepatic cells; hepatocytes (h) which cord-like pattern interrupted by blood vessels and sinusoids (bs). The cords of hepatocytes are arranged around the central vein (cv). The hepatocytes are large in size, polygonal in shape with centrally located nuclei. The hepatocytes have homogenous eosinophilic cytoplasm. The sinusoids are seen as communicating channels occupied by blood cells with Küffer cells (kc) (fig. 4a). Examination of liver sections after exposure to 0.08 mg/l of 4-nonylphenol for 15 days showed degeneration (d) in the form of disintegration in most cytoplasmic contents. Lymphatic aggregations (la), necrosis (n), pyknosis (p), were observed (Fig. 4b). Also, melanomacrophages, pyknosis and rupture of hepatocytes (r) were recorded (fig. 4c). Less damage occurred in liver sections after exposure to 0.05 mg/l of 4-nonylphenol for 15 days (Fig. 4d). As Fig. (5a, b, c, d, e) shows marked severe damage occurred in fishes exposed to 0.1 mg/l of 4-nonylphenol for 15 days. Pyknosis indicated by arrows, fat cell (fc), lymphatic infiltration indicated by arrows, pigments diffusion, aggregation of fibers around central vein and rupture of hepatocytes were recorded. Masson's Trichrome stain indicated this severe damage in liver tissue (Fig. 5f).

3.5 Electron microscope examination of hepatocytes

The fine structure of the hepatocytes shows parallel cisternae of rough endoplasmic reticulum (rer), polygonal centrally located vesicular nuclei (n) with nucleolus (nu), numerous mitochondria (m) with different shapes and sizes and Golgi complex (g) near the nucleus (Fig. 6a).

Hepatocytes of animals exposed to 0.05 mg/l of 4-nonylphenol appeared swollen or hypertrophied with dense bodies (db), karyolysis in nucleus (fig. 6b) and damaged mitochondria (dm), rarified cytoplasm (rc) and vacuoles (v) (fig. 7b). Also, degenerative changes, shrunken and indented nuclei with cytoplasmic fat droplets (fd) were observed (Fig. 8b). In fishes exposed to 0.08 mg/l of 4-nonylphenol the cytoplasm shows tiney vacuoles (cv), the nuclei appeared irregular in shape with nuclear indentation. Some hepatocytes showed signs of karylyosis (Fig. 7a, 9a). Moreover, damaged mitochondria, degenerative rough endoplasmic reticulum (drer), increase in number of lysosomes (ly) were recorded. Some nuclei showed condensation and migration of chromatin at the nuclear periphery with prominent nucleolus with some apoptotic changes in the form of nuclear envelope (Fig. 9a). Some hepatocytes appeared swollen with large rarified areas in the cytoplasm resulting in disorganization and dissociation of cellular organelles (Fig. 7a). Hepatocytes of fishes exposed to 0.1 mg/l of 4-nonylphenol showed similar changes as those exposed to 0.05 and 0.08 mg/l of 4-nonylphenol, however, degenerative changes, hypertrophied, karylyosis, blood sinusoids collapsation, apoptosis and vacuolated hepatocytes (Fig. 10, 11). Mitochondria were swollen with destructive cristae; electron dense materials appeared at the periphery of these mitochondria (em) (Fig. 9b, 10a, 11,a, b). Concentric whorly organization of rough endoplasmic reticulum (cw) with detached ribosomes were also seen (Fig. 11a). Other regions of the reticulum appeared as parallel cisternae with electron lucent cytoplasm between its cisterna also, circular arrays of RER were appeared (Fig. 10a, 11a). An increase in the number of lysosomes and fat drops were seen (Fig. 10b, 11b).

4. Discussion

It has been reported that NP like E2 is estrogenic and affects the histology of developing immune and endocrine organs and those in direct contact with aquatic environment (Yokota *et al.*, 2001; Kang *et al.*, 2003; Seki *et al.*, 2003; Razia *et al.*, 2006). Skin and gills are highly sensitive to pollutants due to their direct contact to aquatic environment. It has been reported that the skin is sensitive to steroid hormone activity (Pottinger and Pickering, 1985). The present results showed severe damage in the skin epithelial cells and necrosis reflecting such sensitivity to NP. NP exposure of rainbow trout resulted in a specific granulation pattern of epidermal mucous cells visible as irregularly shaped and large mucosomes (Burkhardt-Holm *et al.*, 2000). The unique granulation pattern in skin of rainbow trout represents a suitable bioindicator for nonylphenol exposure (Burkhardt-Holm *et al.*, 2000). These latter authors stated that the structural alterations in the skin of the estradiol-injected trout is pointed to a physiological response such as, detached pavement cells, vacuolation of the cytoplasm and severely deformed cell nuclei at a dose of $10 \mu g l^{-1}$ which is lower than those in the present study. The damage occurred in the rainbow trout skin is similar to the quantitative changes of the mucous composition induced by hormones or environmental acidifications (Balm *et al.*, 1995). Schwaiger *et al.*, (1999) reported vitellogenin induction in the liver after exposure to $10 \ \mu g l^{-1}$ nonylphenol.

The present results exhibited severe damage in liver tissue of *C. gariepinus* including necrosis and decrease in the cell number along with vacuolation. Similar results were recorded by Uguz *et al.* (2003) who reported a significant increase in the Küpffer cells after one week of 4-nonylphenol exposure. Hughes *et al.* (2000) have shown NP-induced cell death. Galembeck *et al.* (1998), Hughes *et al.* (2000), and Uguz *et al.* (2003) reported that the disappearance of the cell membranes could be due to the lytic activity of alkylphenols.

In the present study, the liver cell borders disappeared and nuclei became larger after two weeks of exposure to $0.1 mg l^{-1}$ of 4-nonylphenol, this is similar to the findings of Uguz *et al.* (2003) who reported that this may be due to the increase in the DNA/ RNA ratio which was been observed in carcinogenic cells induced by NP (Chiriboga *et al.*, 2000a, b). The increase in the connective tissue with regenerating hepatocytes instead of normal liver tissues recorded in the present work was similar to those of Uguz *et al.* (2003). Such changes can be interpreted as an indication of carcinogenic development in the liver (Chiriboga *et al.*, 2000a, b; Calmak, 2001). Generally, the lesions detected in cells, tissues or organs are represent an integration of cumulative effects of physiological and biochemical stressors and therefore, can be linked to the exposure and subsequent metabolism of chemical contaminants (Adeyemo, 2008).

The gills are the primary initial target of toxicity, and the cytological changes in gill morphology in fish usually occur as a result of contaminant exposure. Gills have an extensive surface area and minimal diffusion distance between dissolved O_2 and blood capillary for efficient gaseous exchange. The fusion occurred in gills of fishes exposed to

4-nonylphenbol in this investigation may cause a drastic reduction in the respiratory surface area. However, very little is known about the toxic impact of 4-nonylphenol on the functional morphology of the gills. The present results indicated such toxic impacts. Increase in the number of mucous cells in gills of fishes exposed to 0.1 mgl^{-1} of 4-nonylphenol was recorded. It has been reported that the immediate morpho-pathological response of the gills to ambient xenobiotics is often manifested by a significant increase in the density of its mucous cells (Dutta, 1997, Hemalatha and Banerjee, 1997). The large quantity of mucous secretion acts as a defense mechanism against several toxic substances (Handy and Eddy, 1991; Mazon *et al.*, 1999). Similar to the findings of Dutta *et al.* (1996), the present results included many alterations such as increase in mucous and chloride cell number and size, necrosis, rupture of epithelium, desquamation, deformed secondary lamellae and oedema.

According to Peuranen *et al.* (1994) any discontinuity of epithelial lining of the gill lead to a negative ion balance and to changes in the haematocrite and mean cellular haemoglobin values of the blood. The number of chloride cells increased in the present study and this is similar to the results of Parashar and Banerjee, (2002). They stated that the number of chloride cells in the epithelial linings of both primary lamellae and secondary lamellae of *Heteropneustes fossilis* increased significantly following exposure to lead nitrate solution. Dutta *et al.* (1996) summarized the increased number of chloride cells in the gills of fishes following exposure to a variety of toxicants.

Increased ion permeability and sodium efflux of gill epithelial cells due to ethoxylate nonylphenol were reported in rainbow trout (Pärt *et al.*, 1985). Similar results in the present work were recorded in the histology of gills under NP-stress and confirmed by the increased NP-induced anion gap.

The kidney of fishes receives the largest proportion of the post-branchial blood and therefore renal lesions might be expected to be good indicators of environmental pollution (Cengiz, 2006). Many studies used histological characteristics of kidney as an indicators of pollution especially nonylphenol (Srivastava *et al.*, 1990; Banerjee and Bhattacharya, 1994; Ortiz *et al.*, 2003; Cengiz, 2006). In the present work, histological changes in the kidney after exposure to 4- nonylphenol were necrosis, hypertrophy of glomerulus, degeneration and dissociation of renal tubules and Bowman's capsule, proliferation in the renal tubule and haemopoieatic tissue, shrinkage of glomerulus, pyknosis, dilated blood vessel, rupture of Bowman's capsule, and obliterated Bowman's space. Similar results were reported in fishes after exposure to other pollutants (Cengiz, 2006; Khidr and Mekkawy, 2008; Abdel-Tawab and Al-Salahy, 2009).

From the results of the current study, it could be suggested that the exposure of adult catfish, Clarias gariepinus to sublethel doses of 4-nonylphenol caused moderate and severe damage to some organs such as gills, skin, kidney, and liver. These adverse effects of NP in gills, skin, kidney and liver were simultaneously correlated with sever biochemical, physiological changes in addition to endocrine disruption (Mekkawy et al., 2011; Mahmoud et al., 2011; Sayed et al., 2011) So, it is concluded that NP works as estrogenic and non-estrogenic factor leadings to general and specific metabolism disruption in different pathway.

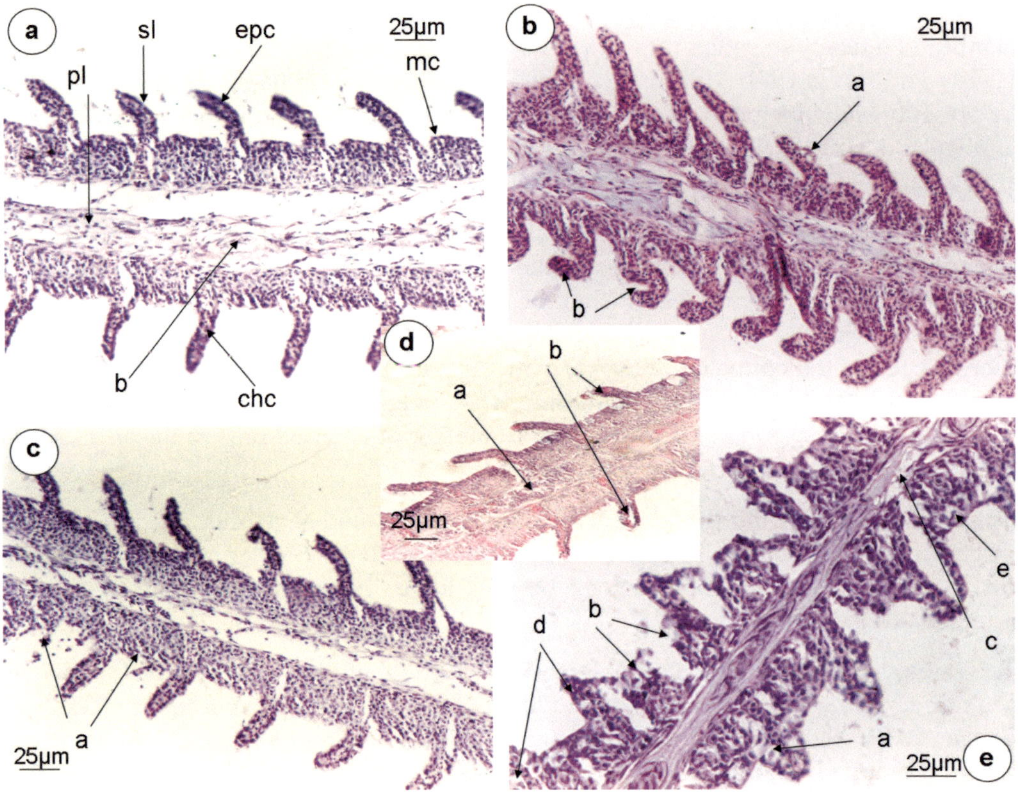

Fig. 1. (a) Gill structure of control adult fish *Clarias gariepinus*. (pl), primary lamellae; (sl) secondary lamellae; (epc) epithelial cell; (mc) mucous cell; (chc) chloride cell. (b) Gill tissue exposed to 0.05 mg/l 4-nonylphenol for 15 days showing a, epithelial lifting and oedema and b, deformed secondary lamellae. (c) Gill tissue exposed to 0.08 mg/l of 4-nonylphenol for 15 days showing a, desquamation and necrosis. (d)) Gill tissue exposed to 0.1 mg/l of 4-nonylphenol for 15 days showing a, degeneration of cartilaginous bar and b, malformed secondary lamellae. (e) Gill tissue exposed to 0.1 mg/l of 4-nonylphenol for 15 days showing a, increase in chloride cell size and number; b, epithelial hyperplasia and diffusion of secondary lamellae; c, degeneration and vacuolation of cartilaginous bar; d, desquamation and necrosis and e, increase number of mucous cells. Stain H& E. Magnification a, b, c, e (400X) and d (100X).

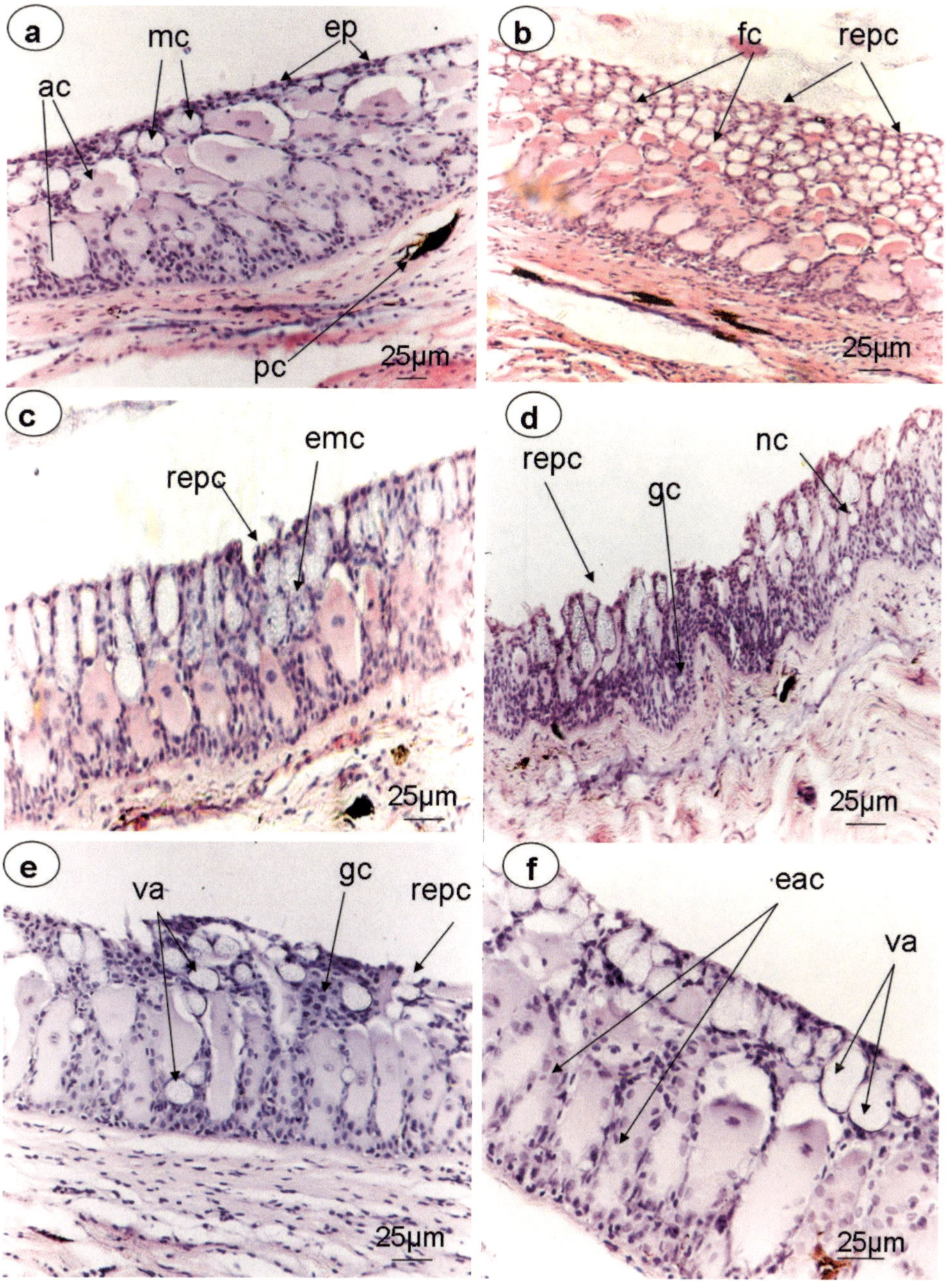

Fig. 2. Vertical sections of the skin of adult catfish *Clarias gariepinus* stained with H&E. (a) skin of control fish showing ac, alarm cell; mc, mucous cell; ep, epithelium; p, pigment cell. (b, d, e) skin of fish exposed to 0.1 mg/l of 4-nonylphenol for 15 days showing fc, fat cell; repc, ruptured epithelial cells; nc, necrotic cells; gc, granuled cells and va, vacuoles.(c) skin of exposed fish to 0.08 mg/l 4-nonylphenol for 15 days showing repc, ruptured epithelial cells and emc, enlarged mucous cells. (f) skin of exposed fish to 0.05 mg/l 4-nonylphenol for 15 days showing eac, enlarged alarm cells and va, vacuoles. Magnification a, b, c, d, e (100X) and f (400X).

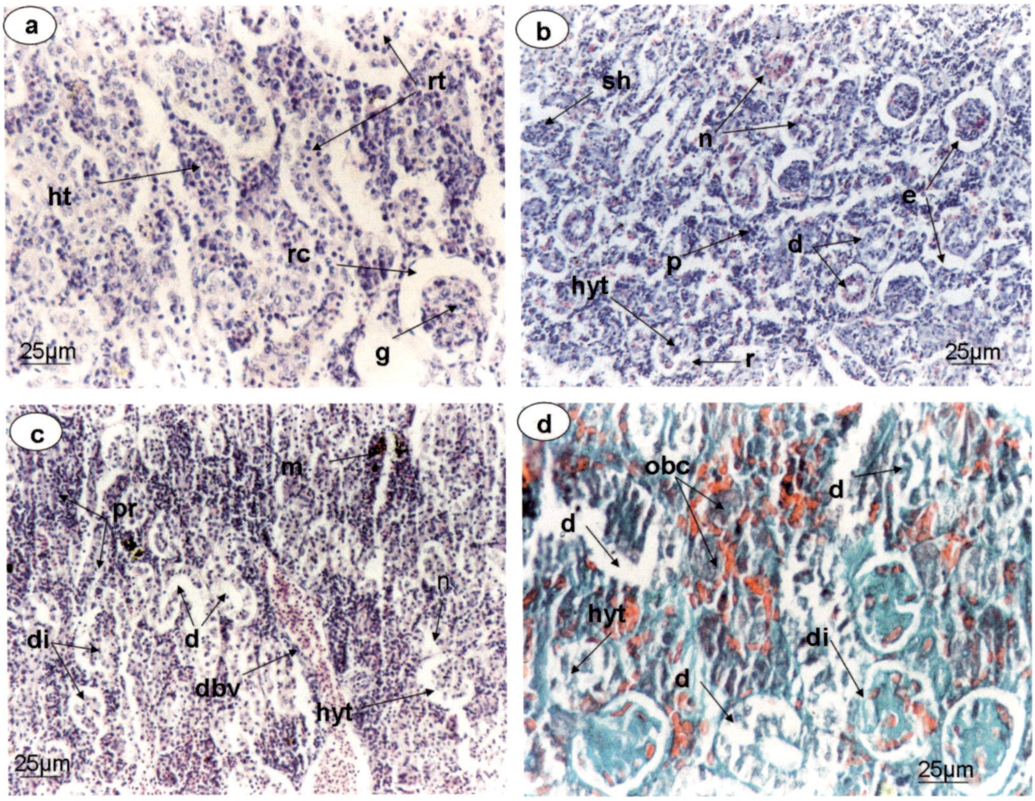

Fig. 3. Transverse sections of kidney of the *C. gariepinus*. (a) Control, (b) fish exposed to 0.08 mg/l of 4-nonylphenol for 15 days, (c) fish exposed to 0.05 mg/l 4-nonylphenol for 15 days, (d) fish exposed to 0.1 mg/l 4-nonylphenol for 15 days. ht, haemopoietic tissue; g, glomerulus; rt, renal tubules; rc, renal corpuscles; n, necrosis; hyt, hypertrophy of glomerulus; d, degeneration; di, dissociation; e, edema of renal tubules and Bowman's capsule; m, melanomacrophages; pr, proliferation in the renal tubules and haemopoieatic tissue; sh, shrinkage of glomerulus; p, pyknosis; dbv, dilated blood vessel; r, rupture of Bowman's capsule; obs, obliterated Bowman's space a, b and c Staind with H&E and d stained with masson's trichrome. Magnification a and d (400X), b and c (200X).

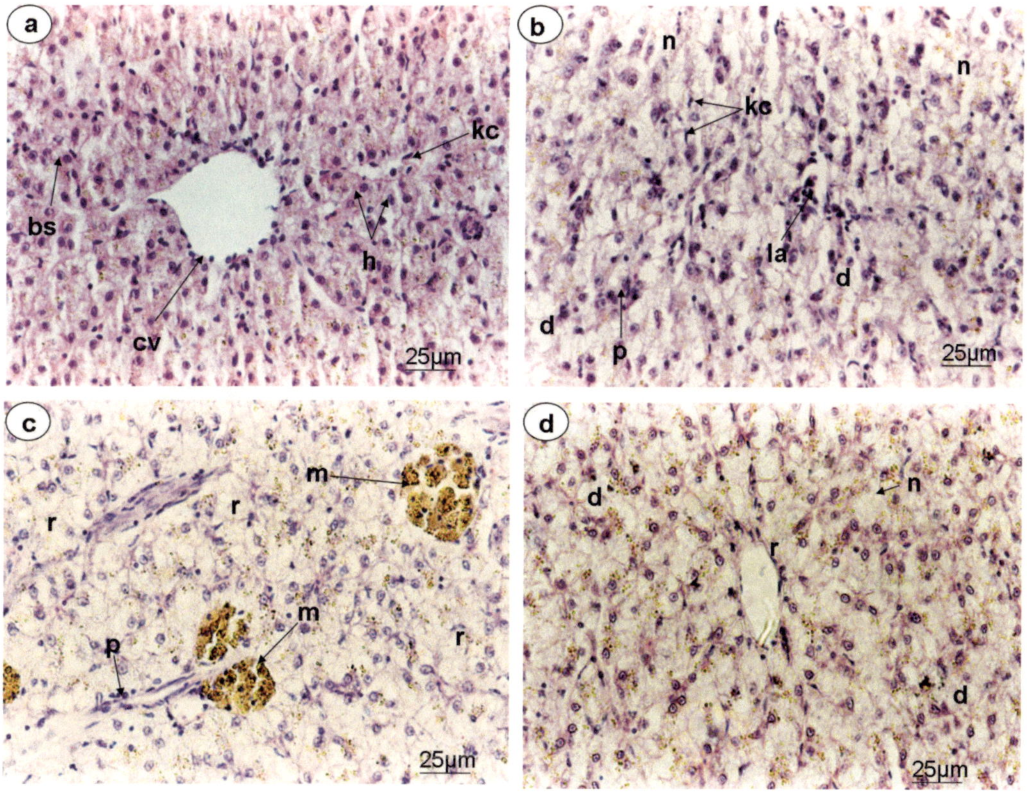

Fig. 4. Sections of liver of the *C. gariepinus.* (a) Control showing cv, central vein; bs, blood sinusoids and h, hepatocyte (b) fish exposed to 0.08 mg/l of 4-nonylphenol for 15 days showing n, necrosis; kc, küpffer cell; la, lymphatic aggregation; d, degeneration and p, pyknosis (c) fish exposed to 0.08 mg/l of 4-nonylphenol for 15 days showing m, melanomacrophages; p, pyknosis; r, rupture of the hepatocytes (d) fish exposed to 0.05 mg/l of 4-nonylphenol for 15 days showing d; d, degeneration; r, rupture of the cell membrane of central vein; n, necrosis. a, b, c and d stained with H&E, Magnification (400X).

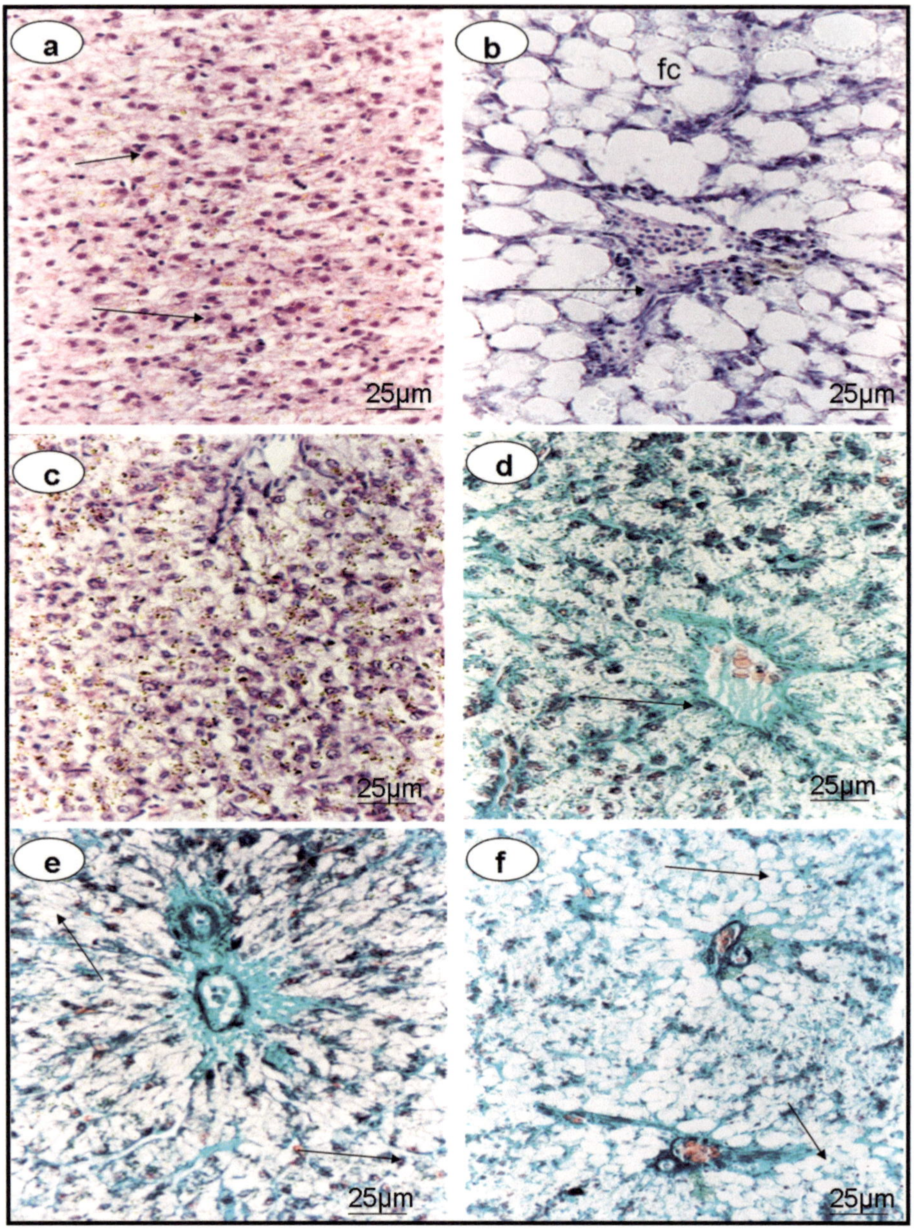

Fig. 5. Sections of liver of the *C. gariepinus.* (a) Fish exposed to 0.1 mg/l of 4-nonylphenol for 15 days showing pyknosis (arrows) (b) fish exposed to 0.1 mg/l of 4-nonylphenol for 15 days showing fc, fat cell and lymphatic infiltration (arrows) (c) fish exposed to 0.1 mg/l of 4-nonylphenol for 15 days showing pigments diffusion (d) fish exposed to 0.1 mg/l of 4-nonylphenol for 15 days showing aggregation of fibres around central vein (arrows) (e) fish exposed to 0.1 mg/l of 4-nonylphenol for 15 days showing rupture of hepatocytes and aggregation of fibres around the central vein (arrows) (f) fish exposed to 0.1 mg/l of 4-nonylphenol for 15 days showing accumulation of fats as fat cells. a, b and c staind with H&E and d, e and f stained with mssson's trichrome. Magnification (400X).

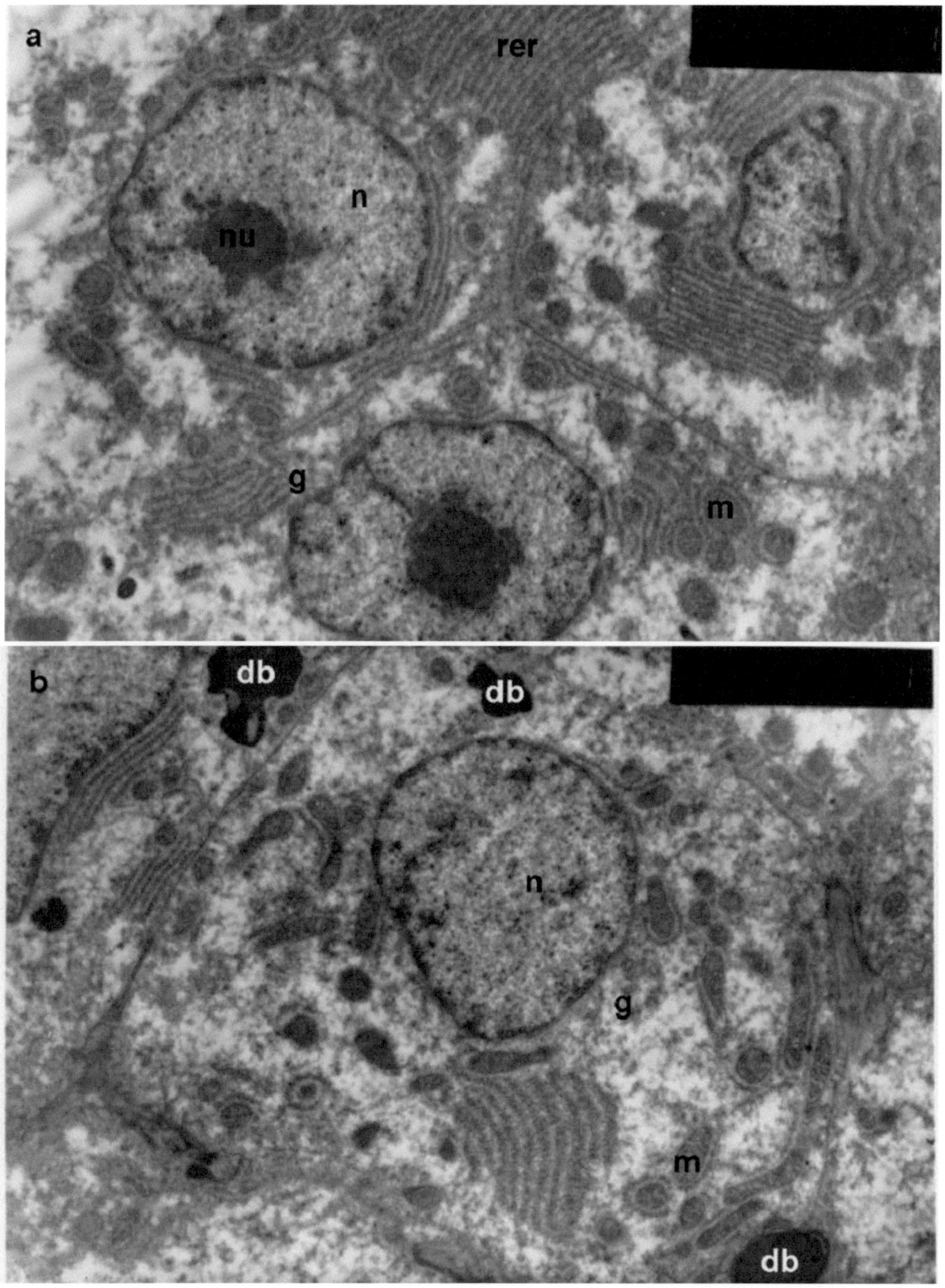

Fig. 6. Transmission electron micrograph of hepatocytes of catfish *Clarias gariepinus*. (a) control (X5000), (b) fish treated with 0.05 mg/l of 4-nonylphenol for 15 days (X5000). (n) nucleus, (nu) nuculeolus, (m) mitochondria, (rer) rough endoplasmic reticulum,(g) Golgi apparatus and (db) dense body.

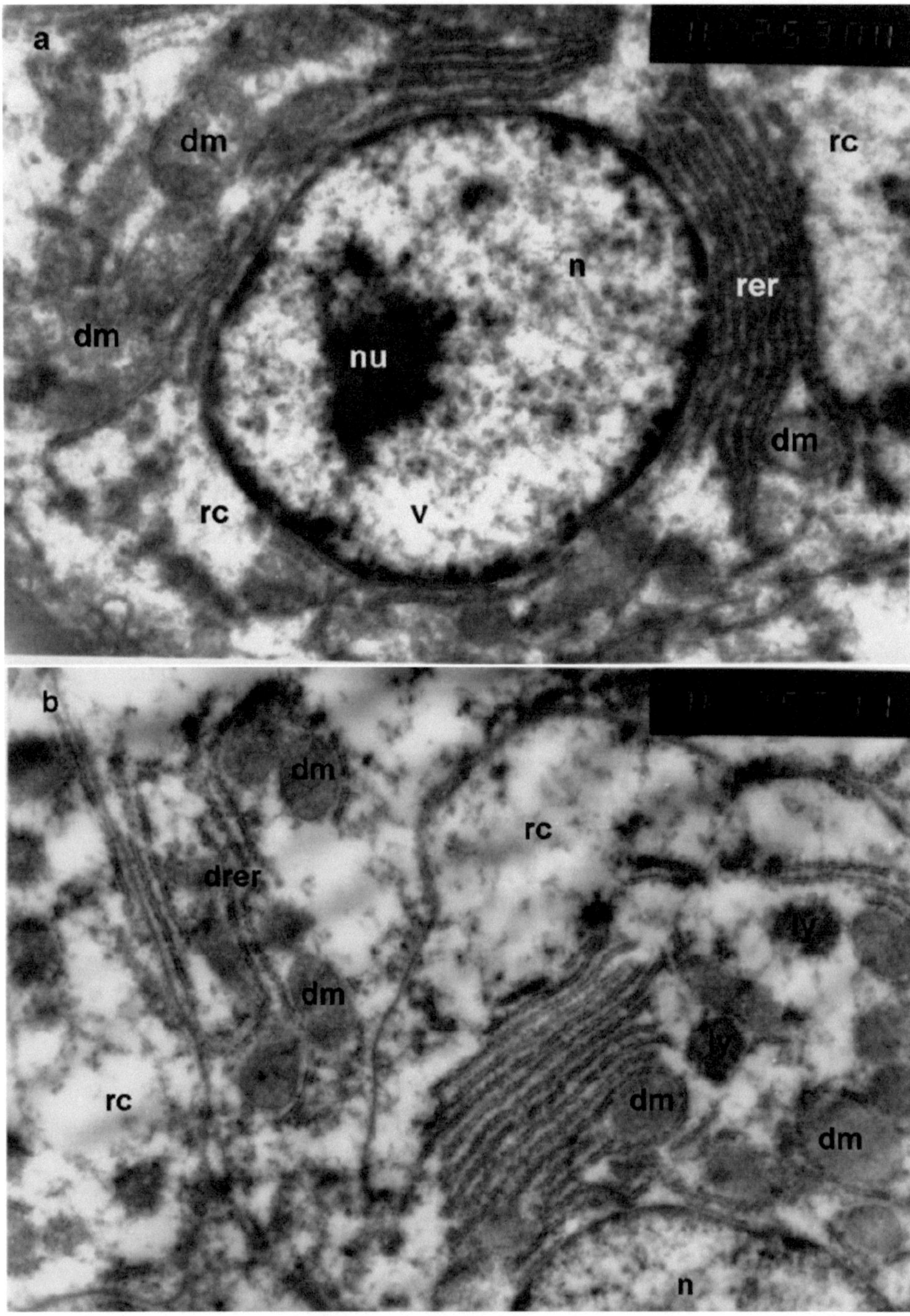

Fig. 7. Transmission electron micrograph of hepatocytes of catfish *Clarias gariepinus* showing marked degeneration of hepatocytes. (a) fish treated with 0.08 mg/l of 4-nonylphenol for 15 days (X8000), (b) fish treated with 0.05 mg/l of 4-nonylphenol for 15 days (X8000). (n) nucleus, (nu) nuculeolus, (dm) damaged mitochondria, (drer) degenerated rough endoplasmic reticulum,(rc) rarfied cytoplasm (v) vacuoles and (ly) lysosomes.

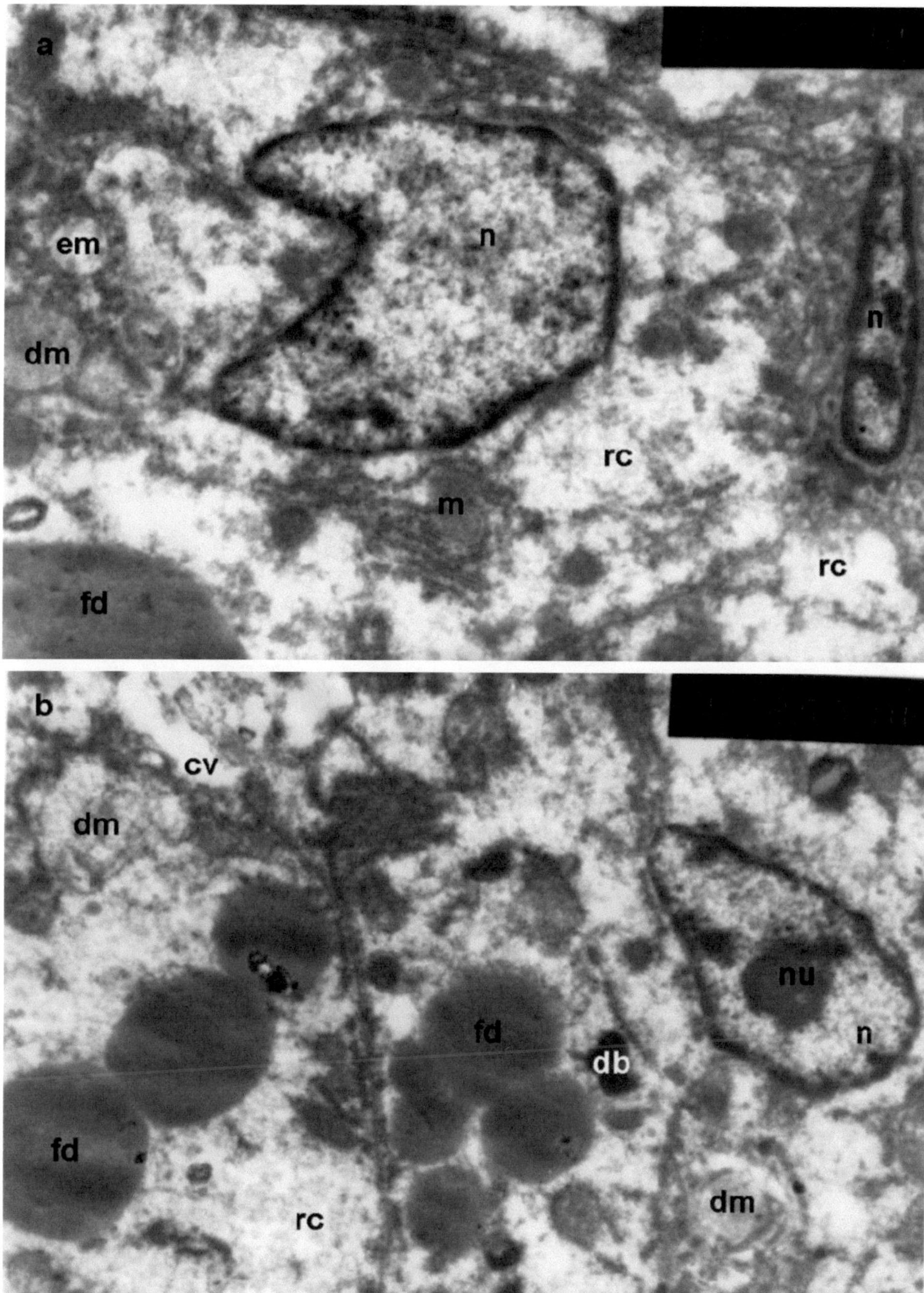

Fig. 8. Transmission electron micrograph of hepatocytes of catfish *Clarias gariepinus* showing marked degeneration of hepatocytes. (a) fish treated with 0.1 mg/l of 4-nonylphenol for 15 days (X5000), (b) fish treated with 0.05 mg/l of 4-nonylphenol for 15 days (X5000). (n) nucleus, (nu) nucleolus , (dm) damaged mitochondria, (em) empty mitochodria, (m) mitochondria, (rc) rarified cytoplasm (fd) fat drops and (db) dense body.

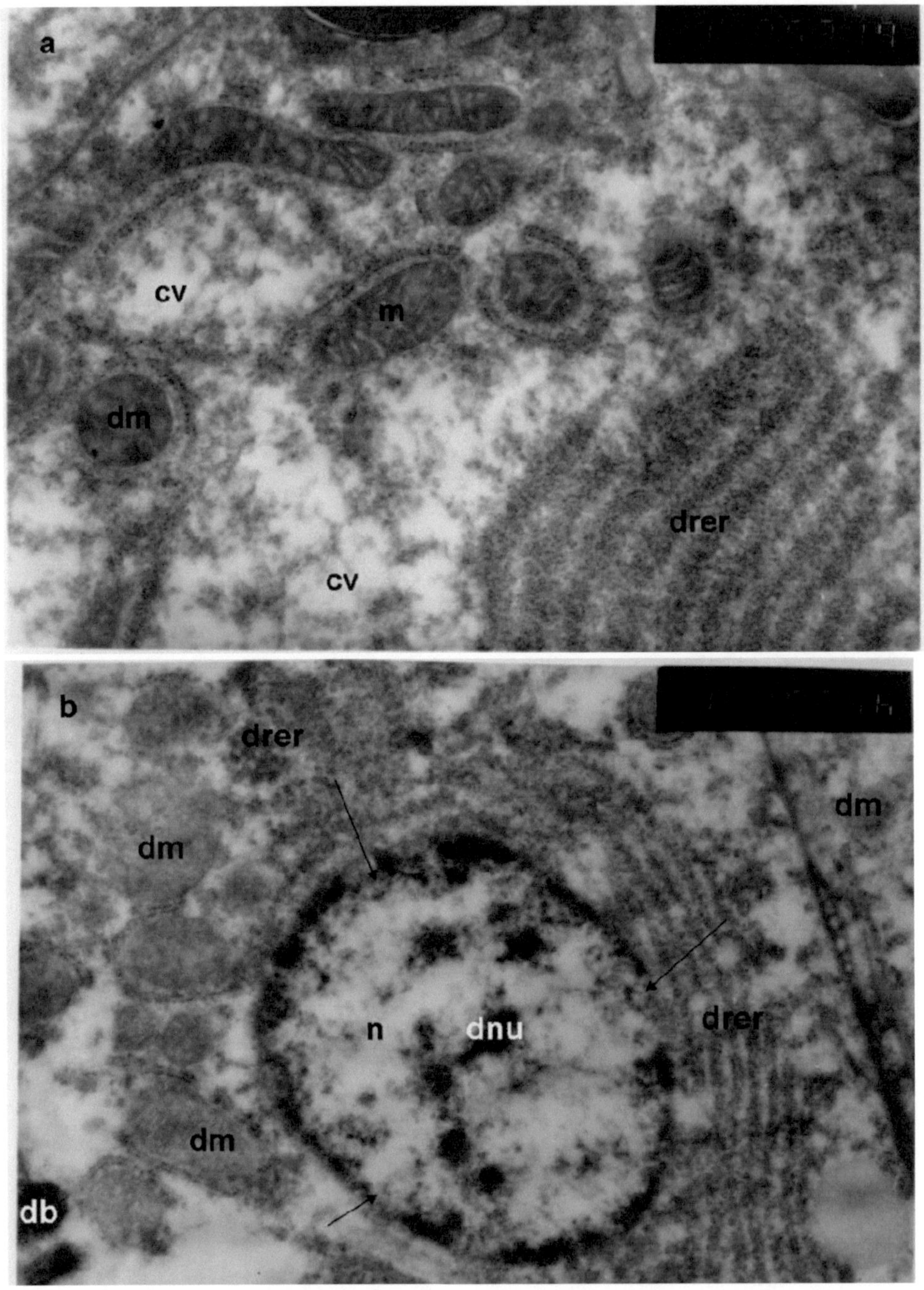

Fig. 9. Transmission electron micrograph of hepatocytes of catfish *Clarias gariepinus* showing marked degeneration of nuclear envelope (arrows) of hepatocytes. (a) fish treated with 0.08 mg/l of 4-nonylphenol for 15 days (X8000), (b) fish treated with 0.1 mg/l of 4-nonylphenol for 15 days (X8000). (n) nucleus, (nu) nuculeols, (dnu) degenerated nucleolus, (dm) damaged mitochondria, (em) empty mitochodria, (cv) cytoplasm vacuoles, (drer) degenerated rough endoplasmic reticulum and (db) dense body.

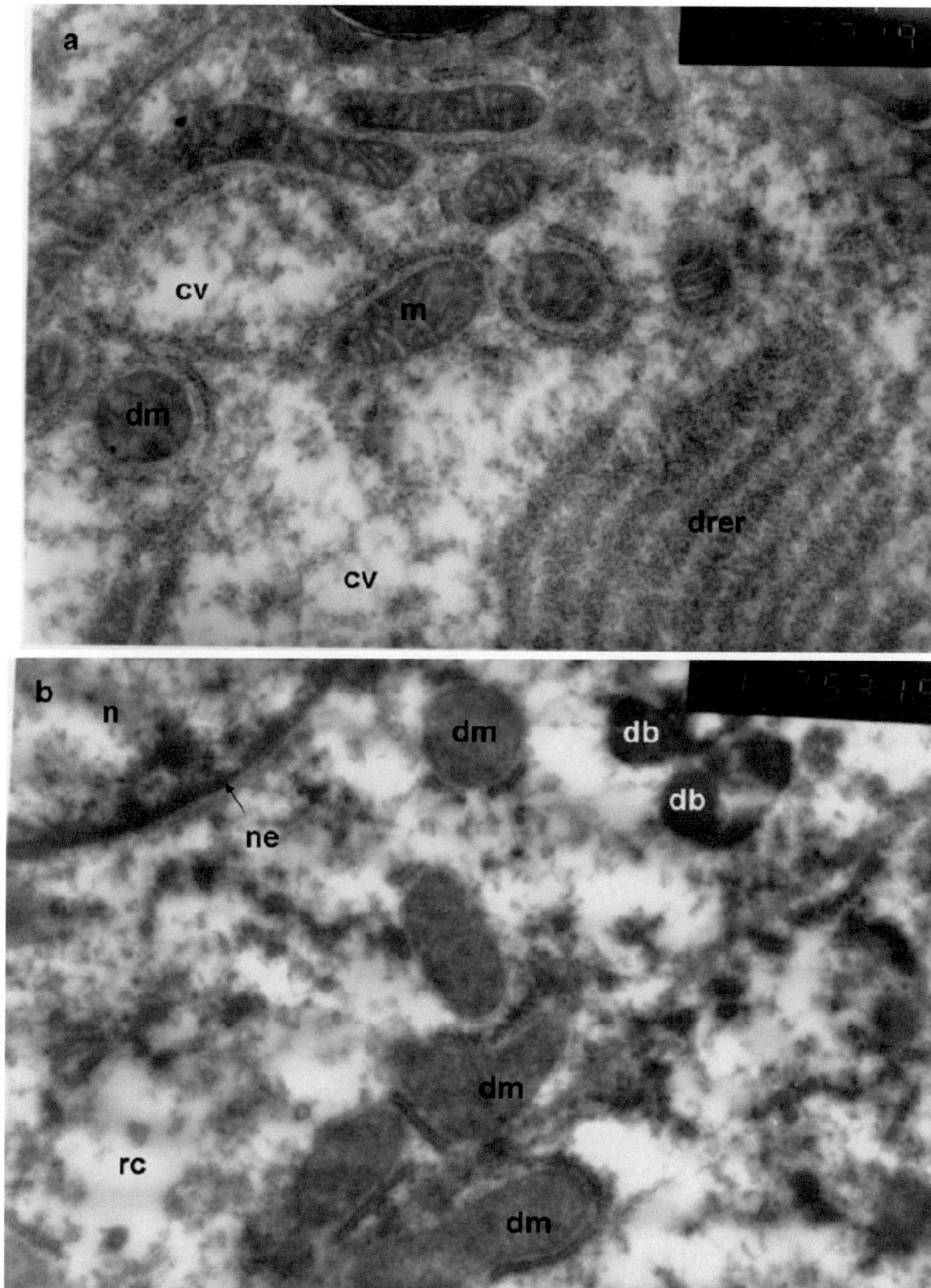

Fig. 10. Transmission electron micrograph of hepatocytes of catfish *Clarias gariepinus* showing marked degeneration of hepatocytes. (a) fish treated with 0.1 mg/l of 4-nonylphenol for 15 days (X14000), (b) fish treated with 0.1 mg/l of 4-nonylphenol for 15 days (X5000). (n) nucleus, (nu) nucleolus, (drer) degenerated rough endoplasmic reticulum, (dm) damaged mitochondria, (m) mitochodria, (rc) rarified cytoplasm, (cv) cytoplasm vacuoles, (fd) fat droplets and (db) dense body.

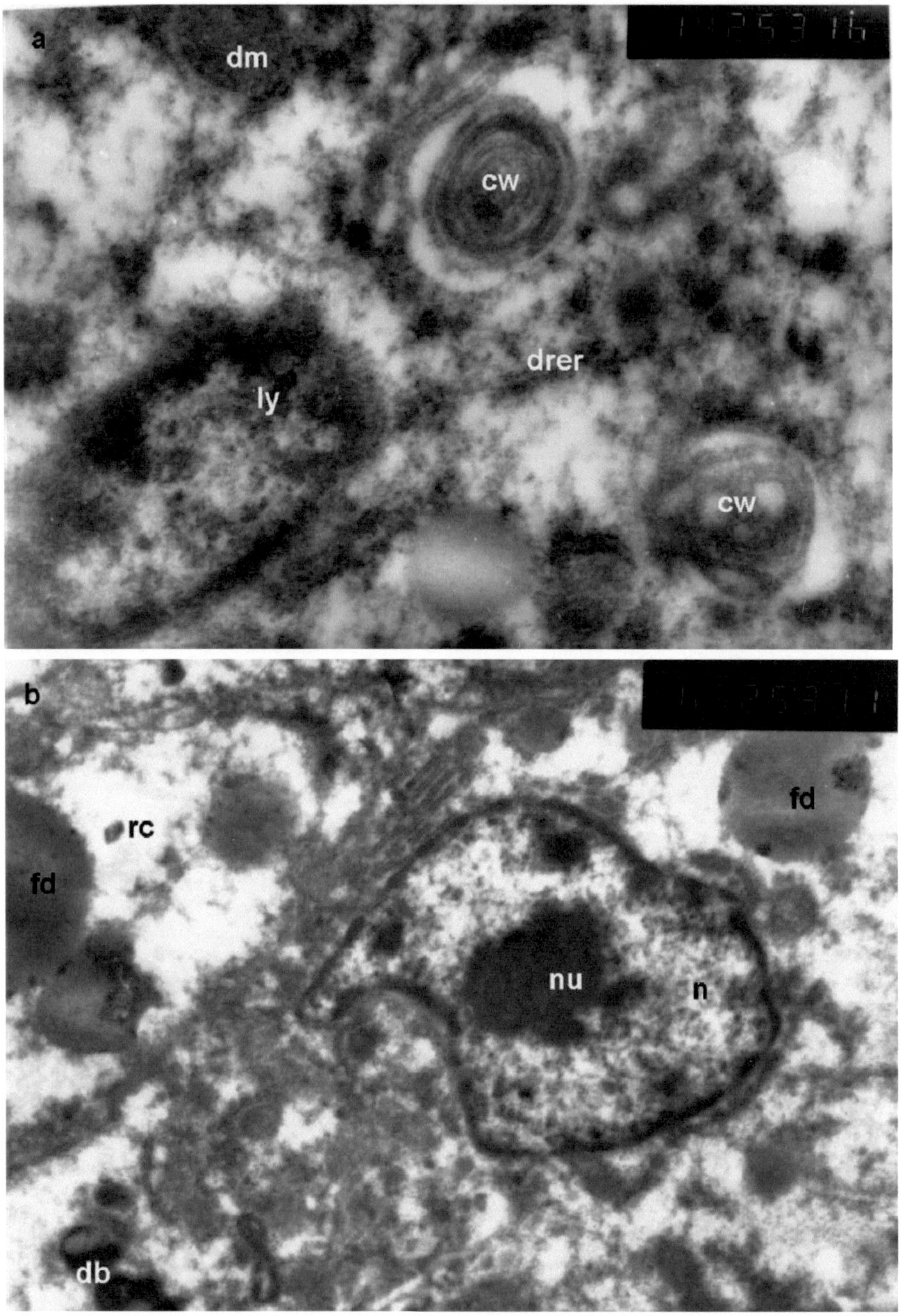

Fig. 11. Transmission electron micrograph of hepatocytes of catfish *Clarias gariepinus* showing marked degeneration of hepatocytes. (a) fish treated with 0.1 mg/l of 4-nonylphenol for 15 days (X14000), (b) fish treated with 0.1 mg/l of 4-nonylphenol for 15 days (X10000). (n) nucleus, (ne) nucleolus envelope, (drer) degenerated rough endoplasmic reticulum, (dm) damaged mitochondria, (cw) concentric whorls appearance of RER, (rc) rarified cytoplasm, (ly) lysosomes and (db) dense body.

5. References

Abdel-Tawab, H.S. & Al-Salahy, M.B. (2010). Biocemical and ultrastructural studies of the effect of garlic juice on the liver of the fish *Clarias gariepinus*. Journal of Egyptian Germany Society of Zoology Vol. 60C: 39-62.

Adewolu, M.A., Adeniji, C.A. & Adejobi, A.B. (2008). Feed utilization, growth and survival of *Clarias gariepinus* (Burchell 1822) fingerlings cultured under different photoperiods. Aquaculture Vol. 283: 64–67.

Adeyemo, O.K. (2008). Histological Alterations Observed in the Gills and Ovaries of *Clarias gariepinus* exposed to environmentally relevant lead concentrations. Journal of Environmental Health Vol. 70 (9): 48-51.

Ahel, M., Giger, W. & Koch, M. (1994). Behaviour of alkylphenol polyethoxylate surfactants in the aquatic environment - I. Occurrence and transformation in sewage treatment. Water Resources Vol. 28: 1131–1142.

Ahel, M., McEvoy, J. & Giger, W. (1993). Bioaccumulation of the lipophilic metabolites of nonionic surfactants in freshwater organisms. Environmental Pollution. Vol. 79: 243-248.

Akinwole, A.O. & Faturoti, E.O. (2007). Biological performance of African Catfish (*Clarias gariepinus*) cultured in recirculating system in Ibadan. Aquaculture Enginering. Vol. 36: 18–23.

Balm, P.H.M., Iger, Y., Prunet, P., Pottinger, T.G. & Wendelaar Bonga, S.E. (1995). Skin ultrastructure in relation to prolactin and MSH function in rainbow trout (*Oncorhynchus mykiss*) exposed to environmental acidification. Cell Tissue Research. Vol. 279: 351-358.

Bancroft, J. & Stevens A, (1982). Theory and Practice of Histological Techniques, 2nd Ed. Churchill-Livingston, NY, pp 131-135.

Banerjee, S. & Bhattacharya, S. (1994). Histopathology of kidney of *Channa punctatus* exposed to chronic nonlethal level of elsan, mercury and ammonia. Ecotoxicology and Environmental Safety. Vol. 29 (3): 65–275.

Bennie, D.T. (1999). Review of the environmental occurrence of alkylphenols and alkylphenol ethoxylates. Water Quality Research Journal of Canada. Vol. 34: 79–122.

Bernet, D., Schmidt, H., Meier, W., Brkhardt-Holm, P. & Wahli, T. (1999). Histopathology in fish: Proposal for a protocol to assess aquatic pollution. Journal of Fish Diseases. Vol. 22: 25–34.

Burkhardt-Holm, P., Escher, M. & Meier, W. (1997). Waste water management plant effluents cause cellular alterations in the skin of brown trout *Salmo trutta*. Journal of Fish Biology. Vol. 50: 744-758.

Burkhardt-Holm, P., Wahli, T. & Meier, W. (2000). Nonylphenol affects the granulation pattern of epidermal mucous cells in rainbow trout, *Oncorhynchus mykiss*. Ecotoxicology and Environmental Safety. Vol. 46: 34–40.

Calmak, G. (2001). Spectroscopic analysis of the livers exposed to nonylphenol in rainbowtrout (*Onchorynchus mykiss*). MS thesis, Biology Department, Middle East Technical University, Turkey.

Cengiz, E. I. (2006). Gill and kidney histopathology in the freshwater fish *Cyprinus carpio* after acute exposure to deltamethrin. Environmental Toxicology Pharmacology. Vol. 22: 200–204.

Chiriboga, L., Yee, H. & Diem, M. (2000a). Infrared spectroscopy of human cells and tissue. Part VI: a comparative study of histology and infrared microspectroscopy of normal, cirrhotic, and cancerous liver tissue. Applied Spectroscopy. Vol. 54 (1): 1–8.

Chiriboga, L., Yee, H. & Diem, M. (2000b). Infrared spectroscopy of human cells and tissue. Part VII: FT-IR microspectroscopy of DNase- and RNase-treated normal, cirrhotic, and neoplastic liver tissue. Applied. Spectroscopy. Vol. 54 (4): 480–485.

Christiansen, T., Korsgaard, B. & Jespersen, A. (1998). Effects of nonylphenol and 17β-oestradiol on vitellogenin synthesis, testicular structure and cytology in male eelpout *Zoarces viviparus*. Journal of Experimental of Biology. Vol. 201: 179-192.

Colborn, T., Vom Saal, F.S. & Soto, A.M. (1993). Developmental effects of endocrine disrupting chemicals in wildlife and humans. Environmental Health Perspectives. Vol. 101: 378–384.

Coldham, N.G., Sivapathasundaram, S., Dave, M., Ashfleld, L.A., Pottinger, T.G., Goodall, C. & Sauer, M.J. (1998). Biotransformation, tissue distribution, and persistence of 4-nonylphenol residues in juvenile rainbow trout (*Oncorhynchus mykiss*). Drug Metabolism Dispososition. Vol. 26: 347-353.

Dutta, H.M. (1997). A composite approach for evaluation of the effect of pesticides on fish, In: Munshi, J. S. D., Dutta, H. M. (Eds.), Fish Morphology, Horizon of new research Science Publisher Inc., USA, pp. 249-277.

Dutta, H.M., Munshi, J.S.D., Roy, P.K., Singh, N.K., Adhikari, S. & Killius, J. (1996). Ultrastructural changes in the respiratory lamellae of the catfish, *Heteropneustes fossilis* after sublethal exposure to mclathion. Environmental Pollution. Vol. 92: 329-341.

Galembeck, E., Alonso, A. & Meirelles, N.C. (1998). Effects of polyoxyethylene chain length on erythrocyte hemolysis induced by poly[oxyethylene(n)nonylphenol] non-ionic surfactants. Chemico-Biollogical Interactions. Vol. 113 (2): 91–103.

Handy, R.D. & Eddy, F.B. (1991). The absence of mucous on the secondary lamellae of unstressed rainbow trout, *Oncorhynchus mykiss*. Journal of Fish Biology. Vol. 38: 153-155.

Hemalatha, S. & Banerjee, T.K. (1997). Histopathological analysis of sublethal toxicity of zinc chloride to the respiratory organs of the air-breathing catfish *Heteropneustes fossilis* (Bloch). Biological Research. Vol. 30: 11-21.

Hinton, D.E., Baumann, P.C., Gardner, G.R., Hawkins, W.E., Hendricks, J.D., Murchelano, R.A. & Okihiro, M.S. (1992). Histopathological biomarkers, In: Huggett, R.J., Kimerle, R.A., Mehrle, P.M., Jr Bergman, H.L., (Eds.), Biomarkers: Biochemical, Physiological, and Histological Markers of Anthropogenic Stress, Lewis, USA, pp 155–196.

Hinton, D.E. & Lauren, D.J. (1990). Integrative histopathological approaches to detecting effects of environmental stressors on fishes. Am. Fish Soc. Sym. Vol. 8: 51–66.

Hughes, P.J., McLellan, H., Lowes, D.A., Khan, S.Z., Bilmen, J.G., Tovey, S.C., Godfrey, R.E., Michell, R.H., Kirk, C.J. & Michelangeli, F. (2000). Estrogenic alkylphenols induce cell death by inhibiting testis endoplasmic reticulum Ca^{2+} pumps. Biochemical and Biophysical Research Communications. Vol. 277: 68–574.

Iger, Y., Balm, P.H., Jenner, H.A. & Wendelaar Bonga, S.E. (1995). Cortisol induces stress-related changes in the skin of rainbow trout (*Oncorhynchus mykiss*). General and Comparative Endocrinology. Vol. 97: 188-198.

Jobling, S., Sheahan, D., Osborne, J.A., Matthiessen, P. & Sumpter, J. P. (1996). Inhibition of testicular growth in rainbow trout (*Oncorhynchus mykiss*) exposed to estrogenic alkylphenolic chemicals. Environmental Toxicology and Chemistry. Vol. 15: 194-202.

Johannessen, J. (1978). Instruction and techniques in Electron Micrscopy in human medicine. Mchgraw-Hill Int. Book Co.

Karami, A., Christianus, A., Ishak, Z., Courtenay S. C., Syed, M. A., Noor Azlina, M. & Noorshinah H. (2010). Effect of triploidization on juvenile African catfish (*Clarias gariepinus*). Aquaculture International. Vol. 18: 851–858.

Khidr, M. B. & Mekkawy, I.A.A. (2008). Effect of separate and combined lead and selenium on the liver of the cichlid fish *Oreochromis niloticus*: ultrastructural study. Egyptian Journal of Zoology. Vol. 50: 89-119.

Lech, J.J., Lewis, S.K. & Ren, L. (1996). In vivo estrogenic activity of nonylphenol in rainbow trout. Fundam. Applied Toxicology. Vol. 30: 229-232.

Lewis, S.K. & Lech, J.J. (1996). Uptake, disposition and persistence of nonylphenol from water in rainbow trout (*Oncorhynchus mykiss*). Xenobiotica. Vol. 26: 813-819.

Malik, R. & Hodgson, H. (2002). The relationship between the thyroid gland and the liver. Quarterly Journal of Medicine. Vol. 95: 559-569.

Marchand, M.J., van Dyk, J.C., Pieterse, G.M., Barnhoorn, I.E.J. & Bornman, M.S., (2008). Histopathological Alterations in the Liver of the Sharptooth Catfish *Clarias gariepinus* from Polluted Aquatic Systems in South Africa. Environmental Toxicology. DOI 10.1002/tox.

Mazon, A.F., Cerqueira, C.C.C., Monteiro, E.A.S. & Fernandes, M.N. (1999). Acute copper exposure in freshwater fish, Morphological and physiological effects. In: Val, A.L., Almeida-Val, V.M.F., (Eds.), Biology of Tropical Fishes, INPA, Manaus, pp. 263-275.

Meyers, T.R. & Hendricks, J.D. (1985). Histopathology. In: Loux, D.B., Dorfman, M., (Eds.), Fundamentals of Aquatic Toxicology: Methods and Applications, Hemisphere USA, pp. 283–330.

Nguyen, L.T.H. & Janssen, C.R. (2002). Embryo-larval toxicity tests with the African catfish (*Clarias gariepinus*): comparative sensitivity of endpoints. Archives of Environmental Contamination and Toxicology. Vol. 42: 256–262.

Nimrod, A.C. & Benson, W.H. (1996). Environmental estrogenic effects of alkylphenol ethoxylates. Critical Reviews in Toxicology. Vol. 26: 335–364.

Ortiz, J.B., De Canales, M.L.G. & Sarasquete, C. (2003). Histopathological changes induced by lindane (gamma-HCH) in various organs of fishes. Scientia Marina. Vol. 67 (1): 53–61.

Parashar, R.S. & Banerjee, T.K. (2002). Toxic impact of lethal concentration of lead nitrate on the gills of air-breathing catfish *Heteropneustes fossilis* (Bloch). Veterinarski Arhiv. Vol. 72 (3): 167-183.

Pärt, P., Svanberg, O. & Bergstrom, E. (1985). The influence of surfactants on gill physiology and cadmium uptake in perfused rainbow trout gills. Ecotoxicology and Environmental Safety. Vol. 9: 135-144.

Peuranen, S., Vuorinen, P. J., Vuorinen, M. & Hollender, A. (1994). The effect of iron, humic acids and low pH on the gills and physiology of brown trout (*Salmo trutta*). Annales Zoologici Fennicii. Vol. 31: 389-396.

Pottinger, T.G. & Pickering, A.D. (1985). Changes in skin structure associated with elevated androgen levels in maturing male brown trout, *Salmo trutta* L. Journal of Fish Biology. Vol. 26: 745-753.

Razia, S., Maegawa, Y., Tamotsu, S. & Oishi, T. (2006). Histological changes in immune and endocrine organs of quail embryos: Exposure to estrogen and nonylphenol. Ecotoxicology and Environmental Safety. Vol. 65: 364–371.

Schwaiger, J., Spieser, O.H., Nardy, E., Kalbfus, W., Braunbeck, Th. & Negele, R. D. (1999). Toxic effects versus endocrine disruption Does the xenoestrogen nonylphenol influence physiological functions in fish? SETAC: 9th Annual Meeting, 25-29 May 1999, Leipzig, Germany.

Servos, M.R. (1999). Review of the aquatic toxicity, estrogenic responses and bioaccumulation of alkylphenols and alkylphenol polyethoxylates. Water Qualality Research Journal of Canada. Vol. 31: 123–177.

Shephard, K.L. (1994). Functions for fish mucus. Reviews in Fish Biology and Fisheries. Vol. 4: 401-429.

Soto, A.M., Chung, K.L. & Sonnenschein, C. (1994). The pesticides endosulfan, toxaphene and dieldrin have estrogenic effects on human estrogen sensitive cells. Environmental Health Perspectives. Vol. 102: 380–385.

Soto, A.M., Lin, T.M., Justicia, H., Silvia, R.M. & Sonnenenschein, C. (1992). An 'in culture' bioassay to assess the estrogenicity of xenobiotics (E-screen), In: Colburn, T.C. (Eds.), Chemically Induced Alterations in Sexual and Functional Development; The Wildlife-human Connection. Princeton Scientific Publishing, Princeton, NJ, USA, pp. 295–309.

Srivastava, S.K., Tiwari, P.R. & Srivastav, A.K. (1990). Effects of chlorpyrifos on the kidney of freshwater catfish. *Heteropneustes fossilis*. Bulletin of Environmental Contamination and Toxicology. Vol. 45: 748–751.

Talmage, S.S. (1994). Environmental and Human Safety of Major Surfactants: Alcohol Ethoxylates and Alkylphenol Ethoxylates. Lewis Publishers, Boca Raton, FL.

Toomey, B.H., Monteverdi, G.H. & Di Giulia, R.T. (1999). Octylphenol induces vitellogenin production and cell death in fish hepatocytes. Environmental Toxicology and Chemistry. Vol. 18: 734–739.

Uguz, C., Iscan, M., Ergu, A., Belgin I.V. & Togan, I. (2003). The bioaccumulation of nonyphenol and its adverse effect on the liver of rainbowtrout (*Onchorynchus mykiss*). Environmental Research. Vol. 92: 262–270.

Wendelaar Bonga, S.E. (1997). The stress response in Fish. Physiological Reviews. Vol. 77: 591-625.

Sayed, A.H., Mekkawy, I.A. & Mahmoud, U.M. (2011). Histopathological alterations in some organs of adults of *Clarias gariepinus* (Burchell, 1822) exposed to 4-nonylphenol. The 19th Conference of Egyptian-German Scocity of Zoology, Bin Sueif University, Egypt.

Mekkawy, I.A., Mahmoud, U.M. & Sayed, A.H. (2011). Effects of 4-nonylphenol on blood cells of the African catfish *Clarias gariepinus* (Burchell, 1822) Tissue Cell, doi:10.1016/j.tice.2011.03.006

Mahmoud, U.M., Sayed, A.H. & Mekkawy, I.A. (2011). Biochemical changes of African catfish *Clarias gariepinus* exposed to 4-nonylphenol. African Journal of Biochemistery,11-053 (Accepted)

Yokota, H., Seki, M., Maeda, M., Oshima, Y., Tadokoro, H., Honjo, T. & Kobayashi, K. (2001). Life-cycle toxicity of 4-nonylphenol to medaka (*Oryzias latipes*), Environmental Toxicology and Chemistry. Vol. 20: 2552-2560.

Kang, I.J., Yokota, H., Oshima, Y., Tsuruda, Y., Hano, T., Maeda, M., Imada, N., Tadokoro, H. & Honjo, T. (2003). Effect of 4-nonylphenol on the reproduction of Japanese medaka, Oryzias latipes, Environmental Toxicology and Chemistery. Vol. 22: 2438-2445.

Seki, M., Yokota, H., Maeda, M., Tadokoro, H. & Kobayashi, K. (2003). Effect of 4-nonylphenol and 4- tert octylphenol on sex differentiation and vitellogenin induction in medaka (*Oryzias Latipes*), Environmental Toxicology and Chemistery. Vol. 22:1507-1516.

9

Apoptosis and Ovarian Follicular Atresia in Mammals

J.K. Bhardwaj and R.K. Sharma
Department of Zoology, Kurukshetra University, Kurukshetra, Haryana, India

1. Introduction

The reproductive performance of any mammalian species can be enhanced by enriching nutrition, regulating environmental factors like photoperiod, temperature, humidity etc., through selective breeding, endocrine manipulations and better management practices (Sharma, 2000; Maillet *et al.*, 2002; Bussiere *et al.*, 2002; Iwata *et al.*, 2004, 2005; Rudolf, 2007; Bhardwaj and Sharma, 2011). The endocrine regulation despite being the most complicated is very effective in improving reproductive output of the species. In females, follicular growth and estrous cyclicity are intricately linked (Craig *et al.*, 2007; Sharma and Batra, 2008). Of the thousands of oocytes and primordial follicles present in neonatal ovary only a small fraction i.e. approximately 0.001% are ovulated throughout the active reproductive life span of mammals (Tabarowski *et al.*, 2005; Sharma and Bhardwaj, 2009). Follicular atresia is a wide spread phenomenon that limits the number of ovulations and thus restricting the full reproductive potential of a species. It results in extensive loss of germ cells during development, prenatal, neonatal, prepubertal, pubertal, estrous cycle, pregnancy, lactation and post reproductive life of mammals (Guraya, 1997, 1998; Sharma, 2000; Manabe *et al.*, 2003, 2004; Bhardwaj and Sharma, 2011). Follicular atresia is a natural fertility regulatory mechanism that can best be exploited for increasing or decreasing the fertility of the species by decreasing or increasing the frequency of atresia. It is, therefore a lot of research papers have been published on morphology, histochemistry, biochemistry and endocrinology of follicular atresia (Williams and Smith, 1993; Guraya, 1998, 1999; Monniaux, 2002; Sharma, 2003; Sharma and Batra, 2008; Sharma and Bhardwaj, 2009; Bhardwaj and Sharma, 2011). However, the mechanism of atresia still needs to be further explored and analysed for its effective implementation in fertility regulation programme. The molecular mechanism of atresia can be best explained on the basis of apoptosis (Palumbo and Yeh, 1994; Sharma, 2000; Yu *et al.*, 2004; Sharma and Bhardwaj, 2009). Apoptosis, a type of physiological cell death is the antithesis of mitosis involved in the regulation of tissue homeostasis (Collins and Lopez, 1993; Schwartzman and Cidlowski, 1993). It affects the single scattered cells in the midest of living tissues without eliciting an inflammatory response and is influenced by growth factors and hormones. Many studies on the morphological changes that occur in granulosa cells and theca-interstitial cells of follicles progressing through atresia have documented that apoptosis is, in all likelihood, the primary mechanism by which cell loss is mediated during follicle degeneration (Tsafriri and Braw, 1984; Hirshfield, 1991; Tilly, 1996). The earliest descriptions of apoptosis, the physiological cell death, in the ovary was made in

1885 on the morphological analysis of granulosa cells during the degeneration of antral follicles in the rabbit ovary (Flemming, 1885). A process referred to as 'Chromatolysis' was proposed as a mechanism by which granulosa cell loss was mediated, and in retrospect these observations closely matched all of the morphological criteria now known to be the hallmarks of apoptosis (Kerr *et al.*, 1972, 1994). In the ovary, the mechanisms underlying decisions of life and death involve cross dialogue between pro-apoptotic and pro-survival molecules (Hussein, 2005). Apoptosis operates in ovarian follicles throughout fetal and adult life. During fetal life, apoptosis is restricted to the oocytes, whereas in adult life, this phenomenon is frequently observed in granulosa cells of secondary and antral follicles (Hussein, 2005).

Apoptosis, a genetically regulated cell suicide is an energy requiring process observed commonly during early embryonic development. Apoptosis permits the safe disposal of cells at the point in time when they have fulfilled their biological functions (Hussein, 2005). The ovary represents the paradigm of effective apoptosis during active reproductive period of the animal. Apoptosis is central to many aspects of the ovary and executed by several molecular pathways. Bcl-2 (B-cell lymphoma 2) family, TNF(Tumor necrosis factor), caspases and Transforming growth factor-β TGF- β proteins constitute well established mechanisms of apoptosis (Hussein, 2005). Morphologically, apoptosis is characterized by cell shrinkage, membrane blebbing and cytoplasmic fragmentation in oocytes (Wu *et al.*, 2000; Chaube *et al.*, 2005), granulosa cells (Sharma and Sawhney, 1999; Sharma, 2003; Sharma and Bhardwaj, 2007, 2009) and corpus luteum (Sharma and Batra, 2005). A unique biochemical event in apoptosis is the activation of a Ca^{+2}/Mg^{+2} dependent endogenous endonuclease. This enzyme cleaves genomic DNA at internucleosomal regions, resulting in chopping of DNA in to fragments of 180-200 base pairs (bp). These DNA fragments can be visualized as a distinct ladder pattern by agarose gel electrophoresis. The presence of this DNA pattern in cells is considered a hallmark of apoptosis (Schwartzman and Cidlowski, 1993; Hsueh *et al.*, 1994; Yang and Rajamahendran, 2000). In addition to the detection of oligonucleosomes in isolated DNA, the occurrence of apoptosis is also inferred from the characteristic morphological appearance of degenerating cells, together with the detection of fragmented DNA in single cells *in situ* through the use of the 3' end-labeling technique (terminal deoxynucleotidyl transferase-mediated dUTP nick end-labeling [TUNEL] (Gavrieli *et al.*, 1992; Palumbo and Yeh, 1994; Sharma and Bhardwaj, 2009). The early development of mammalian ovarian follicles is poorly understood. Although follicle stimulating harmone FSH and luteinizing harmone LH are essential for follicular development from the secondary stage onwards, but the onset of growth of primordial follicles is independent of gonadotrophin hormones (Tilly, 1996). Factors initiating the transition of quiescent primordial follicles to the pool of growing follicles are still unknown. Studies on animals have indicated that several locally produced growth factors are involved in the multifactorial regulation of the growth of primary and secondary follicles (Durlinger, 2000). According to recent findings, the loss of follicles by atresia would be counteracted by the formation of new primordial follicles from germ line stem cells (Johnson *et al.*, 2004), a condition still controversial and is yet to be confirmed (Bristol Gould *et al.*, 2006; Liu *et al.*, 2007). Throughout active reproductive life, follicles leave the resting repository pool to enter the growing pool on a regular basis and pass through subsequent developmental stages under the influence of stage specific subset of intra-ovarian regulators and endocrine factors like growth factors, cytokines and gonadal steroids (Gilchrist *et al.*, 2004, Pangas, 2007). At different developmental stages, follicles behave differently in response to factors promoting follicular cell proliferation, growth, differentiation and apoptosis and very few reach upto ovulation (Tilly *et al.*, 1991; Jiang *et al.*, 2003; Craig *et al.*, 2007).

A large number of follicles undergo atresia in the late pre-antral to early antral stages. During this hormone dependent growth phase, a selection operates whether to allow the growth of the follicles upto the pre-ovulatory stage or not (Durlinger *et al.*, 2000). Gonadotropins are survival factors that prevent atresia (Uilenbroek *et al.*, 1980; Chun *et al.*, 1994). In addition, the growth factors like epidermal growth factor, transforming growth factor-α, basic fibroblast growth factor (Tilly *et al.*, 1992), insulin like growth factor (Chun *et al.*, 1994); and cytokine interleukin (IL)-1β (Chun *et al.*, 1995), also check onset of atresia. The activity of catalase in follicular fluid and granulosa cells exhibited a declining trend from healthy to atretic follicles. Catalase is an ubiquitous enzyme found in all known organisms. It catalyzes the breakdown of hydrogen peroxide to water and molecular oxygen (Aebi, 1984 and Vohra, 2002). The activity of enzyme superoxide dismutase shows a declining trend from healthy to slightly atretic to atretic follicles both in the follicular fluid and granulosa cells. The superoxide dismutase (SOD) family of metalloenzymes are antioxidants that protect cells from the deleterious effects of the oxygen free radical superoxide (Fridovich, 1975, 1986). SOD catalyzes the conversion of superoxide anion in the oxygen and hydrogen peroxide. Hydrogen peroxide is further converted to oxygen and water by catalase and peroxidases. Trace elements have also traditionally been known to play an important role in cellular homeostasis which entails the tight regulation of cell death. Although zinc known as microelement prevents or suppresses apoptosis, several published reports have clearly demonstrated that zinc may actively induce cell injury and cell death (both apoptosis and necrosis) in malignant as well as normal cells (Iitaka *et al.*, 2001; Bae *et al.*, 2006; Wiseman *et al.*, 2006; Rudolf, 2007; Bhardwaj and Sharma, 2011). Zinc has been shown to induce apoptosis as well as necrosis by altering calcium homeostasis via the generation of oxidative stress, with subsequent activation of mitogen activated protein kinases (MAPKs), via p53 – dependent and p53 independent signaling or through its interaction with the cytoskeleton (Feng *et al.*, 2002; Chen *et al.*, 2003). The role of reactive oxygen species (ROS) and anti-oxidants in relation to female reproductive function has, in contrast, received relatively little attention, although there are evidence of both physiological and pathological effects (Guerin *et al.*, 2001; Chaudhary *et al.*, 2004).. Yang *et al.*, (1998) have found high levels of hydrogen peroxide in fragmented embryos and unfertilized oocytes, whilst Paszkowski and Clarke (1996) have reported increased antioxidant consumption (suggesting increased ROS activity) during incubation of poor quality embryos. Recent studies indicate that apoptosis of granulosa cells shows a close relationship between estradiol and progesterone titre in the follicular fluid (Yu *et al.*, 2004; Sharma and Batra, 2008; Sharma *et al.*, 2008). The level of insulin like growth factor-1 IGF-1 but not insulin like growth factor-II IGF-II is the crucial factor in deciding whether a follicle will mature or undergo atresia (Yu *et al.*, 2004). The role of trace elements as pro-apoptotic or anti-apoptotic factors is not well known till date. The cascade of morphological and biochemical alterations need to be studied to rescue the germ cells or somatic cell losses.

A large population of ovarian follicles in mammals is lost through a wide spread phenomenon of follicular atresia that limits the number of ovulations thus restricting the full reproductive potential of a species (Sharma, 2000; Yu *et al.*, 2004; Tabarowski *et al.*, 2005;

Slomczynska *et al.*, 2006; Sharma and Bhardwaj, 2009). Follicle atresia is a wide spread degenerative phenomenon by which follicles lose their structural integrity and oocytes are lost from the ovaries other than the process of ovulation (Guraya, 1997; Sharma, 2000). Follicular atresia decreases the number of ovulations and the follicle wall components transmutate into a steroidogenically functional interstitial gland tissue (Sharma and Guraya, 1992; Manabe *et al.*, 2003) that helps in the endocrine regulation of ovarian physiology. Atresia affects the follicles at all stages of development but extensive loss of germ cells occurs during early development, prenatal, neonatal, prepubertal, pubertal, estrous cycle, pregnancy, lactation and post reproductive life of mammals (Guraya, 1998; Sharma, 2000; Tabarowski *et al.*, 2005). During early development primordial follicle population is the most affected whereas in the active reproductive phase the frequency of atresia is maximum in antral follicles (Danell, 1987; Guraya *et al.*, 1994, Guraya, 2000; Sharma, 2000) The modulation of frequency of atresia can regulate the fertility of the animal (Sharma, 2000). Follicle atresia is characterized by the appearance of pyknotic granulosa cells in intact membrana granulosa; free floating large sized granulosa cells with pyknotic nuclei; hyalinization of granulosa cells; chromolysis of the granulosa cell nuclei and their enucleation; loosening of the intercellular matrix; delamination of intercellular matrix; colloidal, opaque and cloudy follicular fluid; detachment of cumulus-oophorous complex from mural granulosa cells; appearance of RBCs in the follicle; invasion of connective tissue fibres within the follicle and huge accumulation of follicular fluid resulting in the cyst formation (Sharma, 2000; Sharma and Bhardwaj, 2009; Bhardwaj and Sharma, 2011).

2. Apoptosis in normal cycling mature ovary

In prepubertal and mature mammalian ovaries the preantral atretic follicles, the atretic oocyte is characterized by shrinked ooplasmic contents surrounded by a peripheral zone of lipoidal components which is strongly sudanophilic. This zone in a few atretic oocytes is so dense that it occuludens the central mass. (Sharma *et al.*, 1992; Guraya *et al.*, 1994). Initially, pyknosis in granulosa cells was restricted only to specific area which consequently prolonged and large zone of atretic pyknotic granulosa cells was formed, which results in loosening of membrana granulosa (Sharma *et al.*, 1992). These pyknotic granulosa cells give positive test for lipids and 3-β-HSDH, indicating their active role in steroidogenesis (Guraya *et al.*, 1994). Pyknosis of granulosa cells as the first sign of atresia is evocative of the general plan and path of atretogenic changes in mammals (Guraya *et al.*, 1987; Tilly *et al.*, 1996; Burke *et al.*, 2005; Tatone *et al.*, 2008; Sharma and Bhardwaj, 2009; Bhardwaj and Sharma, 2011). The various nuclear and ooplasmic contents are drastically affected in atretic preantral follicles. The organelles showing abnormal morphology and distribution increase in number and proceed towards disintegration and then lead to an accumulation of lipids in the oocytes and degenerating granulosa cells because of liberation of cytosolic membrane bound lipids (Byskov, 1978). In the atretic preantral follicles the granulosa cells develop histochemical characteristics of steroidogenic cells as evidenced by weak 3-β-HSDH activity (Sharma, 2000). In antral follicles, apoptosis is characterized by the presence of pyknotic nucleus in the membrana granulosa, appearance of spaces and release of fragments of pyknotic nuclei in the peripheral zone of antrum in type 1b (Sharma *et al.*, 1992; Guraya *et al.*, 1994; Sharma and Bhardwaj, 2009; Bhardwaj and Sharma,2011). The granulosa cell glyco-conjugates are changed during follicular atresia in the pig and rat ovaries as shown with lectins (Sharma and Guraya 1992; Kimura *et al.*, 1999). The levels of fibronectin, laminin, type IV collagen, proteoglycans, insulin like growth factor II/mannose 6 phosphate receptors, and matrix metaloproteinases 2 and -9 increased whereas 450 aromatase and connexin 43 decreased within the wall of granulosa cells during follicular atresia in sheep (Huet *et al.*, 1997, 1998, Sharma and Guraya 1998a, b). Guraya *et al.*, (1994) have demonstrated that the degenerative signs also appear in cumulus cells of antral atretic follicles, whereas atresia proceeds gradually, membrana granulosa thins off and theca hypertrophy enhances in antral atretic follicles (Sharma *et al.*, 1992; Guraya *et al.*, 1994). The appearance of pyknotic granulosa cells as the first sign of atresia in goat strongly advocate the concept that phenomenon of atresia in bovine species is similar (Zimmerman *et al.*, 1987; Wezel *et al.*, 1999; Hastie and Haresign, 2006). The chromophilic pyknotic granules observed in goat atretic follicles were similar to that of DNA - positive masses observed in sheep (Hay *et al.*, 1976, Hay and Cran 1978; Zhou and Zhang, 2005). However such granular material is negligible in cow (Guraya, 1997). The delamination of mural granulosa cells from the basal lamina leads to disruption causing a decline in oxygen carrying blood transudate that decreases the metabolic pace and results in onset of atresia (Guraya, 1985, 1998; Sharma and Guraya, 1998 b, c). However, Hay *et al.*, (1976) suggested that it is because of decline in estrogen synthesis and release. In advance stage of atresia, there is a tendency towards accumulation and storage of sudanophilic lipids in the atretic oocytes (Guraya, 1973 a). The mural layers acquire more lipids as compared to cumulus cells or peripheral granulosa cells. Similar trends were reported in hamster and rabbit (Guraya and Greenwald, 1964), thus indicating granulosa heterogeneity in morphology and physiology (Parshad and Guraya 1983; Sharma and Guraya, 1990). The NADH and NADPH-dependent tetrazolium reductase activity in goat further endorse this concept. The theca interna of atretic antral follicles undergo hypertrophy and develop lipid droplets largely comprised of phospholipids which are associated with steroidogenesis in mammals (Guraya, 1973 b, 1974 a, b, 1977, 1978 a,b; Nicosia, 1980). This hypertrophied theca interna in atretic follicles of goat finally constitute conspicuous masses of interstitial gland tissue which show strong 3-β-HSDH activity (Guraya, 1971, 1973; Guraya *et al.*, 1994; Sharma and Batra, 2008). These lipid droplets are largely phospholipids and triglycerides which change to cholesterol, its esters and phospholipids, and remain associated with theca type interstitial gland tissue. Various carbohydrates, proteins and enzyme histochemistry were used as an index to indicate whether atresia has initiated or not (Guraya, 1984, 1985; Sharma and Guraya, 1992, 1997; Burke *et al.*, 2005). The follicles undergoing atresia disintegrate oocytes and granulosa cells are reabsorbed while theca interna hypertrophy and acquire morphological and histochemical characteristics of a steroidogenic tissue (Sharma and Batra, 2008).

3. Structure and ultrastructure of the process

Intrafollicular paracrine steroid interactions are dependent on FSH and LH which regulate follicular development and oestrogen synthesis and release (Guraya, 1985, 1998; Ireland, 1987; Richards *et al.*, 1993; Gore-Langton and Armstrong, 1994; Hillier, 1994; Guthrie and Cooper, 1996; Maillet *et al.*, 2002; Burke *et al.*, 2005). Alterations in the production of steroid in the antral fluid are the first biochemical manifestations of atresia. (Guraya, 1985, 1998; Hillier, 1985; Greenwald and Terranova, 1988; Westhof *et al.*, 1991; Gore-Langton and Armstrong, 1994; Greenwald and Roy, 1994; Quirk *et al.*, 2006). Variations in normal and atretic follicular fluid concentration of androgens and that of progesterone are not significantly different (Hillier, 1985; Westhof *et al.*, 1991; Moor, 1977; Gore-Langton and

Armstrong, 1994; Slomczynska *et al.*, 2006). Hormonal profile of normal and atretic follicles depends largely on the stage of advancement of atresia. Normally it is oestrogen to progesterone ratio which determines whether a follicle will mature or undergo atresia (Moor *et al.*, 1978; Ireland and Roche, 1982, 1983 a, b; Guraya, 1997; Yu *et al.*, 2004; Sharma *et al.*, 2008). The estrogen level is highest in the preovulatory healthy follicles of pig, sheep, cow, and goat as compared to atretic ones (Eiler and Nalbandov, 1977; Moor *et al.*, 1978; Carlson *et al.*, 1981; Fortune *et al.*, 1988; Burke *et al.*, 2005; Sharma *et al.*, 2008). Proliferation, migration, differentiation and cellular death constitute the most important stages in the development and growth ovarian follicles and is actively involved in turnover of ovarian tissues. The cellular death basically involves two pathways: necrosis or apoptosis (Wyllie *et al.*, 1980; Kressel and Groscurth, 1994). Necrosis is induced as the result of injuries or environmental pathological influences and produces a series of cellular alterations that begin with the changes in cellular membrane permeability with consequent disruption of cytoplasmic structures and ensuing nuclear degeneration (Pol De *et al.*, 1997). On the other hand, apoptosis is the process of cellular self destruction which also involves active process of intracellular synthesis (Wyllie *et al.*, 1980; Cohen and Duke, 1984) and is controlled by cellular genes. In granulosa cell apoptosis, the primary events observed were the nuclear compaction, the chromatin collapses in to large irregular masses surrounded by nuclear envelope, and plasma membrane introflexes forming deep incisions that confer to the cell a very irregular appearance (Sharma and Bhardwaj 2009; Bhardwaj and Sharma, 2011). Despite this, the cellular organelles maintain their morpho-functional integrity. Subsequently the cell fragments into spheroidal subunits surrounded by membranes (apoptotic bodies) that contain portions of cytoplasm and nucleus that are finally phagocytosed by neighbouring cells or by macrophages. The cellular death in granulosa cells therefore is distinguishable on accounts of histological and ultrastructural morphology, and pattern of changes in cell organelles and the chromatin. In necrosis, chromatin was altered at the end in a disorderly fashion, whereas during apoptosis it was precociously excised in a regular sequence. The apoptotic changes observed in the degenerating granulosa cells from goat ovarian follicles revealed diminished size, withered surface morphology and pyknosis (Hay and Cran, 1978; Hirshfield, 1983; Guraya, 1985; Kaur and Guraya, 1987; Tilly, 1996; Manabe *et al.*, 2003, 2004; Sharma and Bhardwaj, 2009; Bhardwaj and Sharma, 2011). The goat granulsoa cells resemble rat atretic granulosa cells in terms of nuclear condensation and cytoplasmic shrinkage as well as presence of apoptotic bodies (Hurwitz *et al.*, 1996). The undulations of nuclear membrane and pinching off of the apoptotic bodies strongly advocate the concept that the apoptosis is the basic mechanism involved in the phenomenon of atresia of granulosa cells in mammals. The vacuolization of condensed chromatin material within the nucleus was the specific positive indicator of apoptosis (Figs. 1, 2). Various studies, conducted during atresia of ovine and caprine follicles, cumulatively indicate that degenerative changes are restricted only to membrana granulosa layers that lie adjacent to the antral cavity during early phases of atresia (Sharma *et al.*, 1992). The detachment or delamination of mural granulosa cells from the basal lamina in the initial phases induce atresia due to disruption of oxygen and nutritional milieu carying blood transudate to the cells (Guraya, 1985, 1998; Sharma and Guraya, 1998 b, c; Sharma, 2000, 2003). Ultrastructurally, typical apoptotic granulosa cells from the antral follicle of goat show compaction and segregation of chromatin, condensation of the cytoplasm maintaining the integrity of organelles, and subsequent fragmentation in to membrane bounded apoptotic bodies (Figs.3, 4). Necrotic cells are characterized by irregular clumping of chromatin, swelling and dissolution of organelles, rupture of the cell membrane, and finally disintegration of all cellular components. Ultrastructural changes in degenerating granulosa cells of goat ovaries strongly support the morphological hallmarks of apoptotic cells. First, the nucleus and cytoplasm became condensed, then the condensed cells were fragmented but retained the integrity of organelles, i.e. mitochondria, and nuclear and plasma membrane. Since cell debris in atretic follicles increased in number following the appearance of apoptotic bodies and had condensed nuclei similar to apoptotic bodies, the debris is possibly comprised of degraded apoptotic bodies. The investigations on fine morphology of granulosa cells *in vivo* showed similar characteristic features as observed after gonadotrophins and steroid treatment in rat *in vitro*, thus, confirming the role of hormones in structural modification of granulosa cells at different phases of the cycle (Balboni and Zecchi, 1981; Yu *et al.*, 2004; Sharma *et al.*, 2008; Sharma and Batra, 2008).

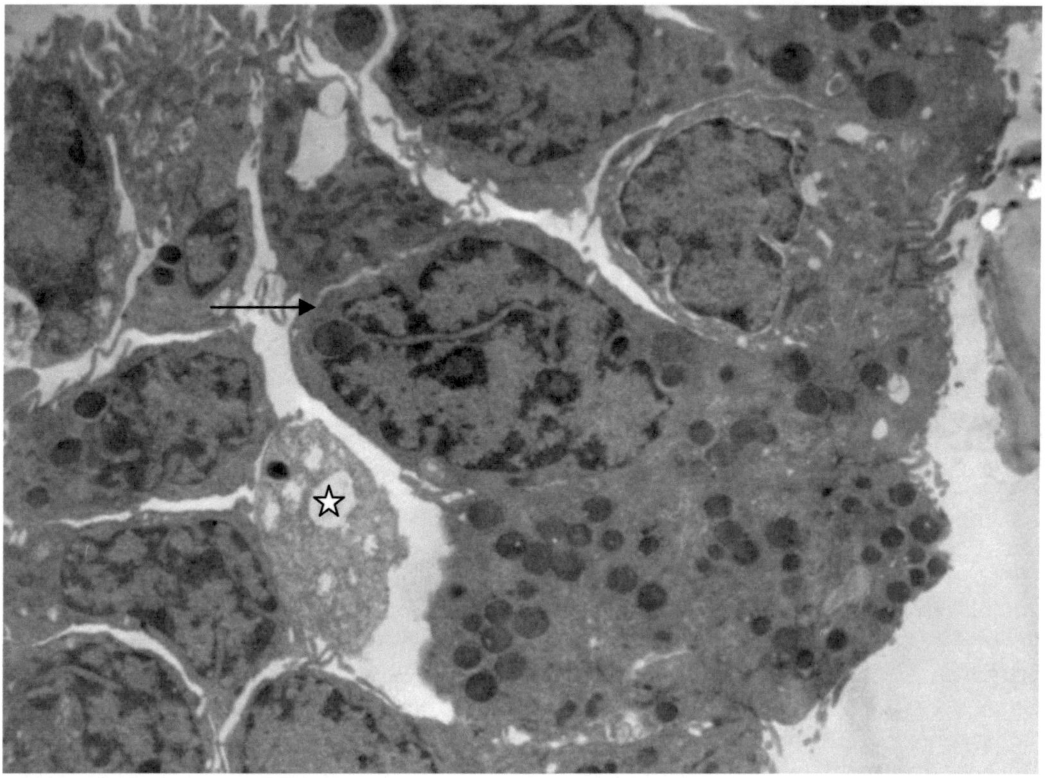

Fig. 1. Electron micrograph of apoptotic granulosa cells showing increased indentation (arrow) and vacuoles (star) of different shapes within condensed chromatin material.

Withdrawal of typical intercellular microvilli or interacting surfaces with the advancement of atresia, the appearance of lysosomal vacuoles and lipid droplets of varied dimensions observed during atresia in goat, further support the histological and histochemical characteristics of follicular atresia in mammals (Motta, 1972; Parshad and Guraya, 1983; Sharma and Bhardwaj, 2009; Bhardwaj and Sharma,2011). The presence of lysosomal vacuoles in the degenerating granulosa cells of goat during advanced stage of atresia further support the findings of Sharma and Guraya (1997) who have reported that lysosomal vacuoles appear in the cytoplasm which subsequently destabilize plasma membrane completely leading to the extrusion of interacellular contents making it more hyaline.

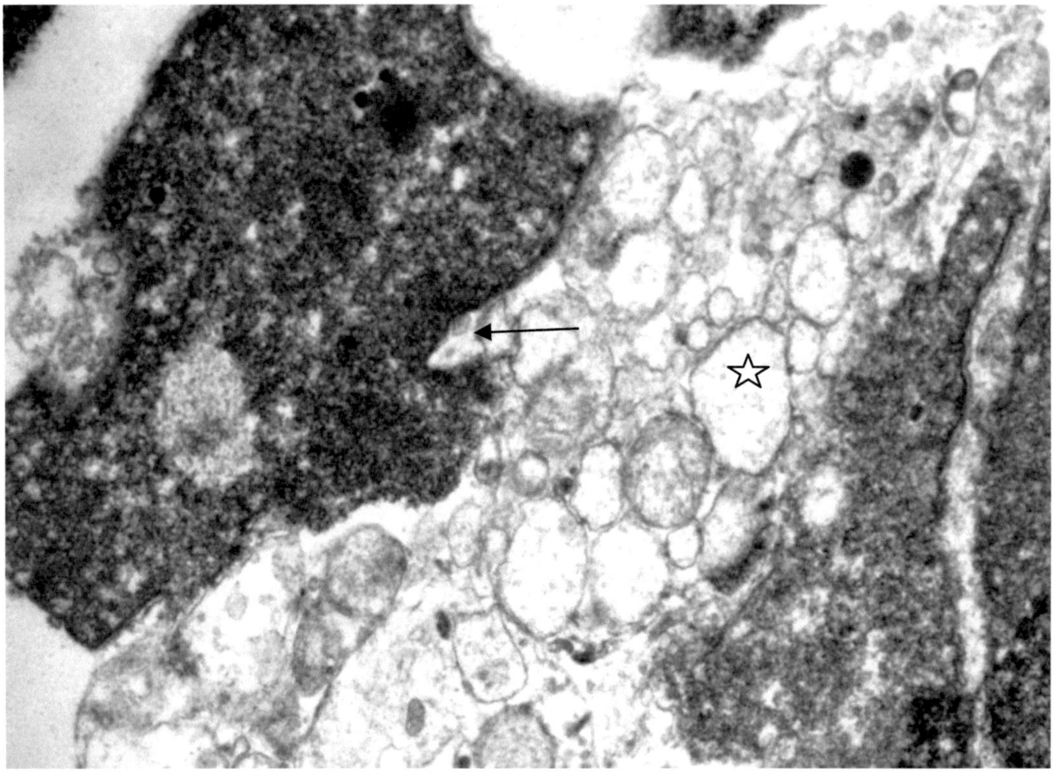

Fig. 2. Electron micrograph of apoptotic granulosa cells showing uneven wavy nuclear envelope (arrow) and increased vacuolization (star) of cytoplasm.

The presence of hyaline granulosa cells in the mural and antral layers of membrana granulosa and free floating in follicular fluid were observed in goat ovary (Sharma and Bhardwaj, 2007). The similar distribution of cells were also observed in primates and other mammals (Byskov, 1974; Balboni and Zecchi, 1981; Bill and Greenwald, 1981; Sharma and Guraya, 1992, 1997). The increased undulation and indentations of the nuclear membrane and pinching off of the nuclear fragments suggests that apoptosis is involved in initiation and execuation of cell death during atresia in goat. The increase in the frequency and dimensions of nuclear pores as well as flattening of the nuclear membrane observed in goat follicles are similar to the earlier findings on ultrastructure of apoptotic granulosa cells in rat and cow (Coucouvanis *et al.*, 1993; Grotowski *et al.*, 1997; Isobe and Yoshimura, 2000; Yang and Rajamahendran, 2000; Inoue *et al.*, 2003), thereby suggesting a common plan of apoptosis in bovine species. The membrane bound pyknotic chromatin material carrying apoptotic vesicles were observed lying within the cytoplasm during the advanced stages of atresia which endorse the concept that apoptotic bodies are formed from condensed chromatin material packed in small vacuoles limited by the nuclear membrane. In a few cells, the presence of condensed cytoplasm in contrast to hyaline one was possibly due to the differential functional impairment of the cytoplasmic membrane. Sharma and Guraya

(1992) have reported changes in glycoconjugates and carbohydrates of atretic granulosa cells in rat and have postulated that changes in histochemical mapping of negatively charged moieties induces uncoupling of membrane interactions subsequently leading to impairment of membrane permeability characteristics that finally lead to atresia or cell death due to apoptosis. The alterations in acidic phospholipids phosphatidyl serine content that acts as apoptosis inducing agent (Krishnamurthy *et al.*, 2000), modulates the membrane chemistry leading to a change in its permeability to water molecules. The cell becomes larger and hyaline if the permeability is enhanced whereas the contents become pyknotic and electron dense if the permeability decreases. Recent studies have demonstrated that apoptosis involves cleaving of DNA in several animal species. Internucleosomal DNA fragmentation has been considered to be characteristic of apoptosis and is one of the earliest event (Schwartzman and Cidlowski, 1993; Okamura *et al.*, 2001; Hastie and Haresign, 2006).

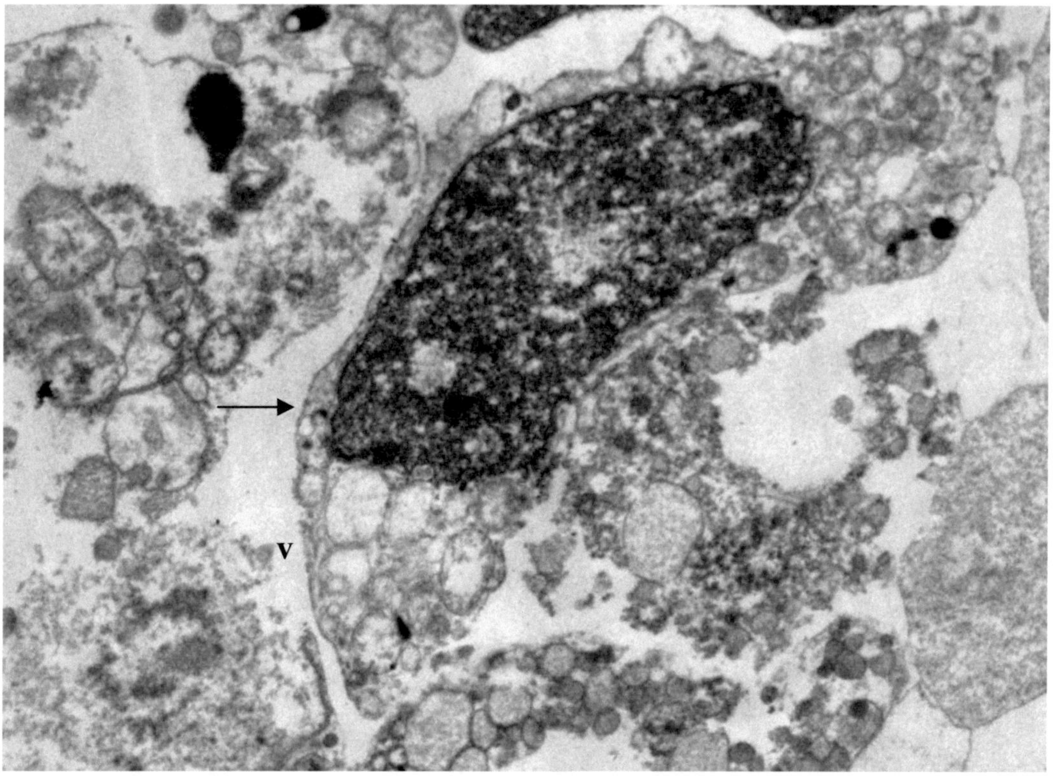

Fig. 3. Electron micrograph of apoptotic granulosa cells showing the pyknotic chromatin material adhering to the periphery of nuclear membrane (arrow) and vacuoles (v) of different shapes and sizes were observed within condensed chromatin material.

In addition to the detection of oligonucleosomes in extracted DNA, the occurrence of apoptosis may also be inferred from the characteristic morphological appearance of degenerating cells, together with the detection of fragment DNA in single cells *in situ* using TUNEL (Gavrieli *et al.*, 1992; Palumbo and Yeh, 1994; Bristol and Gould *et al.*, 2006; Sharma and Bhardwaj, 2009; Bhardwaj and Sharma,2011). Using *in situ* 3' end labeling (TUNEL), which can detect apoptosis precisely at the single cell level without disruption of the tissue morphology (Gavrieli *et al.*, 1992; Palumbo and Yeh, 1994; Liu *et al.*, 2007; Sharma and Bhardwaj, 2009), the specific morphological features of granulosa cell death in follicular atresia (nuclear pyknosis, karyorrhexis, and formation of apoptotic bodies) can be related to the physiological process of apoptosis. The relationship is supported by a combination of biochemical, classic histological evidences, and *in situ* histochemical localization of DNA fragmentation. Different cellular details were observed in atretic and healthy follicles classified by morphological criteria, including cells with a single shrunken and dense nucleus (pyknotic appearance) and cells with marginated chromatin and/or nuclear fragmentation. According to Lussier *et al.*, (1987), non atretic follicles should have intact and normal granulosa layers with the mean pyknotic index per class varying from 0.13 percent to 0.67 percent. However, in another study in cows (Ireland and Roche,1983), pyknotic cells were observed in the granulosa cell layer in 30-60 percent of estrogen-active large follicles.

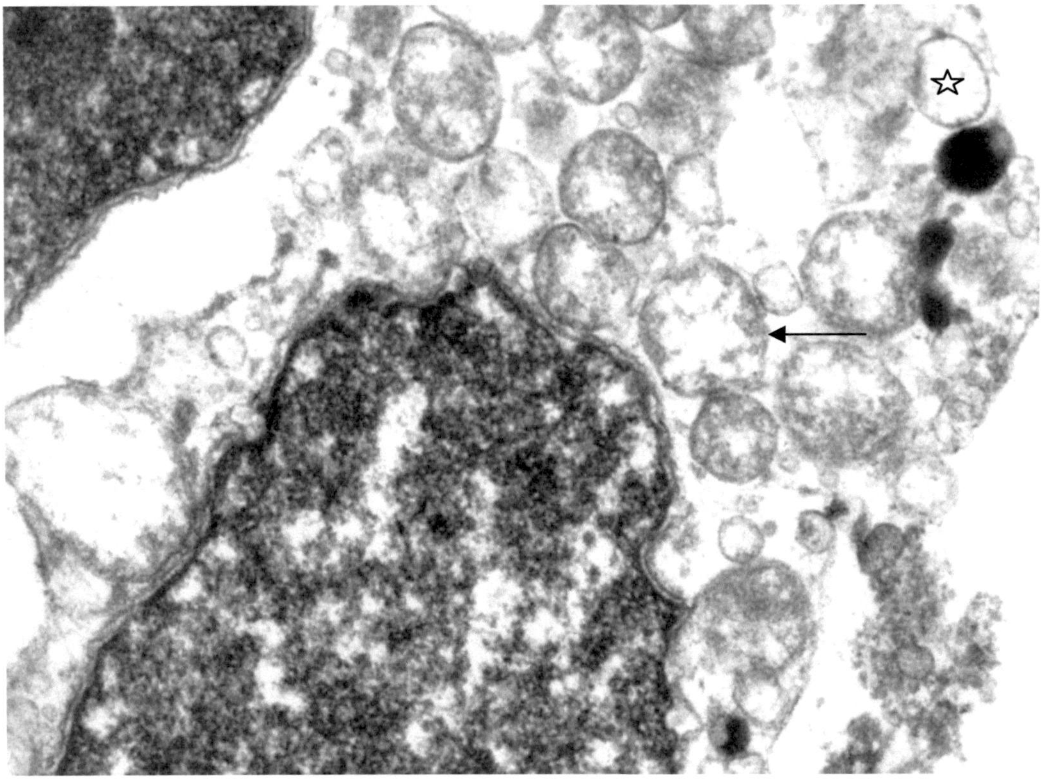

Fig. 4. Electron micrograph of granulosa cells revealing vacuolated cytoplasm (star) and mitochondria (arrow).

Thus, the mere presence of pyknotic cells in the granulosa cell layer does not imply that they are atretic. However, the morphological and biochemical results strongly indicate that apoptosis may occur to a certain level during normal follicle growth and development and that apoptotic death of granulosa cells may be detectable before other morphological and biochemical signs of degeneration in goats. Alkaline phosphatase activity in follicular fluid and granulosa cells exhibited a declining trend from healthy to slightly atretic and atretic follicles. The biochemical estimation of alkaline phosphatase endorse the earlier histochemical mapping of alkaline phsophatase activity opining the possible role of alkaline phsophatase in active transport of nutrients and secretary material across the membrane (Verma and Guraya, 1968; Sangha and Guraya, 1988/89; Sharma, 2000). The association of alkaline phosphatase positive sites with theca interna indicates the involvement of this enzyme in steroid metabolism and transport (Britenecker *et al.,* 1978; Gilchrist *et al.,* 2004). The decline in levels of alkaline phosphatase in follicular fluid and granulosa cells of atretic follicles may be tangibly due to increased vascularity and changed morphology and biochemistry of granulosa cells for steroid hormone synthesis (Guiseppe, 1983; Gilchrist *et al.,* 2004; Pangas, 2007; Tatone *et al.,* 2008). Acid phosphatase activity in follicular fluid and granulosa cells exhibited an increasing trend from healthy to slightly atretic to atretic follicles. The increase in Golgi complex and lysosomes in atretic follicles/cells is possibly attributable to the rise in acid phosphatase activity. The ultrastructural modifications associated with the lysosomal accumulation during atresia which is further increased due to luteinization wherein chief protein synthesizing cells transmutate to steroidogenic cells may be attributable to the increase in acid phosphatase activity in the granulosa cells (Dorrington *et al.* 1975; Armstrong and Dorrington, 1976; Sangha and Guraya, 1988/89). The degenerative/transformative changes involved in reshaping of ovarian subcellular components that facilitate differentiation during follicular development while bringing about lysis and formation of apoptotic vesicles may be responsible for the rise in lysosomal activity (Sangha and Guraya, 1988/89; Sharma, 2000). The increase in acid phosphatase enzyme activity observed in the follicular fluid and granulosa cells of atretic follicles may also be related to some mechanism for the secretion of steroids (Sawyer *et al.,* 1979; Dimino and Elfont, 1980; Pangas, 2007). It has been reported that acid phosphatase activity is higher in active and regressing corpora lutea, provides a lurking possibility that during follicular atresia the rise in acid phosphatase activity may not be associated exclusively with regression but also with the formation of interstitial gland tissue. Catalase is generally associated with superoxide dismutase, constituting a reciprocally protective set, while catalase is inhibited by oxyradicals, and SOD is inhibited by H_2O_2 (Lapluye, 1990). If H_2O_2 produced by SOD, action on oxyradicals is not removed immediately it will react with super oxide radicals (Haber-Weiss reaction) giving rise to highly reactive hydroxyl radicals (Michiels *et al.,* 1994). However, with the increase in H_2O_2 concentration, the catalase contribution for its degradation concomitantly increased (Verkek and Jond Kind, 1992). Michiels *et al.,* (1991) reported a 30 percent increase in survival when catalase was injected in combination with SOD, whereas the survival was only 21 percent and 4 percent for SOD and catalase respectively, when independently injected in human fibroblasts. Singh and Pandey (1994) reported an increased catalase activity in the ovary of metaoestrous rats *Pari pasu* with a decline in H_2O_2 production in the mitochondria and microsomal fraction. In addition to its effects on oxygen free radical metabolism, SOD has been shown to influence cell functions by increasing the levels of the second messenger cGMP (Ignarro *et al.,* 1987; Schmidt *et al.,* 1993; Burke *et al.,* 2005). There are three known forms of SOD with specific subcellular and extracellular distributions (Ravindranath and Fridovich, 1975; Crouch *et al.,* 1984; Redmond *et al.,* 1984; Tibell *et al.,* 1987). The manganese-associated form of SOD is localized in mitochondria of cells, whereas the copper/zinc associated form is found in the cytoplasm. Furthermore, there is an extracellular form of SOD that is secreted from cells. All three forms of SOD are expressed in the ovary (Laloraya *et al.,* 1988; Shiotani *et al.,* 1991; Sato *et al.,* 1992; Hesla *et al.,* 1992, Tilly and Tilly, 1995) and the pattern of expression appears to be related to gonadotropin induced follicular development and luteal steroidogenesis and regression (Laloraya *et al.*, 1988; Hesla *et al.*, 1992; Sato *et al.*, 1992; Tilly and Tilly, 1995; Pangas, 2007). Furthermore, SOD effectively blocks gonadotropin-induced ovulation (Miyazaki *et al.*, 1991). The activity of enzyme glutathione peroxidase in follicular fluid and granulosa cells shows a declining trend from healthy to slightly atretic to atretic follicles. It has been reported that glutathione peroxidase (GPx) catalyzes the breakdown of H_2O_2 with much more affinity than the catalase (Gul *et al.*, 2000). The major protection against both lipid peroxide and H_2O_2 is reported to be achieved by GPx (Halliwel and Gutteridge, 1985). Land and Verdetti (1989) reported decreased level of GPx with age in kidney and liver in rats, whereas Imre *et al.*, (1984) and Hazelton and Lang (1985) reported a decrease in GPx activity with age in liver and kidney and many other tissues of mice. It has been reported that GPx activity remained constant in the caudate putamen and the temporal cortex but decreased in the substantial nigra and the thalamus in rats (Benzi *et al.*, 1989).

4. Conclusion

Thus, the importance of understanding the mechanistic machinery of apoptosis is vital because programmed cell death is a component of both health and disease, being initiated by various physiologic and pathologic stimuli. Moreover, the widespread involvement of apoptosis in the pathophysiology of disease lends itself to therapeutic intervention at many different checkpoints. Therefore, understanding the mechanisms of apoptosis and other variants of programmed cell death, at the molecular level provides deeper insight into various disease processes and may thus influence therapeutic strategy.

5. References

Aebi, H. (1984). Catalase *in vitro.* In : Methods in Enzymology (I. Packer, ed.). Vol. 105, Academic Press, NY, pp. 121-126.

Armstrong, D.T., and Dorrington, J.H. (1976). Androgen augment FSH-induced progesterone secretion by cultured rat granulosa cells. *Endocrinol.* 99, 1411-1414.

Bae, S.N., Lee, Y.S., Kim, M.Y., Kim, J.D., Park, L.O. (2006). Antiproliferative and apoptotic effects of zinc-citrate compound on human epithelial ovarian cancer cell line, OVCAR-3. *Gynecol. Oncol.* 103, 127-136.

Balboni, G.C., and Zecchi, S. (1981). On the structural changes of granulosa cells cultured *in vitro.* Histochemical, ultrastructural and stereological observations. *Acta Anat.* 110, 136-145.

Benzi, G., Marzatico, F., Pastoris, O., Villa, R.F. (1989). Relationship between ageing, drug treatment and the cerebral enzymatic antioxidant system. *Exp. Gerontol.* 24, 469-479.

Bhardwaj, J.K. and Sharma, R.K. (2011). Changes in trace elements during follicular atresia in goat (Capra hircus) ovary. *Biol Trace Elem Res* 140, 291-298.

Bhardwaj, J.K. and Sharma, R.K. (2011). Scanning electron microscopic changes in granulosa cells during follicular atresia in caprine ovary. *J. Scanning.* 33, 21-24.

Bill, C.H., and Greenwald, G.S. (1981). Acute gonadotropin deprivation. A model for the study of follicular atresia. *Biol Reprod.* 24, 913-921.

Brietenecker, G., Friedrich, F., Kemeter, P. (1978). Further investigation on the maturation and degeneration of human ovarian follicles and their oocytes. *Fertil. Steril.* 29, 336-341.

Bristol-Gould, S.K., Kreeger, P.K., Selkirk, C.G., Kilen, S.M., Mayo, K.E., Shea, L.D., Woodruff, T.K. (2006). Fate of the initial follicle pool: empirical and mathematical evidence supporting its sufficiency for adult fertility. *Dev. Biol.* 298, 149-154.

Burke, C.R., Cardenas, H., Mussard, M.L., Day, M.L. (2005). Histological and steroidogenic changes in dominant ovarian follicles during oestradiol-induced atresia in heifers-*Reproduction.* 129, 611-620.

Bussiere, F.I., Zimowska, W., Gueux, E., Rayssiguier, Y., Mazur, A. (2002). Stress protein expression cDNA array study supports activation of neutrophils during acute magnesium deficiency in rats. *Magnes. Res.* 15, 37-42.

Byskov, A.G. (1978). *The Vertebrate Ovary*: Follicular Atresia. In: Jones, R.E. (ed.). Plenum press, New York, pp. 533-562.

Chaube, S.K., Prasad, P.V., Thakur, S.C., and Srivastava, T.G. (2005). Hydrogen peroxide modulates meiotic cell cycle and induces morphological features characteristics of apoptosis in rat oocytes cultured *in vitro. Apoptosis.* 10, 863-875.

Chaudhary, A., Sharma, R.K., Saini, K. (2004). Antioxidant effect of Mandukparni on cerebellum and spinal cord of ageing rats. *J. Tiss. Res.* 4, 113-115.

Chen, W., Wang, Z., Zhang, Y. (2003). The effect of zinc on the apoptosis of cultured human retinal pigment epithelial cells. *J. Huazhong Univ. Sci Technology. Med Sci.* 23, 414-417.

Chun, S.Y., Eisenhauer, K.M., Kubo, M., Hsueh, A.J. (1995). Interleukin-1β suppresses apoptosis in rat ovarian follicles by increasing nitric oxide production. *Endocrinology.* 136, 3120-3127.

Cohen, J.J., and Duke, R.C. (1984). Glucocorticoid activation of a calcium dependent endonuclease in thymocyte nuclei leads to cell death. *J. Immunol.* 132, 38-42.

Coucouvanis, E.C., Sherwood, S.W., Carswell-Crumpton, C., Spack, E.G., Jones, P.P. (1993). Evidence that the mechanism of prenatal germ cell death in mouse is apoptosis. *Exp. Cell Res.* 209, 238-247.

Craig, J., Orisaka, M., Wang, H., Orisaka, S., Thompson, W., Zhu, C., Kotsuji, F., Tsang, B.K. (2007). Gonadotropin and intra-ovarian signals regulating follicle development and atresia: the delicate balance between life and death. *Front Biosci.* 12, 3628-3639.

Crouch, R.K., Gandy, S., Patrick, J., Reynolds, S., Buse, M.G., Simson, J.A. (1984). Localization of copper-zinc superoxide dismutase in the endocrine pancreas. *Exp Mol Pathol.* 41, 377-383.

Danell, B. (1987). 'Oestrous behaviour, ovarian morphology and cyclical variations in follicular system and endocrine pattern in water buffalo heifers'. Ph.D. Thesis, pp. 1-124. Swedish Univ. Agricultural Sci., Uppsala, Sweden.

Dimino, M.J., and Elfont, E.A. (1980). The role of lysosomes in ovarian physiology. In: *Biology of the Ovary.* (Eds.) Motta, P.M. and Hafez, E.S.E., Nijhoff, The Hague/Boston/London ed. Chap. XV, 196-201.

Dorrington, J.H., Moon, Y.S., Armstrong, D.T. (1975). Estradiol 17 β biosynthesis in cultured granulosa cells from hypophysectomized immature rats. Stimulation by follicle-stimulating hormone. *Endocrinol.* 97, 1328-1331.

Durlinger, A.L.L., Krammer, P., Karels, B., Grootegoed, J.A., Vilenbroek, J., Themmen, A.P.N. (2000). Apoptotic and proliferative changes during induced atresia of pre-ovulatory follicles in the rat. *Hum. Reprod.* 15, 2504-2511.

Feng, P., Li, T.L., Guan, Z.X., Franklin, R.B., Costello, L.C. (2002). Direct effect of zinc on mitochondrial apoptogenesis in prostate cells. *Prostate.* 52, 311-318.

Flemming, W. (1885). Uber die bildung von richtungsfiguren in sauge thiereiern beim untergang graafscher follikel. *Arch Anat Entwickl.* 221-244.

Fridovich, I. (1975). Superoxide dismutases. *Annu Rev Biochem.* 44, 147-159.

Fridovich, I. (1986). Biological effects of the superoxide radical. *Arch Biochem Biophys.* 247, 1-11.

Gavrieli, Y., Sherman, Y., Bensasson, S.A. (1992). Identification of programmed cell death *in situ* via specific labeling of nuclear DNA fragmentation. *J. Cell. Biol.* 119, 493-501.

Gilchrist, R.B., Ritter, L.J. Armstrong, D.T. (2004). Oocyte-somatic cell interactions during follicle development in mammals. *Anim. Reprod. Sci.* 82-83, 431-446.

Gore-Langton, R.E., and Armstrong, D.T. (1994). Follicular steroidogenesis and its control. *The physiology of Reproduction,* Vol. I, 2nd Edn., Knobil, E and Neil, J.D., Raven Press, New York, pp. 571-628.

Greenwald, G.S. and Terranova, P.F. (1988). Follicular selection and its control In: *The Physiology of Reproduction.* (Eds.) Knobil, E. and Neil, J.D. Vol. 1, 387-446. Raven Press, New York.

Greenwald, G.S., and Roy, S.K. (1994). Follicular development and its control. *The Physiology of Reproduction.* Knobil, E. and Neil, J.D. Raven Press, New York, pp. 629-724.

Grotowski, W., Lecybyl, R., Warenik-Szymankiewicz, A., Trzeciak, W.H. (1997). The role of apoptosis in granulosa cells in follicular atresia. *Ginekol. Pol.* 68, 317-326.

Guerin, P., El Mouatassim, S., Menezo, Y. (2001). Oxidative stress and protection against reactive oxygen species in the pre-implantation embryo and its surroundings. *Hum. Reprod. Update.* 7, 175-189.

Guiseppe, C.B. (1983). Structural changes: Ovulation and luteal phase. In: *The Ovary.* (Ed.) Serra, G.B. Raven Press, New York, Vol. I, Chap-VIII, 123-142.

Gul, M., Kutay, F.Z., Temocin, S., Hannien, O. (2000). Cellular and Clinical implications of glutathione. *I.J. Exp. Biol.* 38, 325-634.

Guraya, S.S. (1971). Morphology, histochemistry and biochemistry of human ovarian compartments and steroid hormone synthesis. *Physiol. Rev.* 51, 785-807.

Guraya, S.S. (1973a). Morphology, histochemistry and biochemistry of follicular growth and atresia. *In: Proceedings of the Symposium-Ovogenesis-Folliculogenesis.* Nouzilly, *Ann. Biol. Anim. Biochem. Biophys.* 13, 229-240.

Guraya, S.S. (1973b). Interstitial gland tissue of mammalian ovary. *Acta endocr.* 72, 1-27.

Guraya, S.S. (1974a). Morphology, histochemistry, and biochemistry of human oogenesis and ovulation. *Int. Rev. Cytol.* 37, 121-151.

Guraya, S.S. (1974b). Gonadotrophins and functions of granulosa and thecal cells *in vivo* and *in vitro.* Gonadotrophins and Gonadal Functions. (Ed.) Moudgal N.R. Academic press, New York, pp. 280-337.

Guraya, S.S. (1985). Biology of Ovarian Follicles in Mammals. Springer Verlag, Heidelberg-Berlin, New York.

Guraya, S.S. (1997). Comparative biology of corpus luteum cellular and molecular regulatory mechanisms. In : *frontiers in Environmental and Metabolic Endocrinology* (Ed.) Maitra, S.K. pp. 31-58, University of Burdwan, India.

Guraya, S.S. (1998). Cellular and Molecular Biology of Gonadal Development and Maturation in Mammals : Fundamentals and Biomedical Implications. Narosa Publishing House, New Delhi.

Guraya, S.S. (1998). Cellular and Molecular Biology of Gonadal Development and Maturation in Mammals : Fundamentals and Biomedical Implications. Narosa Publishing House, New Delhi.

Guraya, S.S. (2000). Comparative Cellular and Molecular Biology of Ovary in Mammals: Fundamentals and Applied Aspects: Science Publishers, INC, PO Box 699, Enfield, USA.

Guraya, S.S., and Greenwald, G.S. (1964). A comparative histochemical study of interstitial tissue and follicular atresia in the mammalian ovary. *Anat. Rec.* 149, 411-434.

Guthrie, H.D., and Cooper, B.S. (1996). Follicular atresia, follicular fluid hormones, and circulating hormones during the mid luteal phase of the estrous cycle in pigs. *Biol. Reprod.* 55, 543-547.

Halliwell, B., and Gutteridge, J.M.C. (1985). Lipid peroxidation: A radical chain reaction. In B. Halliwell and J.M.C. Gutteridge eds: *Free radicals in biology and medicine,* Clarend press, Oxford.

Hastie, P.M., and Haresign, W. (2006). Expression of mRNAs encoding insulin-like growth factor (IGF) ligands, IGF receptors and IGF binding proteins during follicular growth and atresia in the ovine ovary throughout the oestrous cycle. *Anim. Reprod. Sci.* 92, 284-299.

Hay, M.F., and Cran, D.G. (1978). Differential response of the components of sheep Graafian follicle to atresia. *Annl. Biol. Anim. Biochem. Biophys.* 18, 453-460.

Hay, M.F., Cran, D.G., and Moor, R.M. (1976). Structural changes occurring during atresia in sheep ovarian follicles. *Cell Tissue Res.* 169, 515-529.

Hazeleton, G.A., and Lang, C.A. (1985). Glutathione reductase and peroxidase activity in aging mouse. *Mech. Aging Dev.* 29, 71.

Hesla, J.S., Miyazaki, T., Dasko, L.M., Wallach, E.E., Dharmarajan, A.M. (1992). Superoxide dismutase activity, lipid peroxide production and corpus luteum steroidogenesis during natural luteolysis and regression induced by oestradiol deprivation of the ovary in pseudopregnant rabbits. *J. Reprod Fertil.* 95, 915-924.

Hillier, S.G. (1985). Sex steroid metabolism and follicular development in the ovaries. *Oxford Rev. Reprod. Biol.* 7, 168-222.

Hillier, S.G. (1994). Hormonal control of folliculogenesis and luteinization. *Molecular Biology of Female Reproductive System.* Academic Press, San Diego, pp. 1-37.

Hirshfield, A.N. (1991). Development of follicles in the mammalian ovary. *Int. Rev. Cytol.* 124, 43-101.

Ho, J.S., Gargano, M., Cao, J., Bronson, R.T., Heimler, I., Hutz, R.J. (1998). Reduced fertility in female mice lacking copper-zinc superoxide dismutase. *J. Biol. Chem.* 273, 7765-7769.

Hsueh, A.J.W., Billig, H., Tsafriri, A. (1994). Ovarian follicle atresia: a hormonally controlled apoptotic process. *Endocr Rev.* 15, 707-724.

Huet, C., Monget, P., Pisselet, C., Hennequet, C., Locatelli, A., Monniaux, D. (1998). Chronology of events accompanying follicular atresia in hypophysectomized ewes. Changes in levels of steroidogenic enzymes, connexin 43, insulin-like growth factor II/mannose 6 phosphate receptor, extracellular and matrix metaloproteinases. *Biol. Reprod.* 58, 175-185.

Huet, C., Monget, P., Pisselet, C., Monniaux, D. (1997). Changes in extracellular matrix components and steroidogenic enzymes during growth and atresia of antral ovarian follicles in sheep. *Biol. Reprod.* 56, 1025-1034.

Hughes, F.M.J.R., and Gorospe, W.C. (1991). Biochemical identification of apoptosis (Programmed cell death) in granulosa cells: evidence for a potential mechanism underlying follicular atresia. *Endocrinology.* 129, 2415-2422.

Hurwitz, A., Ruutiainen-Altman, K., Marzella, L., Botero, L., Dushnik, M., Adashi, E.Y. (1996). Follicular atresia as an apoptotic process: atresia-associated increased in the ovarian expression of the putative apoptotic marker sulfated glycoprotein-2, *J. Soc. Gynecol. Investig.* 3, 199-208.

Hussein, M.R., (2005). Apoptosis in the ovary: molecular mechanisms. *Hum. Reprod.* 11, 162-178.

Ignarro, L.J., Byrns, R.E., Buga, G.M., Wood, K.S. (1987). Endothelium derived relaxing factor from pulmonary artery and vein possesses pharmacologic and chemical properties identical to those of nitric oxide radical. *Circ Res.* 61, 866-879.

Iitaka, M., Kakinuma, S., Fujimaki, S., Posuta, I., Fujita, T., Yamanaka, K., Wada, S., Katayama, S. (2001). Induction of apoptosis and necrosis by zinc in human thyroid cancer cell lines. *J. Endocrinol.* 169, 417-424.

Imre, S., Toth, F., Fachet, J. (1984). Superoxide dismutase, catalase and lipid peroxidation in liver mice of different ages. *Mech. Aging Dev.* 28, 297.

Inoue, N., Manabe, N., Matsui, T., Maeda, A., Nakagawa, S., Wada, S., Miyamoto, H. (2003). Role of tumor necrosis factor-related apoptosis-inducing ligand signaling pathway in granulosa cell apoptosis during atresia in pig ovaries. *J. Reprod. Dev.* 49, 313-321.

Ireland, J.J. (1987). Control of follicular growth and development. *J. Reprod. Fertil.* 34, 39-54.

Ireland, J.J., and Roche, J.F. (1982). Development of antral follicles in cattle after prostaglandin-induced luteolysis. Changes in serum, hormones, steroids in follicular fluid and gonadotrophin receptors. *Endocrinology.* 111, 2077-2086.

Ireland, J.J., and Roche, J.F. (1983). Growth and differentiation of large antral follicles after spontaneous luteolysis in heifers: changes in concentration of hormones in follicular fluid and specific binding of gonadotropins to follicles. *J. Anim. Sci.* 57, 157-167.

Ireland, J.J., and Roche, J.F. (1983a). Development of nonovulatory antral follicles after spontaneous luteolysis in heifers: changes in the concentration of hormones in follicular fluid and specific binding of gonadotrophins to follicles. *J. Anim. Sci.* 57, 157-167.

Ireland, J.J., and Roche, J.F. (1983b). Development of nonovulatory antral follicles in heifer: changes in the steroids in follicular fluid and receptors for gonadotrophins. *Endocrinology.* 112, 150-156.

Isobe, N., and Yoshimura, Y. (2000). Localization of apoptotic cells in the cystic ovarian follicles of cows: a DNA-end labeling histochemical study. *Theriogenology.* 53, 897-904.

Iwata, H., Hashimoto, S., Ohota, M., Kimura, K., Shibano, K., Miyake, M. (2004). Effects of follicle size and electrolytes and glucose in maturation medium on nuclear maturation and developmental competence of bovine oocytes. *Reproduction.* 127, 159-164.

Iwata, H., Hayashi, T., Sato, H., Kimura, K., Kuwayama, T., Manju, Y. (2005). Modification of ovary stock solution with magnesium and raffinose improves the developmental competence of oocytes after long preservation. *Zygote.* 13, 303-308.

Jiang, J.Y., Cheung, C.K., Wang, Y., Tsang, B.K. (2003). Regulation of cell death and cell survival gene expression during ovarian follicular development and atresia. *Front Biosci.* 8, 222-237.

Johnson, N.C., Dan, H.C., Cheng, J.Q., Kruk, P.A. (2004). BRCAI 185detAG mutation inhibits Akt-dependent, IAP-mediated caspase-3 inactivation in human ovarian surface epithelial cells. *Exp. Cell Res.* 298, 9-18.

Kaur, P., and Guraya, S.S. (1987). Ovarian characteristics of the Indian mole rat *Bandicota bongalensis. Proc. Ind. Natl Acad Sci.* 96, 667.

Kerr, J.F., Winterford, C.M., and Harman, B.V. (1994). Apoptosis. Its significance in cancer and cancer therapy. *Cancer.* 73, 2013-26.

Kerr, J.F.R., Wyllie, A.H., Currie, A.R. (1972). Apoptosis: a basic biological phenomenon with wide ranging implications in tissue kinetics. *Br. J. Cancer.* 26, 239-257.

Kimura, Y., Manabe, N., Nishihara, S., Matsushita, H., Tajima, C., Wada, S., Miyamoto, H. (1999). Up-regulation of the alpha 2,6-sialyl- transferase messenger ribonucleic acid increased glycoconjugation containing alpha 2,6-linked sialic acid residues in granulosa cells during follicular atresia of porcine ovaries. *Biol. Reprod.* 60, 1475-1482.

Kressel, M., Groscurth, P. (1994). Distinction of apoptotic and necrotic cell death by *in situ* labeling of fragmented DNA. *Cell Tiss. Res.* 278, 549-556.

Krishnamurthy, K.V., Krishnaraj, R., Chozhavendan, R., Samuel, C.F. (2000). The program of cell death in plants and animals - A comparison. *Curr. Sci.* 79, 1169-1181.

Laloraya, M., Kumar, G.P., Laloraya, M.M. (1988). Changes in the levels of superoxide anion radical and superoxide dismutase during the estrous cycle of *Rattus norvegicus* and induction of superoxide dismutase in rat ovary by lutropin. *Biochem Biophys Res Commun.* 157, 146-153.

Lapluye, G. (1990). SOD mimicking properties of copper (II) complexes: health side effects. In: Emerit I., Packer, L., Auclair, C. Antioxidants in therapy and preventive medicine. Plenum Press, New York, 59.

Liu, Y., Wu, C., Lyu, Q., Yang, D., Albertini, D.F., Keefe, D.L., Liu, L. (2007). Germline stem cells and neo-oogenesis in the adult human ovary. *Dev. Biol.,* 306, 112-120.

Lussier, J.G., Matton, P., Dufour, J.J., (1987). Growth rates of follicles in the ovary of the cow. *J. Reprod. Fertil.* 81, 301-307.

Maillet, G., Breard, E., Benhaim, A., Leymarie, P., Feral, C. (2002). Hormonal regulation of apoptosis in rabbit granulosa cells *in vitro:* evaluation by flow cytometric detection of plasma membrane phosphatidylserine externalization. *Reproduction.* 123, 243-251.

Manabe, N., Inoue, N., Miyano, T., Sakamaki, K., Sugimoto, M., Miyamoto, H. (2003). Ovarian follicle selection in mammalian ovaries : regulatory mechanism of granulosa cell apoptosis during follicular atresia. In : *The Ovary*. Leung PK, Adashi E (Eds.), Academic press/Elsevier Science Publishers, Amsterdam, pp. 369-385.

Manabe, N., Inoue, N., Miyano, T., Sakamaki, K., Sugimoto, M., Miyamoto, H. (2003). Ovarian follicle selection in mammalian ovaries : regulatory mechanism of granulosa cell apoptosis during follicular atresia. In : *The Ovary*. Leung PK, Adashi E (Eds.), Academic press/Elsevier Science Publishers, Amsterdam, pp. 369-385.

Matzuk, M.M., Dionne, L., Guo, Q., Kumar, T.R., Lebovitz, R.M. (1998). Ovarian function in superoxide dismutase 1 and 2 knockout mice. *Endocrinology*. 139, 4008-4011.

Michiels, C., Raes, M., Haubion, A., Remacle, J. (1991). Association of antioxidant system in the protection of human fibroblasts against oxygen derived free radicals. *Free Rad. Res. Comm.* 14, 323.

Michiels, C., Raes, M., Joussaint, O., Remacle, J. (1994). Importance of Se-glutathione peroxidase, catalase and Cu/Zn-SOD for cell survival against oxidative stress. *Free Rad. Biol. Med.* 17, 235.

Monniaux, D., (2002). Oocyte apoptosis and evolution of ovarian reserve. *Gynecol Obstet Fertil.* 30, 822-826.

Moor, R.M. (1977). Sites of steroid production in ovine Graafian follicles in culture. *J. Endocrinol.* 73, 143-150.

Moor, R.M., Hay, M.F., Dott, H.M., and Cran, D.G. (1978). Macroscopic identification and steroidogenic functions of atretic follicles in sheep. *Endocrinology*. 77, 309-318.

Motta, P. (1972). Histochemical evidence of early atretic follicles in different mammals. *J. Cell Biol.* 55, 18.

Nicosia, S.V. (1980). Endocrine Physiopathology of the ovary. Tozzini RI, Reeves G. and Pineda R.L. (Eds.) Amsterdam, New York, pp. 43-62.

Okamura, Y., Miyamoto, A., Manabe, N., Tanaka, N., Okamura, H., Fukomoto, M. (2001). Protein tyrosine kinase expression in the porcine ovary. *Mol Human Reprod.* 7, 723-729.

Osman, P. (1985). Rate and Course of atresia during follicular development in the adult cycling rat. *J. Reprod. Fertil.* 73, 261-270.

Palumbo, A., and Yeh, J. (1994). *In situ* localization of apoptosis in rat ovary during follicular atresia. *Biol Reprod.* 51, 888-895.

Pangas, S.A. (2007). Growth factors in ovarian development. *Semin Reprod Med.* 25, 225-234.

Parshad, V.R., and Guraya, S.S. (1983). Histochemical distribution of tetrazolium reductases, dehydrogenases and lipids in the follicular wall of normal and atretic follicles in the ovary of the Indian gerbil, *Tatera indica* (Muridea: Rodentia). *Proc. Ind. Acad. Sci.* 92, 121-128.

Paszkowski, T., Clarke, R.N. (1996). Antioxidant capacity of preimplantation embryo culture medium declines following the incubation of poor quality embryos. *Hum Reprod.* 11, 2493-2495.

Paszkowski, T., Traub, A.I., Robinson, S.Y., McMaster, D. (1995). Selenium dependent glutathione peroxidase activity in human follicular fluid. *Clin. Chim. Acta.* 236, 173-180.

Pol De, A., Vaccina, F., Forabosco, A., Cavazzuti, E., Marzono, L. (1997). Apoptosis of germ cells during human prenatal oogenesis. *Hum Reprod.* 12, 2235-2241.

Quirk, S.M., Cowan, R.G., and Harman, R.M. (2006). The susceptibility of granulosa cells to apoptosis is influenced by oestradiol and the cell cycle. *J. Endocrinol.* 189, 441-453.

Rajakoshi, E. (1960). The ovarian follicular system in sexually mature heifers with special reference to seasonal, cyclical and left right variations. *Acta endocr.* 34, 7-68

Ravindranath, S.D., and Fridovich, I. (1975). Isolation and Characterization of a manganese-containing superoxide dismutase from yeast. *J. Biol Chem.* 250, 6107-6112.

Redmond, T.M., Duke, E.J., Coles, W.H., Simson, J.A., Crouch, R.K. (1984). Localization of corneal superoxide dismutase by biochemical and histocytochemical techniques. *Exp. Eye Res.* 38, 369-378.

Richards, J.S., Sirois, J., Natraj, V., Morris, J.K., Fiotzpatrick, S.L., Clemens, J.W. (1993). Molecular regulation of genes involved in ovulation and luteinization. Ovarian cell interaction. Genes to physiology. Hsueh AJW and Schomberg DW (eds.), W. Springs-Verlag, New York, pp. 125-133.

Rudolf, E. (2007). Depletion of ATP and oxidative stress underlie zinc-induced cell injury. *Acta Medica.* 50, 43-49.

Sangha, G.K., and Guraya, S.S. (1988/89). Histochemical changes in acid and alkaline phosphatase activities in the growing follicles and corpora lutea of the rat ovary. *Acta Morphol. Neerl. Scand.* 26, 43-49.

Sato, T., Irie, S., Krajeueski, S., Reed, J.C. (1994). Cloning and sequencing of a cDNA encoding the rat Bcl-2 protein. *Gene.* 140, 291-292.

Sawyer, H.R., Abel, J.H., McClellan, M.C., Schwitz, M., Niswender, G.D. (1979). Secretory granules and progesterone secretion by ovine corpora lutea *in vitro. Endocrinol.* 104. 476-486.

Schmidt, H.H., Lohmann, S.M., Walter, U. (1993). The nitric oxide and cGMP signal transduction system: regulation and mechanism of action. *Biochim Biophys Acta.* 1178, 153-175.

Schwartzman, R.A., and Cidlowski, J.A. (1993). Apoptosis: the biochemistry and molecular biology of programmed cell death. *Endocr Rev.* 14, 133-151.

Sharma, R.K. (2000). Follicular atresia in goat: A review. *Ind. J. Anim. Sci.* 70, 1035-1046.

Sharma, R.K. (2003). Structural analysis of cumulus and corona cells of goat antral follicles: possible functional significance. *Indian J. Anim. Sci.* 73, 28-32.

Sharma, R.K. and Batra, S. (2005). Ultrastructure of the regressing corpus luteum in the goat ovary. *Indian J. Anim. Sci.* 75, 936-937.

Sharma, R.K. and Bhardwaj, J.K. (2009). Ultrastructural characterization of apoptotic granulosa cells in caprine ovary. *J. Microsc.* 236, 236-242.

Sharma, R.K., and Batra, S. (2008). Changes in the steroidogenic cells of the ovaries in small ruminants. *Indian J. Anim. Sci.*, 78, 584-596.

Sharma, R.K., and Bhardwaj, J.K. (2007). Granulosa cell apoptosis *in situ* in caprine ovary. *J. Cell Tissue Res.* 7, 1111-1114.

Sharma, R.K., and Bhardwaj, J.K. (2009). *In situ* evaluation of granulosa cells during apoptosis in caprine ovary. *Int. J. Integ Biol.* 5, 58-61.

Sharma, R.K., and Guraya, S.S. (1990). Granulosa heterogeneity: A histo-chemical lectin staining and scanning electron microscopic study on *Rattus rattus* ovary. *Acta Embr. Morph. Exp.* 11, 107-129.

Sharma, R.K., and Guraya, S.S. (1992). Lectin staining studies on follicular atresia in house rat (*Rattus rattus*). *Acta Morphol. Hung.* 40, 25-34.

Sharma, R.K., and Guraya, S.S. (1992). Lectin staining studies on follicular atresia in house rat (*Rattus rattus*). *Acta Morphol. Hung.* 40, 25-34.

Sharma, R.K., and Guraya, S.S. (1998a). Distribution, histochemistry and biochemistry of carbohydrates in the mammalian oocytes during folliculogenesis. *J. Agri. Rev.* 19, 73-85.

Sharma, R.K., and Guraya, S.S. (1998b). Carbohydrate histochemistry of atretic follicles in house rat ovary. *J. Anim. Morphol. Physiol.* 45, 112-117.

Sharma, R.K., and Guraya, S.S. (1998c). Simultaneously histochemical and lectin staining studies and electron microsopic observations on interstitial gland tissue in the rat ovary. *J. Anim. Morphol. Physiol.* 45, 112-117.

Sharma, R.K., Khajuria, M., and Guraya, S.S. (1992). Morphology of normal and atretic follicles of goat during anoestrous. *Int. J. Anim. Sci.* 6, 81-85.

Sharma, R.K., Sharma, M.B., and Bhardwaj, J.K. (2008). Gonadotropins titre and apoptosis in caprine granulosa cells (Abstract). In: Proceedings of International conference on *Molecular and Clinical aspects of gonadal and nongonadal actions of gonadotropins.* Feb. 7-9. AIIMS, New Delhi, p. 51.

Shiotani, M., Noda, Y., Narimoto, K., Imai, K., Mori, T., Fujimoto, K., Ogawa, K. (1991). Immunohistochemical localization of superoxide dismutase in the human ovary. *Hum Reprod.* 6, 1349-1353.

Singh, D., and Pandey, R.S. (1994). Changes in the catalase activity and hydrogen peroxide production in the rat ovary during estrous cycle. XVI. *Int. Cong. Biochem. Mol. Bio.* 364.

Slomczynska, M., Tabarowski, Z., Duda, M., Burek, M., Knapczyk, K. (2006). Androgen receptor in early apoptotic follicles in the porcine ovary at pregnancy. *Folia Histochem. Cytobiol.* 44, 185-188.

Tabarowski, Z., Szaltys, M., Bik, M., Slomczynska, M. (2005). Atresia of large ovarian follicles of the rat. *Folia Histochem. Cytobiol.* 43, 43-55.

Tatone, C., Amicarelli, F., Carbone, M.C., Monteleone, P., Caserta, D., Marci, R., Artini, P.G., Piomboni, P., Focarelli, R., (2008). Cellular and molecular aspects of ovarian follicle ageing. *Human Reprod.* 14, 131-142.

Tibell, L., Hjalmarsson, K., Edlund, T., Skogman, G., Engstrom, A., Marklund, S.C. (1987). Expression of human extracellular superoxide dismutase in Chinese hamster ovary cells and characterization of the product. *Proc Natl Acad Sci USA.* 84, 6634-6638.

Tilly, J.L. (1996). Apoptosis and ovarian function. *Rev Reprod.* 1, 162-172.

Tilly, J.L. (1996). The molecular basis of ovarian cell death during germ cell attrition, follicular atresia, and luteolysis. *Front Biosci.* 1, d1-11.

Tilly, J.L., and Tilly, K.I. (1995). Inhibitors of oxidative stress mimic the ability of follicle-stimulating hormone to suppress apoptosis in cultured rat ovarian follicles. *Endocrinology.* 136, 242-252.

Tilly, J.L., Kowalski, K.I., Johnson, A.L., Huseh, A.J.W. (1991). Involvement of apoptosis in ovarian follicular atresia and postovulatory regression. *Endocrinology.* 129, 2799-2801.

Tsafriri, A. and Braw, R. (1984). Experimental approaches to atresia in mammals. *Oxford Review in Reproductive Biology.* 6, 226-265.

Uilenbroek, J.M.J., Wouterson, P.J., Van der Schoot, P. (1980). Atresia of preovulatory follicles: gonadotropin binding and steroidogenic activity. *Biol. Reprod.* 23, 219-229.

Verkek, A., and Jondkind, J.F. (1992). Vascular cells under peroxide induced stress: a balance study on *in vitro* peroxide handling by vascular endothelial and smooth muscle cells. *Free Rad. Res. Comm.* 17, 121.

Verma, S.K., and Guraya, S.S. 1968. The localization and functional significance of alkaline phosphatase in the vertebrate ovary. *Experientia.* 24, 398-399.

Vohra, B.P.S., James, T.J., Sharma, S.P., Kansal, V.K., Chaudhary, A., Gupta, S.K., (2002). Dark neurons in the ageing cerebellum: their mode of formation and effect of Maharishi Amrit Kalash. *Biotechnology,* 3, 347-354.

Westhof, G., Westhof, K.F., Braendle, W.L., Dizerega, G.S. (1991). Differential steroid secretion and gonadotrophin response by individual tertiary porcine follicles *in vitro*. Possible physiological role of atretic follicles. *Biol. Reprod.* 44, 461-468.

Wezel, I.L.V., Dharmarajan, A.M., Lavranos, T.C., Rodgers, R.J. (1999). Evidence for alternative pathways of granulosa cell death in healthy and slightly atretic bovine antral follicles. *Endocrinology.* 140, 2602-2612.

Williams, G.T., and Smith, C.A. (1993). Molecular regulation of apoptosis: genetic controls on cell death. *Cell.* 74, 777-779.

Wiseman, D.A., Wells, S.M., Wilham, J., Hubbard, M., Welker, J.E., Boack, S.M. (2006). Endothelial response to stress from exogeneous Zn^{+2} resembles that of NO-mediated nitrosative stress, and is protected by MT-1 overexpression. *Am. J. Physiol Cell Physiol.* 291, 555-568.

Wu, Ji., Zhang, L. and Wang, X. (2000). Maturation and apoptosis of human oocytes *in vitro* are age related. *Fertil steril.* 74, 1137-1141.

Wyllie, A.H., Kerr, J.F.R., Currie, A.R. (1980). Cell death: the significance of apoptosis. *Int. Rev. Cytol.* 68, 251-306.

Yang, H.W., Hwang, K.J., Kwon, H.C., Kim, H.S., Choi, K.W., Oh, K.S. (1998). Detection of reactive oxygen species (ROS) and apoptosis in human fragmented embryos. *Hum. Reprod.* 13, 998-1002.

Yang, M.Y., and Rajamahendran, R. (2000). Morphological and biochemical identification of apoptosis in small, medium, and large bovine follicles and the effects of follicle-stimulating hormone and insulin-like growth factor-1 on spontaneous apoptosis in cultured bovine granulosa cells. *Biol Reprod.* 2000, 62, 1209-1217.

Yu, Y.S., Sui, H.S., Han, Z.B., Li, W., Luo, M.J., Tan, J.H. (2004). Apoptosis in granulosa cells during follicular atresia : relationship with steroids and insulin like growth factors. *Cell Research.* 14, 341-346.

Zhou, H., and Zhang, Y. (2005). Effect of growth factors on *in vitro* development of caprine preantral follicle oocytes. *Ann. Reprod. Sci.* 90, 265-272.

Zimmerman, R.C., Westhof, G., Peukert-Adam, I., Hoedemaker, M., Grunert, E., Braendle, W. (1987). *In vitro* steroid secretion of tertiary atretic bovine follicles in a superfusion system correlated to their histological features. *Human Reprod.* 2, 457-461.